Student Solutions Manual

to accompany

Physics

Second Edition

Alan Giambattista
Cornell University

Betty McCarthy Richardson
Cornell University

Robert C. Richardson
Cornell University

Prepared by
William Fellers
Fellers Math & Science

 **Higher Education**

Boston   Burr Ridge, IL   Dubuque, IA   New York   San Francisco   St. Louis
Bangkok   Bogotá   Caracas   Kuala Lumpur   Lisbon   London   Madrid   Mexico City
Milan   Montreal   New Delhi   Santiago   Seoul   Singapore   Sydney   Taipei   Toronto

The McGraw·Hill Companies

 **Higher Education**

Student Solutions Manual to accompany
PHYSICS, SECOND EDITION
ALAN GIAMBATTISTA, BETTY MCCARTHY RICHARDSON, AND ROBERT C. RICHARDSON

Published by McGraw-Hill Higher Education, an imprint of The McGraw-Hill Companies, Inc., 1221 Avenue of the Americas, New York, NY 10020. Copyright © 2010, 2008 by The McGraw-Hill Companies, Inc. All rights reserved.

This book is printed on recycled, acid-free paper containing 10% post consumer waste.

3 4 5 6 7 8 9 0 QDB/QDB 14 13 12

ISBN: 978-0-07-334892-6
MHID: 0-07-334892-9

www.mhhe.com

Student Solutions Manual
to accompany
PHYSICS
Second Edition

Table of Contents

Chapter 1

INTRODUCTION

Conceptual Questions

1. Knowledge of physics is important for a full understanding of many scientific disciplines, such as chemistry, biology, and geology. Furthermore, much of our current technology can only be understood with knowledge of the underlying laws of physics. In the search for more efficient and environmentally safe sources of energy, for example, physics is essential. Also, many study physics for the sense of fulfillment that comes with learning about the world we inhabit.

5. Scientific notation eliminates the need to write many zeros in very large or small numbers. Also, the appropriate number of significant digits is unambiguous when written this way.

9. The kilogram, meter, and second are three of the base units used in the SI system.

13. Trends in a set of data are often the most interesting aspect of the outcome of an experiment. Such trends are more apparent when data is plotted graphically rather than listed in numerical tables.

Problems

1. **Strategy** The new fence will be $100\% + 37\% = 137\%$ of the height of the old fence.

   Solution Find the height of the new fence.
   $$1.37 \times 1.8 \text{ m} = \boxed{2.5 \text{ m}}$$

3. **Strategy** Relate the surface area S to the radius r using $S = 4\pi r^2$.

   Solution Find the ratio of the new radius to the old.
   $S_1 = 4\pi r_1^2$ and $S_2 = 4\pi r_2^2 = 1.160 S_1 = 1.160(4\pi r_1^2)$.
   $$4\pi r_2^2 = 1.160(4\pi r_1^2)$$
   $$r_2^2 = 1.160 r_1^2$$
   $$\left(\frac{r_2}{r_1}\right)^2 = 1.160$$
   $$\frac{r_2}{r_1} = \sqrt{1.160} = 1.077$$
   The radius of the balloon increases by $\boxed{7.7\%}$.

5. **Strategy** The surface area S and the volume V are given by $S = 6s^2$ and $V = s^3$, respectively.

   Solution Find the ratio of the surface area to the volume.
   $$\frac{S}{V} = \frac{6s^2}{s^3} = \boxed{\frac{6}{s}}$$

9. **Strategy** Use a proportion.

 Solution Find Jupiter's orbital period.

 $T^2 \propto R^3$, so $\dfrac{T_J^2}{T_E^2} = \dfrac{R_J^3}{R_E^3} = 5.19^3$. Thus, $T_J = 5.19^{3/2} T_E = \boxed{11.8 \text{ yr}}$.

11. **Strategy** The area of the poster is given by $A = \ell w$. Let the original and final areas be $A_1 = \ell_1 w_1$ and $A_2 = \ell_2 w_2$, respectively.

 Solution Calculate the percentage reduction of the area.

 $A_2 = \ell_2 w_2 = (0.800 \ell_1)(0.800 w_1) = 0.640 \ell_1 w_1 = 0.640 A_1$

 $\dfrac{A_1 - A_2}{A_1} \times 100\% = \dfrac{A_1 - 0.640 A_1}{A_1} \times 100\% = \boxed{36.0\%}$

13. **(a) Strategy** Rewrite the numbers so that the power of 10 is the same for each. Then add and give the answer with the number of significant figures determined by the less precise of the two numbers.

 Solution Perform the operation with the appropriate number of significant figures.

 $3.783 \times 10^6 \text{ kg} + 1.25 \times 10^8 \text{ kg} = 0.03783 \times 10^8 \text{ kg} + 1.25 \times 10^8 \text{ kg} = \boxed{1.29 \times 10^8 \text{ kg}}$

 (b) Strategy Find the quotient and give the answer with the number of significant figures determined by the number with the fewest significant figures.

 Solution Perform the operation with the appropriate number of significant figures.

 $(3.783 \times 10^6 \text{ m}) \div (3.0 \times 10^{-2} \text{ s}) = \boxed{1.3 \times 10^8 \text{ m/s}}$

17. **Strategy** Multiply and give the answer in scientific notation with the number of significant figures determined by the number with the fewest significant figures.

 Solution Solve the problem.

 $(3.2 \text{ m}) \times (4.0 \times 10^{-3} \text{ m}) \times (1.3 \times 10^{-8} \text{ m}) = \boxed{1.7 \times 10^{-10} \text{ m}^3}$

21. **Strategy** There are approximately 39.37 inches per meter.

 Solution Find the thickness of the cell membrane in inches.

 $7.0 \times 10^{-9} \text{ m} \times 39.37 \text{ inches/m} = \boxed{2.8 \times 10^{-7} \text{ inches}}$

25. **(a) Strategy** There are 60 seconds in one minute, 5280 feet in one mile, and 3.28 feet in one meter.

 Solution Express 0.32 miles per minute in meters per second.

 $\dfrac{0.32 \text{ mi}}{1 \text{ min}} \times \dfrac{1 \text{ min}}{60 \text{ s}} \times \dfrac{5280 \text{ ft}}{1 \text{ mi}} \times \dfrac{1 \text{ m}}{3.28 \text{ ft}} = \boxed{8.6 \text{ m/s}}$

 (b) Strategy There are 60 minutes in one hour.

 Solution Express 0.32 miles per minute in miles per hour.

 $\dfrac{0.32 \text{ mi}}{1 \text{ min}} \times \dfrac{60 \text{ min}}{1 \text{ h}} = \boxed{19 \text{ mi/h}}$

29. **Strategy** There are 1000 grams in one kilogram and 100 centimeters in one meter.

    Solution Find the density of mercury in units of g/cm^3.

    $$\frac{1.36\times10^4\,kg}{1\,m^3}\times\frac{1000\,g}{1\,kg}\times\left(\frac{1\,m}{100\,cm}\right)^3 = \boxed{13.6\ g/cm^3}$$

33. (a) **Strategy** There are 12 inches in one foot, 2.54 centimeters in one inch, and 60 seconds in one minute.

    Solution Express the snail's speed in feet per second.

    $$\frac{5.0\,cm}{1\,min}\times\frac{1\,min}{60\,s}\times\frac{1\,in}{2.54\,cm}\times\frac{1\,ft}{12\,in} = \boxed{2.7\times10^{-3}\ ft/s}$$

    (b) **Strategy** There are 5280 feet in one mile, 12 inches in one foot, 2.54 centimeters in one inch, and 60 minutes in one hour.

    Solution Express the snail's speed in miles per hour.

    $$\frac{5.0\,cm}{1\,min}\times\frac{60\,min}{1\,h}\times\frac{1\,in}{2.54\,cm}\times\frac{1\,ft}{12\,in}\times\frac{1\,mi}{5280\,ft} = \boxed{1.9\times10^{-3}\ mi/h}$$

37. **Strategy** Replace each quantity in $T^2 = 4\pi^2 r^3/(GM)$ with its dimensions.

    Solution Show that the equation is dimensionally correct.

    T^2 has dimensions $[T]^2$ and $\dfrac{4\pi^2 r^3}{GM}$ has dimensions $\dfrac{[L]^3}{\dfrac{[L]^3}{[M][T]^2}\times[M]} = \dfrac{[L]^3}{[M]}\times\dfrac{[M][T]^2}{[L]^3} = [T]^2$.

    Since $\boxed{[T]^2 = [T]^2}$, the equation is dimensionally correct.

41. **Strategy** Approximate the distance from your eyes to a book held at your normal reading distance.

    Solution The normal reading distance is about 30-40 cm, so the approximate distance from your eyes to a book you are reading is $\boxed{30\text{-}40\ cm}$.

45. Strategy (Answers will vary.) In this case, we use San Francisco, CA for the city. The population of San Francisco is approximately 750,000. Assume that there is one automobile for every two residents of San Francisco, that an average automobile needs three repairs or services per year, and that the average shop can service 10 automobiles per day.

Solution Estimate the number of automobile repair shops in San Francisco.

If an automobile needs three repairs or services per year, then it needs $\dfrac{3 \text{ repairs}}{\text{auto} \cdot \text{y}} \times \dfrac{1 \text{ y}}{365 \text{ d}} \approx \dfrac{0.01 \text{ repairs}}{\text{auto} \cdot \text{d}}$.

If there is one auto for every two residents, then there are $\dfrac{1 \text{ auto}}{2 \text{ residents}} \times 750{,}000 \text{ residents} \approx 4 \times 10^5 \text{ autos}$.

If a shop requires one day to service 10 autos, then the number of shops-days per repair is

$1 \text{ shop} \times \dfrac{1 \text{ d}}{10 \text{ repairs}} = \dfrac{0.1 \text{ shop} \cdot \text{d}}{\text{repair}}$.

The estimated number of auto shops is $4 \times 10^5 \text{ autos} \times \dfrac{0.01 \text{ repairs}}{\text{auto} \cdot \text{d}} \times \dfrac{0.1 \text{ shop} \cdot \text{d}}{\text{repair}} = \boxed{400 \text{ shops}}$.

Checking the phone directory, we find that there are approximately 463 automobile repair and service shops in

San Francisco. The estimate is off by $\dfrac{400 - 463}{400} \times 100\% = \boxed{-16\%}$. The estimate was 16% too low, but in the ball park!

47. Strategy One story is about 3 m high.

Solution Find the order of magnitude of the height in meters of a 40-story building.

$(3 \text{ m})(40) \sim \boxed{100 \text{ m}}$

49. Strategy The plot of temperature versus elapsed time is shown. Use the graph to answer the questions.

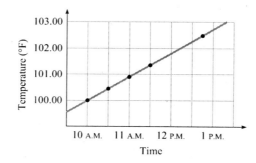

Solution

(a) By inspection of the graph, it appears that the temperature at noon was $\boxed{101.8°F}$.

(b) Estimate the slope of the line.

$m = \dfrac{102.6°F - 100.0°F}{1:00 \text{ P.M.} - 10:00 \text{ A.M.}} = \dfrac{2.6°F}{3 \text{ h}} = \boxed{0.9°F/h}$

(c) In twelve hours, the temperature would, according to the trend, be approximately
$T = (0.9 \ °F/h)(12 \text{ h}) + 102.5°F = 113°F$.

The patient would be dead before the temperature reached this level. So, the answer is $\boxed{\text{no}}$.

53. Strategy Put the equation that describes the line in slope-intercept form, $y = mx + b$.

$$at = v - v_0$$
$$v = at + v_0$$

Solution

(a) v is the dependent variable and t is the independent variable, so $\boxed{a}$ is the slope of the line.

(b) The slope-intercept form is $y = mx + b$. Find the vertical-axis intercept.
$v \leftrightarrow y$, $t \leftrightarrow x$, $a \leftrightarrow m$, so $v_0 \leftrightarrow b$.

Thus, $\boxed{+v_0}$ is the vertical-axis intercept of the line.

57. Strategy For parts (a) through (d), perform the calculations.

Solution

(a) $186.300 + 0.0030 = \boxed{186.303}$

(b) $186.300 - 0.0030 = \boxed{186.297}$

(c) $186.300 \times 0.0030 = \boxed{0.56}$

(d) $186.300 / 0.0030 = \boxed{62,000}$

(e) **Strategy** For cases (a) and (b), the percent error is given by $\dfrac{0.0030}{Actual\ Value} \times 100\%$.

Solution Find the percent error.

Case (a): $\dfrac{0.0030}{186.303} \times 100\% = \boxed{0.0016\%}$

Case (b): $\dfrac{0.0030}{186.297} \times 100\% = \boxed{0.0016\%}$

> For case (c), ignoring 0.0030 causes you to multiply by zero and get a zero result. For case (d), ignoring 0.0030 causes you to divide by zero.

(f) **Strategy** Make a rule about neglecting small values using the results obtained above.

Solution

> You can neglect small values when they are added to or subtracted from sufficiently large values. The term "sufficiently large" is determined by the number of significant figures required.

59. Strategy Assuming that the cross section of the artery is a circle, we use the area of a circle, $A = \pi r^2$.

Solution

$A_1 = \pi r_1^2$ and $A_2 = \pi r_2^2 = \pi (2.0 r_1)^2 = 4.0 \pi r_1^2$.

Form a proportion.

$$\frac{A_2}{A_1} = \frac{4.0 \pi r_1^2}{\pi r_1^2} = 4.0$$

The cross-sectional area of the artery increases by a factor of $\boxed{4.0}$.

61. Strategy If s is the speed of the molecule, then $s \propto \sqrt{T}$ where T is the temperature.

Solution Form a proportion.

$$\frac{s_{\text{cold}}}{s_{\text{warm}}} = \frac{\sqrt{T_{\text{cold}}}}{\sqrt{T_{\text{warm}}}}$$

Find s_{cold}.

$$s_{\text{cold}} = s_{\text{warm}} \sqrt{\frac{T_{\text{cold}}}{T_{\text{warm}}}} = (475 \text{ m/s}) \sqrt{\frac{250.0 \text{ K}}{300.0 \text{ K}}} = \boxed{434 \text{ m/s}}$$

65. Strategy Use the metric prefixes n (10^{-9}), μ (10^{-6}), m (10^{-3}), or M (10^6).

Solution

(a) M (or mega) is equal to 10^6, so $6 \times 10^6 \text{ m} = \boxed{6 \text{ Mm}}$.

(b) There are approximately 3.28 feet in one meter, so $6 \text{ ft} \times \dfrac{1 \text{ m}}{3.28 \text{ ft}} = \boxed{2 \text{ m}}$.

(c) μ (or micro) is equal to 10^{-6}, so $10^{-6} \text{ m} = \boxed{1 \text{ μm}}$.

(d) n (or nano) is equal to 10^{-9}, so $3 \times 10^{-9} \text{ m} = \boxed{3 \text{ nm}}$.

(e) n (or nano) is equal to 10^{-9}, so $3 \times 10^{-10} \text{ m} = \boxed{0.3 \text{ nm}}$.

67. Strategy The circumference of a viroid is approximately 300 times 0.35 nm. The diameter is given by $C = \pi d$, or $d = C/\pi$.

Solution Find the diameter of the viroid in the required units.

(a) $d = \dfrac{300(0.35 \text{ nm})}{\pi} \times \dfrac{10^{-9} \text{ m}}{1 \text{ nm}} = \boxed{3.3 \times 10^{-8} \text{ m}}$

(b) $d = \dfrac{300(0.35 \text{ nm})}{\pi} \times \dfrac{10^{-3} \text{ μm}}{1 \text{ nm}} = \boxed{3.3 \times 10^{-2} \text{ μm}}$

(c) $d = \dfrac{300(0.35 \text{ nm})}{\pi} \times \dfrac{10^{-7} \text{ cm}}{1 \text{ nm}} \times \dfrac{1 \text{ in}}{2.54 \text{ cm}} = \boxed{1.3 \times 10^{-6} \text{ in}}$

69. Strategy The volume of the blue whale can be found by dividing the mass of the whale by its average density.

Solution Find the volume of the blue whale in cubic meters.
$$V = \frac{m}{\rho} = \frac{1.9 \times 10^5 \text{ kg}}{0.85 \text{ g/cm}^3} \times \frac{1000 \text{ g}}{1 \text{ kg}} \times \left(\frac{1 \text{ m}}{100 \text{ cm}} \right)^3 = \boxed{2.2 \times 10^2 \text{ m}^3}$$

73. Strategy There are 2.54 cm in one inch and 3600 seconds in one hour.

Solution Find the conversion factor for changing meters per second to miles per hour.
$$\frac{1 \text{ m}}{1 \text{ s}} \times \frac{100 \text{ cm}}{1 \text{ m}} \times \frac{1 \text{ in}}{2.54 \text{ cm}} \times \frac{1 \text{ ft}}{12 \text{ in}} \times \frac{1 \text{ mi}}{5280 \text{ ft}} \times \frac{3600 \text{ s}}{1 \text{ h}} = \boxed{2.24 \text{ mi/h} = 1 \text{ m/s}}$$
So, $\boxed{\text{for a quick, approximate conversion, multiply by 2}}$.

77. Strategy The SI base unit for mass is kg. Replace each quantity in $W = mg$ with its SI base units.

Solution Find the SI unit for weight.
$$\text{kg} \cdot \frac{\text{m}}{\text{s}^2} = \boxed{\frac{\text{kg} \cdot \text{m}}{\text{s}^2}}$$

81. (a) Strategy There are 7.0 leagues in one pace and 4.8 kilometers in one league.

Solution Find your speed in kilometers per hour.
$$\frac{120 \text{ paces}}{1 \text{ min}} \times \frac{7.0 \text{ leagues}}{1 \text{ pace}} \times \frac{4.8 \text{ km}}{1 \text{ league}} \times \frac{60 \text{ min}}{1 \text{ h}} = \boxed{2.4 \times 10^5 \text{ km/h}}$$

(b) Strategy The circumference of the earth is approximately 40,000 km. The time it takes to march around the Earth is found by dividing the distance by the speed.

Solution Find the time of travel.
$$40{,}000 \text{ km} \times \frac{1 \text{ h}}{2.4 \times 10^5 \text{ km}} \times \frac{60 \text{ min}}{1 \text{ h}} = \boxed{10 \text{ min}}$$

85. Strategy The dimensions of k and m are mass per time squared and mass, respectively. Dividing either quantity by the other will eliminate the mass dimension.

Solution The square root of k/m has dimensions of inverse time, which is correct for frequency.
So, $f = \sqrt{k/m}$. Find k.
$$f_1 = \sqrt{\frac{k}{m_1}}, \text{ so } f_1^2 = \frac{k}{m_1}, \text{ or } k = m_1 f_1^2.$$
Find the frequency of the chair with the 75-kg astronaut.
$$f_2 = \sqrt{\frac{k}{m_2}} = \sqrt{\frac{m_1 f_1^2}{m_2}} = f_1 \sqrt{\frac{m_1}{m_2}} = (0.50 \text{ s}^{-1}) \sqrt{\frac{62 \text{ kg} + 10.0 \text{ kg}}{75 \text{ kg} + 10.0 \text{ kg}}} = \boxed{0.46 \text{ s}^{-1}}$$

Chapter 2

MOTION ALONG A LINE

Conceptual Questions

1. Distance traveled is a scalar quantity equal to the total length of the path taken in moving from one point to another. Displacement is a vector quantity directed from the initial point towards the final point with a magnitude equal to the straight line distance between the two points. The magnitude of the displacement is always less than or equal to the total distance traveled.

5. The area under the curve of an a_x versus time graph is equal to the change in the x-component of the velocity.

9. (a) $a_x > 0$ and $v_x < 0$ means you are moving south and slowing down.

   (b) $a_x = 0$ and $v_x < 0$ means you are moving south at a constant speed.

   (c) $a_x < 0$ and $v_x = 0$ means you are momentarily at rest but speeding up in a southward direction.

   (d) $a_x < 0$ and $v_x < 0$ means you are moving south and speeding up.

   (e) As can be seen from our answers above, it is not a good idea to use the term "negative acceleration" to mean slowing down. In parts (c) and (d), the acceleration is negative, but the bicycle is speeding up. Also, in part (a), the acceleration is positive, but the bicycle is slowing down.

Problems

1. **Strategy** Let east be the $+x$-direction.

   Solution Draw a vector diagram; then compute the sum of the three displacements.
   The vector diagram:

   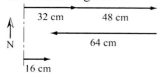

   The sum of the three displacements is $(32 \text{ cm} + 48 \text{ cm} - 64 \text{ cm})$ east = $\boxed{16 \text{ cm east}}$.

5. **Strategy** Let south be the $+x$-direction.

   Solution Draw vector diagrams for each situation; then find the displacements of the car.

   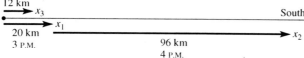

(a)

$$x_3 - x_1$$

$$x_3$$

$$-x_1$$

$x_3 - x_1 = 12 \text{ km} - 20 \text{ km} = -8 \text{ km}$

The displacement of the car between 3 P.M. and 6 P.M. is $\boxed{8 \text{ km north of its position at 3 P.M.}}$

(b)

$$x_1 \qquad x_2$$

$$x_1 + x_2$$

$x_1 + x_2 = 20 \text{ km} + 96 \text{ km} = 116 \text{ km}$

The displacement of the car from the starting point to the location at 4 P.M. is $\boxed{116 \text{ km south of the starting point.}}$

(c)

$$-(x_1 + x_2) \qquad\qquad x_3$$

$$x_3 - (x_1 + x_2)$$

$x_3 - (x_1 + x_2) = 12 \text{ km} - 116 \text{ km} = -104 \text{ km}$

The displacement of the car between 4 P.M. and 6 P.M. is $\boxed{104 \text{ km north of its position at 4 P.M.}}$

9. Strategy Jason never changes direction, so the direction of the average velocity is due west. Find the average speed by dividing the total distance traveled by the total time.

Solution The distance traveled during each leg of the trip is given by $\Delta x = v_{av}\Delta t$.

$$v_{av} = \frac{(35.0 \text{ mi/h})(0.500 \text{ h}) + (60.0 \text{ mi/h})(2.00 \text{ h}) + (25.0 \text{ mi/h})(10.0/60.0 \text{ h})}{0.500 \text{ h} + 2.00 \text{ h} + 10.0/60.0 \text{ h}} = 53.1 \text{ mi/h}$$

So, the average velocity is $\boxed{53.1 \text{ mi/h due west}}$.

13. Strategy Use the graph to answer the questions. The slope of the graph at any instant represents the speed at that instant.

Solution

(a) The section of the graph with the largest magnitude (steepest) slope represents the highest speed, $\boxed{DE}$.

(b) The slope changes from positive to negative at D, and from negative to positive at E, so the object reverses its direction of motion at times $\boxed{4 \text{ s and } 5 \text{ s}}$.

(c) During the time interval $t = 0$ s to $t = 2$ s, the speed of the object is

$$v = v_{av} = \frac{20 \text{ m} - 0}{2 \text{ s}} = 10 \text{ m/s, and from } t = 2 \text{ s to } t = 3 \text{ s it is } v = v_{av} = \frac{20 \text{ m} - 20 \text{ m}}{3 \text{ s} - 2 \text{ s}} = 0.$$

Therefore, the distance traveled is $(10 \text{ m/s})(2 \text{ s}) = \boxed{20 \text{ m}}$.

17. Strategy Use the area under the curve to find the displacement of the skateboard.

Solution The displacement of the skateboard is given by the area under the v vs. t curve. Under the curve for $t = 3.00$ s to $t = 8.00$ s, there are 16.5 squares and each square represents $(1.0 \text{ m/s})(1.0 \text{ s}) = 1.0$ m; so the board moves $\boxed{16.5 \text{ m}}$.

19. Strategy The slope of the x vs. t curve is equal to v_x. Use the definition of average speed.

Solution Compute the average speed at $t = 2.0$ s.
$$v_x = \frac{6.0 \text{ m} - 4.0 \text{ m}}{3.0 \text{ s} - 1.0 \text{ s}} = \boxed{1.0 \text{ m/s}}$$

21. (a) Strategy Let the positive direction be to the right. Draw a diagram.

Solution Find the chipmunk's total displacement.

$$80 \text{ cm} - 30 \text{ cm} + 90 \text{ cm} - 310 \text{ cm} = -170 \text{ cm}$$
The total displacement is $\boxed{170 \text{ cm to the left}}$.

(b) Strategy The average speed is found by the dividing the total distance traveled by the elapsed time.

Solution Find the total distance traveled.
$$80 \text{ cm} + 30 \text{ cm} + 90 \text{ cm} + 310 \text{ cm} = 510 \text{ cm}$$
Find the average speed.
$$\frac{510 \text{ cm}}{18 \text{ s}} = \boxed{28 \text{ cm/s}}$$

(c) Strategy The average velocity is found by dividing the displacement by the elapsed time.

Solution Find the average velocity.
$$\vec{v}_{av} = \frac{\Delta\vec{r}}{\Delta t} = \frac{170 \text{ cm to the left}}{18 \text{ s}} = \boxed{9.4 \text{ cm/s to the left}}$$

25. Strategy Use the definition of average acceleration.

Solution
$$\vec{a}_{av} = \frac{\Delta\vec{v}}{\Delta t} = \frac{\vec{v}_f - \vec{v}_i}{\Delta t}$$
$$= \frac{0 - 28 \text{ m/s in the direction of the car's travel}}{4.0 \text{ s}} = \boxed{7.0 \text{ m/s}^2 \text{ in the direction opposite the car's velocity}}$$

29. Strategy The magnitude of the acceleration is the absolute value of the slope of the graph at $t = 7.0$ s.

Solution
$$a_x = \left|\frac{\Delta v_x}{\Delta t}\right| = \left|\frac{0 - 20.0 \text{ m/s}}{12.0 \text{ s} - 4.0 \text{ s}}\right| = \boxed{2.5 \text{ m/s}^2}$$

33. (a) Strategy The graph will be a line with a slope of 1.20 m/s^2.

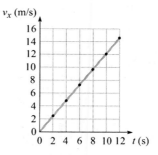

   Solution $v_x = 0$ when $t = 0$. The graph is shown.

(b) Strategy Use Eq. (2-15).

   Solution Find the distance the train traveled.
$$\Delta x = v_{ix}\Delta t + \frac{1}{2}a_x(\Delta t)^2 = (0)\Delta t + \frac{1}{2}a_x(\Delta t)^2 = \frac{1}{2}a_x(\Delta t)^2$$
$$= \frac{1}{2}(1.20 \text{ m/s}^2)(12.0 \text{ s})^2 = \boxed{86.4 \text{ m}}$$

(c) Strategy Use Eq. (2-12).

   Solution Find the final speed of the train.
$$v_{fx} - v_{ix} = v_{fx} - 0 = a_x\Delta t, \text{ so } v_{fx} = a_x\Delta t = (1.20 \text{ m/s}^2)(12.0 \text{ s}) = \boxed{14.4 \text{ m/s}}.$$

(d) Strategy Refer to Figure 2.16, which shows a motion diagram.

   Solution The motion diagram is shown.

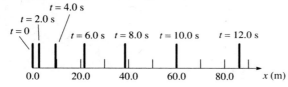

37. Strategy Use Eq. (2-12).

   Solution

(a) Since the acceleration is constant, we have $\Delta v_x = a_x\Delta t = (2.0 \text{ m/s}^2)(12.0 \text{ s} - 10.0 \text{ s}) = \boxed{4.0 \text{ m/s}}$.

(b) The speed when the stopwatch reads 12.0 s is the sum of the speed at 10.0 s plus the change in speed, so
$$v(t = 10.0 \text{ s}) + \Delta v_x = 1.0 \text{ m/s} + 4.0 \text{ m/s} = \boxed{5.0 \text{ m/s}}.$$

41. Strategy Refer to the figure. Analyze graphically and algebraically.

   Solution Graphical analysis:
The displacement of the object is given by the area under the v_x vs. t curve between $t = 9.0$ s and $t = 13.0$ s. The area is a triangle, $A = \frac{1}{2}bh$.

$$\Delta x = \frac{1}{2}(13.0 \text{ s} - 9.0 \text{ s})(40 \text{ m/s}) = 80 \text{ m}$$

Algebraic solution:
Use the definition of average velocity.
$$\Delta x = v_{av,x}\Delta t = \frac{v_{ix} + v_{fx}}{2}\Delta t = \frac{40 \text{ m/s} + 0}{2}(13.0 \text{ s} - 9.0 \text{ s}) = 80 \text{ m}$$

The object goes $\boxed{80 \text{ m}}$.

45. Strategy Use Eq. (2-16).

Solution Find the final speed of the penny.

$v_{fy}^2 - v_{iy}^2 = v_{fy}^2 - 0 = -2g\Delta y$, so $v_{fy} = \sqrt{-2g\Delta y} = \sqrt{-2(9.80 \text{ m/s}^2)(0 \text{ m} - 369 \text{ m})} = 85.0 \text{ m/s}$.

Therefore, $\bar{v} = \boxed{85.0 \text{ m/s down}}$.

49. Strategy Use Eq. (2-15) to find the time it takes for the coin to reach the water. Then, find the time it takes the sound to reach Glenda's ear. Add these two times. Let $h = 7.00$ m.

Solution Find the time elapsed between the release of the coin and the hearing of the splash.

$h = v_{iy}\Delta t + \frac{1}{2}a_y(\Delta t)^2 = 0 + \frac{1}{2}g(\Delta t_1)^2$, so $\Delta t_1 = \sqrt{\dfrac{2h}{g}}$. $h = v_s \Delta t_2$, so $\Delta t_2 = \dfrac{h}{v_s}$.

Therefore, the time elapsed is $\Delta t = \Delta t_1 + \Delta t_2 = \sqrt{\dfrac{2h}{g}} + \dfrac{h}{v_s} = \sqrt{\dfrac{2(7.00 \text{ m})}{9.80 \text{ m/s}^2}} + \dfrac{7.00 \text{ m}}{343 \text{ m/s}} = \boxed{1.22 \text{ s}}$.

51. Strategy Use Eqs. (2-15), (2-16), and (2-12). Let the $+y$-direction be down.

Solution

(a) Ignoring air resistance, the lead ball falls

$\Delta y = \frac{1}{2}g(\Delta t)^2 = \frac{1}{2}(9.80 \text{ m/s}^2)(3.0 \text{ s})^2 = \boxed{44 \text{ m}}$.

(b) The lead ball is initially at rest. Find the speed of the ball after it has fallen 2.5 m.

$v_{fy}^2 = 2g\Delta y$, so $v_{fy} = \sqrt{2g\Delta y} = \sqrt{2(9.80 \text{ m/s}^2)(2.5 \text{ m})} = \boxed{7.0 \text{ m/s}}$.

(c) After 3.0 s, the lead ball is falling at a speed of $v_y = g\Delta t = (9.80 \text{ m/s}^2)(3.0 \text{ s}) = \boxed{29 \text{ m/s}}$.

(d) Find the change in height of the ball when $\Delta t = 2.42$ s.

$\Delta y = v_{iy}\Delta t - \frac{1}{2}g(\Delta t)^2 = (4.80 \text{ m/s})(2.42 \text{ s}) - \frac{1}{2}(9.80 \text{ m/s}^2)(2.42 \text{ s})^2 = -17.1 \text{ m}$

The ball will be $\boxed{17.1 \text{ m below the top of the tower}}$.

53. **(a) Strategy** Use Eq. (2-15) and the quadratic formula to find the time it takes the rock to reach Lois. Then, use Eq. (2-15) again to find Superman's required constant acceleration.

Solution Solve for Δt using the quadratic formula.

$$\Delta y = v_{iy}\Delta t + \frac{1}{2}a_y(\Delta t)^2, \text{ so } \frac{1}{2}a_y(\Delta t)^2 + v_{iy}\Delta t - \Delta y = 0.$$

$$\Delta t = \frac{-v_{iy} \pm \sqrt{v_{iy}^2 + 2a_y\Delta y}}{a_y} = \frac{-(-2.8 \text{ m/s}) \pm \sqrt{(-2.8 \text{ m/s})^2 + 2(-9.80 \text{ m/s}^2)(-14.0 \text{ m})}}{-9.80 \text{ m/s}^2} = -2 \text{ s or } 1.4286 \text{ s}$$

Since $\Delta t > 0$, it takes about 1.43 s for the rock to reach Lois. Find Superman's required acceleration.

$$\Delta x = v_{ix}\Delta t + \frac{1}{2}a_x(\Delta t)^2 = 0 + \frac{1}{2}a_x(\Delta t)^2 = \frac{1}{2}a_x(\Delta t)^2, \text{ so } a_x = \frac{2\Delta x}{(\Delta t)^2} = \frac{2(120 \text{ m})}{(1.43 \text{ s})^2} = 120 \text{ m/s}^2.$$

Superman must accelerate at $\boxed{120 \text{ m/s}^2 \text{ toward Lois}}$ to save her.

(b) Strategy Use Eq. (2-12).

Solution Find Superman's speed when he reaches Lois.
$$v_x = a_x\Delta t = (120 \text{ m/s}^2)(1.43 \text{ s}) = \boxed{170 \text{ m/s}}$$

57. **(a) Strategy** Use Eqs. (2-3) and (2-13) since the acceleration is constant.

Solution Find the distance traveled.
$$\Delta x = v_{av,\,x}\Delta t = \frac{v_{fx} + v_{ix}}{2}\Delta t = \frac{27.3 \text{ m/s} + 17.4 \text{ m/s}}{2}(10.0 \text{ s}) = \boxed{224 \text{ m}}.$$

(b) Strategy Use the definition of average acceleration.

Solution Find the magnitude of the acceleration.
$$a = \frac{\Delta v}{\Delta t} = \frac{27.3 \text{ m/s} - 17.4 \text{ m/s}}{10.0 \text{ s}} = \boxed{0.99 \text{ m/s}^2}$$

61. **Strategy** Each car has traveled the same distance Δx in the same time Δt when they meet.

Solution Using Eq. (2-15), we have
$$\Delta x = v_i\Delta t + \frac{1}{2}a(\Delta t)^2 = 0 + \frac{1}{2}a(\Delta t)^2 = v\Delta t, \text{ so } \Delta t = \frac{2v}{a}. \text{ The speed of the police car is } v_p = a\Delta t = a(2v/a) = \boxed{2v}.$$

65. **Strategy** Use the definitions of displacement, average velocity, and average acceleration.

Solution
$$\Delta\vec{r} = \vec{r}_f - \vec{r}_i = 185 \text{ mi north} - 126 \text{ mi north} = \boxed{59 \text{ mi north}}$$

$$\vec{v}_{av} = \frac{\Delta\vec{r}}{\Delta t} = \frac{59 \text{ mi north}}{37 \text{ min}}\left(\frac{60 \text{ min}}{1 \text{ h}}\right) = \boxed{96 \text{ mi/h north}}$$

$$\vec{a}_{av} = \frac{\Delta\vec{v}}{\Delta t} = \frac{105.0 \text{ mi/h north} - 112.0 \text{ mi/h north}}{37 \text{ min}} = \frac{-7.0 \text{ mi/h north}}{37 \text{ min}}\left(\frac{60 \text{ min}}{1 \text{ h}}\right) = \boxed{11 \text{ mi/h}^2 \text{ south}}$$

69. (a) Strategy Use Eq. (2-15).

    Solution Find the rocket's altitude when the engine fails.

$$\Delta y = \frac{1}{2} a (\Delta t_1)^2 = \frac{1}{2} (20.0 \ \text{m/s}^2)(50.0 \ \text{s})^2 = \boxed{25.0 \ \text{km}}$$

  (b) Strategy v_{iy} = the speed when the engine fails = $a\Delta t_1$; $v_y = v_{iy} - g\Delta t = 0$ at maximum height.

    Solution Find the time elapsed from the engine failure to maximum height.

$$0 = v_{iy} - g\Delta t = a\Delta t_1 - g\Delta t, \text{ so } \Delta t = \frac{a}{g}\Delta t_1 = \frac{20.0 \ \text{m/s}^2}{9.80 \ \text{m/s}^2}(50.0 \ \text{s}) = 102 \ \text{s}.$$

    The time to maximum height from lift off is $\Delta t + \Delta t_1 = 102 \ \text{s} + 50.0 \ \text{s} = \boxed{152 \ \text{s}}$.

  (c) Strategy Use Eq. (2-15).

    Solution Find the maximum height reached by the rocket.

$$y_f = y_i + v_{iy}\Delta t - \frac{1}{2}g(\Delta t)^2 = y_i + (a\Delta t_1)\left(\frac{a}{g}\Delta t_1\right) - \frac{1}{2}g\left(\frac{a}{g}\Delta t_1\right)^2 = y_i + \frac{a^2(\Delta t_1)^2}{g} - \frac{a^2(\Delta t_1)^2}{2g} = y_i + \frac{a^2(\Delta t_1)^2}{2g}$$

$$= 25.0 \ \text{km} + \frac{(20.0 \ \text{m/s}^2)^2(50.0 \ \text{s})^2}{2(9.80 \ \text{N/kg})} = \boxed{76.0 \ \text{km}}$$

  (d) Strategy Use Eq. (2-16). $v_{iy} = 0$ at the maximum height.

    Solution Find the final velocity.

$$v_{fy}^2 - v_{iy}^2 = v_{fy}^2 - 0 = 2a_y\Delta y = -2g\Delta y, \text{ so } v_{fy} = \sqrt{-2g\Delta y} = \sqrt{-2(9.80 \ \text{N/kg})(0 - 76.0\times10^3 \ \text{m})} = 1220 \ \text{m/s}.$$

    Thus, $\vec{\mathbf{v}} = \boxed{1220 \ \text{m/s downward}}$.

73. (a) Strategy Use the definition of average speed.

    Solution

$$v_{av} = \frac{\Delta x}{\Delta t} = \frac{100\times10^{-9} \ \text{m}}{0.10\times10^{-3} \ \text{s}} = \boxed{1.0 \ \text{mm/s}}$$

  (b) Strategy Find the time it takes the pain signal to travel the length of a 1.0-m long neuron. Then, add the times of travel across synapses and neurons.

    Solution

$$t_n = \frac{x}{v} = \frac{1.0 \ \text{m}}{100 \ \text{m/s}} = 10 \ \text{ms}$$

    Find the total time to reach the brain.

$$t_n + t_{syn} + t_n + t_{syn} = 2t_n + 2t_{syn} = 2(t_n + t_{syn}) = 2(10 \ \text{ms} + 0.10 \ \text{ms}) = \boxed{20 \ \text{ms}}$$

  (c) Strategy Use the definition of average speed.

    Solution

$$v_{av} = \frac{\Delta x}{\Delta t} = \frac{2.0 \ \text{m} + 2(100\times10^{-9} \ \text{m})}{20\times10^{-3} \ \text{s}} = \boxed{100 \ \text{m/s}}$$

Chapter 3

MOTION IN A PLANE

Conceptual Questions

1. No; to be equal they must also have the same direction. If the magnitudes are different, they cannot be equal.

5. The trajectory is sometimes parabolic in another reference frame that moves with a constant velocity with respect to the first. The only possibility other than parabolic is a straight-line trajectory.

9. The average speed and the magnitude of the average velocity of an object are equal if and only if the object travels along a straight line path without changing direction. In all other cases, the average speed is greater than the magnitude of the average velocity because the total distance traveled must be greater than the straight-line distance between the starting and ending points.

13. (a) A vector of magnitude $1L$ may be obtained by adding two vectors with lengths $3L$ and $4L$ aligned in opposite directions.

    (b) A vector of magnitude $7L$ may be obtained by adding two vectors with lengths $3L$ and $4L$ aligned in the same direction.

    (c) A vector of magnitude $5L$ may be obtained as the hypotenuse of a right triangle with sides composed of vectors with magnitudes $3L$ and $4L$.

17. The demonstration works when the hunter is aiming either up or down at the monkey and coconut. In the absence of gravity, either case will result in the arrival of a bullet at the position occupied by the coconut. Gravity alters the vertical motion of the coconut and the bullet identically—in a given time interval, both objects will fall an equal distance from the trajectory they would have followed in the absence of gravity. Thus, either case concludes with the result that the bullet and coconut arrive at the same position.

Problems

1. **Strategy** Let east be the $+x$-direction.

   Solution Draw vector diagrams; then find the magnitudes and directions of the vectors.

   (a)

   $(2.56 \text{ km} + 7.44 \text{ km}) \text{ west} = \boxed{10.00 \text{ km west}}$

   (b)

   $(2.56 \text{ km} - 7.44 \text{ km}) \text{ west} = -4.88 \text{ km west} = \boxed{4.88 \text{ km east}}$

   (c)

   $(7.44 \text{ km} - 2.56 \text{ km}) \text{ west} = \boxed{4.88 \text{ km west}}$

5. **Strategy** Sketch the displacement vectors using graph paper, ruler, and protractor. Then find the vector sum by sketching a graphical addition of the displacement vectors.

   Solution The combined sketches are shown.

   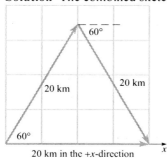

9. **Strategy** Draw the displacement vectors. Use the diagram to answer the questions.

   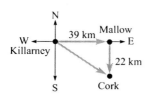

   Solution

   (a) The magnitude of the displacement from Killarney to Cork is the hypotenuse of a right triangle with legs 22 km and 39 km.
   $$\Delta r = \sqrt{(22\text{ km})^2 + (39\text{ km})^2} = \boxed{45\text{ km}}$$

   (b) The distance along Michaela's chosen route is $39\text{ km} + 22\text{ km} = 61\text{ km}$, so the additional distance traveled is $61\text{ km} - 45\text{ km} = \boxed{16\text{ km}}$.

13. **Strategy** The vector makes an angle of 60.0° counterclockwise from the *y*-axis. So, the angle from the positive *x*-axis is 90.0° + 60.0° = 150.0°.

   Solution Find the components of the vector.
   x-comp = $(20.0\text{ m})\cos(150.0°) = \boxed{-17.3\text{ m}}$ and *y*-comp = $(20.0\text{ m})\sin(150.0°) = \boxed{10.0\text{ m}}$.

   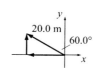

15. (a) **Strategy** Since each vector is directed along a different axis, each component of the vector sum is equal to the vector that lies along that component's axis. Use the Pythagorean theorem.

   Solution Find the magnitude of $\vec{\mathbf{A}} + \vec{\mathbf{B}}$.
   $$\left|\vec{\mathbf{A}} + \vec{\mathbf{B}}\right| = \sqrt{[(A+B)_x]^2 + [(A+B)_y]^2} = \sqrt{(-1.0)^2 + (\sqrt{3.0})^2} = 2.0\text{ units}$$
   Find the direction.
   $$\theta = \tan^{-1}\frac{\sqrt{3.0}}{-1.0} = 30°\text{ CCW from the }+y\text{-axis, so}$$
   $$\vec{\mathbf{A}} + \vec{\mathbf{B}} = \boxed{2.0\text{ units at }30°\text{ CCW from the }+y\text{-axis}}.$$

(b) Strategy Subtract a vector by adding its opposite. Use the Pythagorean theorem.

Solution Find the magnitude.

$$\left|\vec{A}-\vec{B}\right| = \sqrt{[(A-B)_x]^2 + [(A-B)_y]^2} = \sqrt{1.0^2 + \left(\sqrt{3.0}\right)^2} = 2.0 \text{ units}$$

Find the direction.

$$\theta = \tan^{-1}\frac{\sqrt{3.0}}{1.0} = 60° \text{ CW from the } -x\text{-axis, so } \vec{A}-\vec{B} = \boxed{2.0 \text{ units at } 30° \text{ CW from the } +y\text{-axis}}.$$

(c) Strategy The vectors lie on the axes.

Solution Find the components of $\vec{B}-\vec{A}$.

$$x\text{-comp} = B_x = \boxed{-1.0 \text{ unit}} \text{ and } y\text{-comp} = -A_y = \boxed{-\sqrt{3.0} \text{ units}}.$$

17. Strategy Use the fact that $|\vec{A}| = |\vec{B}|$, symmetry, and the component method to find the magnitude of $\vec{C}$.

Solution By symmetry, $\vec{C}$ points downward, since the horizontal components cancel when $\vec{A}$ and $\vec{B}$ are added. The downward components of each vector have the same magnitude, $A_y = B_y = (4.0 \text{ cm})\sin 10° = 0.69 \text{ cm}$. So, the magnitude of $\vec{C}$ is $\boxed{1.4 \text{ cm}}$.

21. Strategy Use the Pythagorean theorem to find the magnitude of each vector. Give the angle with respect to the axis to which it lies closest.

Solution Find the magnitude and direction of each vector.

(a) $A = \sqrt{(-5.0 \text{ m/s})^2 + (8.0 \text{ m/s})^2} = \boxed{9.4 \text{ m/s}}$ and $\theta = \tan^{-1}\frac{8.0}{-5.0} = \boxed{32° \text{ CCW from the } +y\text{-axis}}$.

(b) $B = \sqrt{(120 \text{ m})^2 + (-60.0 \text{ m})^2} = \boxed{130 \text{ m}}$ and $\theta = \tan^{-1}\frac{-60.0}{120} = \boxed{27° \text{ CW from the } +x\text{-axis}}$.

(c) $C = \sqrt{(-13.7 \text{ m/s})^2 + (-8.8 \text{ m/s})^2} = \boxed{16.3 \text{ m/s}}$ and $\theta = \tan^{-1}\frac{-8.8}{-13.7} = \boxed{33° \text{ CCW from the } -x\text{-axis}}$.

(d) $D = \sqrt{(2.3 \text{ m/s}^2)^2 + (6.5\times10^{-2} \text{ m/s}^2)^2} = \boxed{2.3 \text{ m/s}^2}$ and

$$\theta = \tan^{-1}\frac{0.065}{2.3} = \boxed{1.6° \text{ CCW from the } +x\text{-axis}}.$$

25. Strategy Add the displacement from Jerry's dorm to the fitness center to the displacement from Cindy's apartment to Jerry's dorm to find the total displacement from Cindy's apartment to the fitness center.

Solution Add the displacements.

1.50 mi east + 2.00 mi north + 3.00 mi east = 4.50 mi east + 2.00 mi north
Let north be along the $+y$-axis and east be along the $+x$-axis. Then, the components of the total displacement are $\Delta x = 4.50$ mi and $\Delta y = 2.00$ mi.
Find the magnitude.

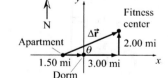

$$\Delta r = \sqrt{(\Delta x)^2 + (\Delta y)^2} = \sqrt{(4.50 \text{ mi})^2 + (2.00 \text{ mi})^2} = \boxed{4.92 \text{ mi}}$$

Find the direction.

$$\theta = \tan^{-1}(2.00/4.50) = \boxed{24.0° \text{ north of east}}$$

27. **Strategy** Draw a diagram and use the component method. Let north be $+y$ and east be $+x$.

    Solution The diagram is shown (with the answer included).

    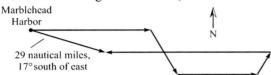

    Find the position of the sailboat.
    $$\Delta x = 45 \text{ n.m.} + (20.0 \text{ n.m.})\cos 300° + 30.0 \text{ n.m.} + (10.0 \text{ n.m.})\cos 60° - 62 \text{ n.m.} = 28 \text{ n.m.}$$
    $$\Delta y = (20.0 \text{ n.m.})\sin 300° + (10.0 \text{ n.m.})\sin 60° = -8.7 \text{ n.m.}$$
    $$\Delta r = \sqrt{(\Delta x)^2 + (\Delta y)^2} = \sqrt{(28 \text{ n.m.})^2 + (-8.7 \text{ n.m.})^2} = 29 \text{ n.m.}$$
    $$\theta = \tan^{-1}\frac{-8.7}{28} = 17° \text{ south of east}$$
    So, $\Delta \vec{\mathbf{r}} = \boxed{29 \text{ nautical miles at } 17° \text{ south of east}}$.

29. **Strategy** Draw diagrams of the situation. Use the definitions of average speed and average velocity.

    Solution

    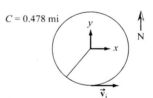

    (a) Find the runner's average speed.
    $$v_{av} = \frac{\Delta x}{\Delta t} = \frac{0.750 \text{ mi}}{4.00 \text{ min}} \times \frac{1609 \text{ m}}{\text{mi}} \times \frac{1 \text{ min}}{60 \text{ s}} = \boxed{5.03 \text{ m/s}}$$

    (b) Find the location of the runner on the track.
    $\frac{0.750}{0.478} = 1.569$, so the runner has gone around once plus 0.569 times.
    Find the angle θ shown in the diagram.
    $$0.569 \times 360° - 180° = 24.84°$$
    Find the radius of the track.
    $$C = 2\pi r, \text{ so } r = \frac{C}{2\pi}.$$
    Find $\Delta \vec{\mathbf{r}}$.
    $$\Delta r = \sqrt{(r_i + r_f \cos\theta)^2 + (r_f \sin\theta)^2} = \sqrt{(r + r\cos\theta)^2 + (r\sin\theta)^2} = r\sqrt{(1 + \cos\theta)^2 + (\sin\theta)^2}$$
    $$= r\sqrt{1 + 2\cos\theta + \cos^2\theta + \sin^2\theta} = r\sqrt{1 + 2\cos\theta + 1} = \frac{C}{2\pi}\sqrt{2(1 + \cos\theta)} = \frac{0.478 \text{ mi}}{2\pi}\sqrt{2(1 + \cos 24.84°)}$$
    $$= 0.1486 \text{ mi}$$
    $$\phi = \tan^{-1}\frac{r\sin\theta}{r + r\cos\theta} = \tan^{-1}\frac{\sin 24.84°}{1 + \cos 24.84°} = 12.4°$$
    Find the runner's average velocity.
    $$\left|\vec{\mathbf{v}}_{av}\right| = \frac{\Delta r}{\Delta t} = \frac{0.1486 \text{ mi}}{4.00 \text{ min}} \times \frac{1609 \text{ m}}{\text{mi}} \times \frac{1 \text{ min}}{60 \text{ s}} = 0.996 \text{ m/s, so } \vec{\mathbf{v}}_{av} = \boxed{0.996 \text{ m/s at } 12.4° \text{ west of north}}.$$

33. Strategy Use the definition of average velocity. Draw a diagram.

Solution Let east be in the $+x$-direction and north be in the $+y$-direction.
Find the magnitude of $\Delta \vec{r}$.

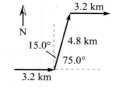

$$\left| \Delta \vec{r} \right| = \sqrt{[3.2 \text{ km} + (4.8 \text{ km})\cos 75.0° + 3.2 \text{ km}]^2 + [(4.8 \text{ km})\sin 75.0°]^2} = 8.9 \text{ km}$$

Find the direction of $\Delta \vec{r}$.

$$\theta = \tan^{-1} \frac{4.6 \text{ km}}{7.6 \text{ km}} = 31° \text{ north of east}$$

So, $\left| \vec{v}_{av} \right| = \dfrac{\Delta r}{\Delta t} = \dfrac{8.94 \text{ km}}{0.10 \text{ h} + 0.15 \text{ h} + 0.10 \text{ h}} = 26 \text{ km/h}$ and $\vec{v}_{av} = \dfrac{\Delta \vec{r}}{\Delta t} = \boxed{26 \text{ km/h at } 31° \text{ north of east}}$.

37. Strategy Draw diagrams. Use the definitions of average speed and average velocity.

Solution

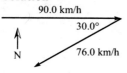

(a) Find the displacement.

$$\begin{aligned} \Delta \vec{r} &= \vec{r}_f - \vec{r}_i \\ &= (90.0 \text{ km/h})(80.0 \text{ min})[1 \text{ h}/(60 \text{ min})] \text{ east} - (76.0 \text{ km/h})(45.0 \text{ min})[1 \text{ h}/(60 \text{ min})] \text{ south of west} \\ &= 120 \text{ km east} - 57.0 \text{ km at } 30.0° \text{ south of west} \end{aligned}$$

Find the distance.

$$\Delta r = \sqrt{[120 \text{ km} - (57.0 \text{ km})\cos 30.0°]^2 + [-(57.0 \text{ km})\sin 30.0°]^2} = \boxed{76.2 \text{ km}}$$

(b) Find the magnitude of the average velocity.

$$v_{av} = \frac{\Delta r}{\Delta t} = \frac{76.2 \text{ km}}{45.0 \text{ min}} \times \frac{60 \text{ min}}{1 \text{ h}} = 102 \text{ km/h}$$

Find the direction.

$$\theta = \tan^{-1} \frac{(57.0 \text{ km})\sin 30.0°}{-120 \text{ km} + (57.0 \text{ km})\cos 30.0°} = 22.0° \text{ north of west}$$

The average velocity on the third leg is $\boxed{102 \text{ km/h at } 22.0° \text{ north of west}}$.

(c) Compute the time.
$$80.0 \text{ min} + 15.0 \text{ min} + 45.0 \text{ min} = 140.0 \text{ min}$$
Find the magnitude of the average velocity.

$$v_{av} = \frac{\Delta r}{\Delta t} = \frac{76.17 \text{ km}}{140.0 \text{ min}} \times \frac{60 \text{ min}}{1 \text{ h}} = 32.6 \text{ km/h}$$

The direction is opposite that found in part (b), so the average velocity during the first two legs is
$\boxed{32.6 \text{ km/h at } 22.0° \text{ south of east}}$.

(d) Since the displacement is zero, the average velocity over the entire trip is $\boxed{0}$.

(e) Compute the total time.

80.0 min + 15.0 min + 45.0 min + 45.0 min + 55.0 min = 240.0 min, or 4.000 h

Compute the total distance.

120 km + 57.0 km + 76.2 km = 253.2 km

The average speed during the entire trip is $\dfrac{253.2 \text{ km}}{4.000 \text{ h}} = \boxed{63.3 \text{ km/h}}$.

41. Strategy The magnitude of the velocity is constant, but the direction changes. Recall that the circumference of a circle is given by $2\pi r$.

Solution

(a) Find the car's speed.

$$v = \frac{C/4}{\Delta t} = \frac{C}{4\Delta t} = \frac{2\pi r}{4\Delta t} = \frac{\pi r}{2\Delta t} = \frac{\pi(10.0 \text{ m})}{2(1.60 \text{ s})} = \boxed{9.82 \text{ m/s}}$$

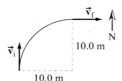

(b) Let east be the $+x$-direction and north the $+y$-direction.

$$\Delta\vec{v} = \vec{v}_f - \vec{v}_i = v \text{ east} - v \text{ north}$$

$$\left| \Delta\vec{v} \right| = \sqrt{v^2 + (-v)^2} = \sqrt{2v^2} = v\sqrt{2}$$

$$\theta = \tan^{-1}\frac{-v}{v} = \tan^{-1}(-1) = 45° \text{ south of east (SE)}$$

$$\Delta\vec{v} = \sqrt{2}(9.82 \text{ m/s}) \text{ southeast} = \boxed{13.9 \text{ m/s southeast}}$$

(c) $\vec{a}_{av} = \dfrac{\Delta\vec{v}}{\Delta t} = \dfrac{13.88 \text{ m/s southeast}}{1.60 \text{ s}} = \boxed{8.68 \text{ m/s}^2 \text{ southeast}}$

45. Strategy Use the component method. Solve for the time. Let north be in the $+y$-direction and east be in the $+x$-direction.

Solution

$v_x = 60$ m/s and $v_y = a_y \Delta t$.

Use the Pythagorean theorem.

$$v^2 = v_x{}^2 + v_y{}^2 = v_x{}^2 + (a_y\Delta t)^2, \text{ so } \Delta t = \frac{\sqrt{v^2 - v_x{}^2}}{a_y} = \frac{\sqrt{(100 \text{ m/s})^2 - (60 \text{ m/s})^2}}{100 \text{ m/s}^2} = \boxed{0.8 \text{ s}}.$$

47. Strategy Use equations of motion with constant acceleration to determine the vertical and horizontal positions of the clay after 1.50 s have elapsed.

Solution Find the position of the clay.

$$x_f = v_{ix}\Delta t = (20.0 \text{ m/s})(1.50 \text{ s}) = 30.0 \text{ m}$$

$$y_f = y_i + v_{iy}\Delta t - \frac{1}{2}g(\Delta t)^2 = 8.50 \text{ m} + 0 - \frac{1}{2}(9.80 \text{ m/s}^2)(1.50 \text{ s})^2 = -2.53 \text{ m}$$

The clay cannot pass through the ground, so it hit and stuck prior to 1.50 s. Find the time it took for the clay to land.

$$y_f = 0 = y_i + v_{iy}\Delta t - \frac{1}{2}g(\Delta t)^2 = y_i - \frac{1}{2}g(\Delta t)^2, \text{ so}$$

$$\Delta t = \sqrt{\frac{2y_i}{g}} = \sqrt{\frac{2(8.50 \text{ m})}{9.80 \text{ m/s}^2}} = 1.317 \text{ s, and } x_f = v_{ix}\Delta t = (20.0 \text{ m/s})(1.317 \text{ s}) = 26.3 \text{ m}.$$

The clay hits and $\boxed{\text{it is on the ground after 1.32 s, so the horizontal distance along the ground is 26.3 m}}$.

49. Strategy Use Eqs. (3-13) and (3-14). Set $v_{fy} = 0$, since the vertical component of the velocity is zero at the maximum height.

Solution

(a) Find the maximum height.

$$v_{fy}^2 - v_{iy}^2 = 0 - v_{iy}^2 = -2g\Delta y, \text{ so } \Delta y = \frac{v_{iy}^2}{2g} = y_f - y_i \text{ and}$$

$$y_f = \frac{v_i^2 \sin^2 \theta}{2g} + y_i = \frac{(19.6 \text{ m/s})^2 \sin^2 30.0°}{2(9.80 \text{ m/s}^2)} + 1.0 \text{ m} = \boxed{5.9 \text{ m}}.$$

(b) At the ball's highest point, $v_{fy} = 0$, so the speed v equals v_x.

$$v = v_x = v_{ix} = v_i \cos \theta = (19.6 \text{ m/s}) \cos 30.0° = \boxed{17.0 \text{ m/s}}$$

53. (a) Strategy At the maximum height of the cannonball's trajectory, $v_{fy} = 0$. Use Eq. (3-13).

Solution Find the maximum height reached by the cannonball.

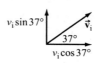

$$v_{fy}^2 - v_{iy}^2 = 0 - (v_i \sin \theta)^2 = 2a_y \Delta y = -2g(y_f - y_i), \text{ so}$$

$$y_f = y_i + \frac{v_i^2 \sin^2 \theta}{2g} = 7.0 \text{ m} + \frac{(40 \text{ m/s})^2 \sin^2 37°}{2(9.80 \text{ m/s}^2)} = \boxed{37 \text{ m}}.$$

(b) Strategy Solve $\Delta x = v_x \Delta t$ for the time and substitute the result into Eq. (3-12). Then, solve for Δx to find the horizontal distance from the release point.

Solution When the cannonball hits the ground, $\Delta y = -7.0$ m.

$$\Delta x = v_x \Delta t = (v_i \cos \theta) \Delta t, \text{ so } \Delta t = \frac{\Delta x}{v_i \cos \theta}. \text{ Substitute.}$$

$$\Delta y = y_f - y_i = v_{iy} \Delta t + \frac{1}{2} a_y (\Delta t)^2 = (v_i \sin \theta) \frac{\Delta x}{v_i \cos \theta} - \frac{1}{2} g \frac{(\Delta x)^2}{v_i^2 \cos^2 \theta} = \Delta x \tan \theta - \frac{g(\Delta x)^2}{2v_i^2 \cos^2 \theta}, \text{ so}$$

$$0 = \frac{g}{2v_i^2 \cos^2 \theta} (\Delta x)^2 - (\tan \theta) \Delta x + \Delta y. \text{ Use the quadratic formula.}$$

$$\Delta x = \frac{\tan \theta \pm \sqrt{\tan^2 \theta - \frac{4g\Delta y}{2v_i^2 \cos^2 \theta}}}{\frac{2g}{2v_i^2 \cos^2 \theta}} = \frac{\tan 37° \pm \sqrt{\tan^2 37° - \frac{2(9.80 \text{ m/s}^2)(-7.0 \text{ m})}{(40 \text{ m/s})^2 \cos^2 37°}}}{\frac{9.80 \text{ m/s}^2}{(40 \text{ m/s})^2 \cos^2 37°}} = 170 \text{ m or } -9 \text{ m}$$

Since the catapult doesn't fire backward, -9 m is extraneous. So, the cannonball lands $\boxed{170 \text{ m}}$ from its release point.

(c) **Strategy** The x-component is the same as the initial value. Find the y-component using Eq. (3-13).

Solution The x-component of the velocity is $v_{fx} = v_{ix} = v_i \cos\theta = (40 \text{ m/s})\cos 37° = \boxed{32 \text{ m/s}}$.
Find the y-component of the velocity.

$$v_{fy}^2 - v_{iy}^2 = v_{fy}^2 - v_i^2 \sin^2\theta = 2a_y \Delta y = -2g\Delta y, \text{ so}$$

$$v_{fy} = \pm\sqrt{v_i^2 \sin^2\theta - 2g\Delta y} = \pm\sqrt{(40 \text{ m/s})^2 \sin^2 37° - 2(9.80 \text{ m/s}^2)(-7.0 \text{ m})} = \boxed{-27 \text{ m/s}},$$

where the negative sign was chosen because the cannonball is on its way down.

57. (a) **Strategy** Solve for the time using Eq. (3-12).

Solution

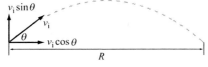

$$\Delta y = 0 = v_{iy}\Delta t + \frac{1}{2}a_y(\Delta t)^2 = v_{iy}\Delta t - \frac{1}{2}g(\Delta t)^2 = v_{iy} - \frac{1}{2}g\Delta t, \text{ so } \Delta t = \frac{2v_{iy}}{g} = \boxed{\frac{2v_i \sin\theta}{g}}.$$

(b) **Strategy** Use $\Delta x = v_x \Delta t = v_{ix}\Delta t$ and $\Delta y = v_{iy}\Delta t - \frac{1}{2}g(\Delta t)^2$ to find the range. Use the trigonometric identity $\sin 2\theta = 2\sin\theta\cos\theta$.

Solution Solve for the time.

$$\Delta x = v_{ix}\Delta t, \text{ so } \Delta t = \frac{\Delta x}{v_{ix}}.$$

$$\Delta x = R, \text{ so } \Delta t = \frac{R}{v_{ix}}. \text{ Find } \Delta t \text{ in terms of } v_{iy}.$$

$$\Delta y = 0 = v_{iy}\Delta t - \frac{1}{2}g(\Delta t)^2 = v_{iy} - \frac{1}{2}g\Delta t, \text{ so } \Delta t = \frac{2v_{iy}}{g} = \frac{R}{v_{ix}}. \text{ Therefore, the range is}$$

$$R = \frac{2v_{iy}v_{ix}}{g} = \frac{2v_i^2 \sin\theta\cos\theta}{g} = \frac{v_i^2 \sin 2\theta}{g}.$$

(c) **Strategy and Solution** The maximum value of $\sin 2\theta$ occurs when $2\theta = 90°$ or $\theta = \boxed{45°}$.

Therefore, $R_{max} = \dfrac{v_i^2 \sin 90°}{g} = \boxed{\dfrac{v_i^2}{g}}$.

61. Strategy The skater must be up the ramp far enough for their speed at the end of the horizontal section to be just great enough so that the skater travels a horizontal distance of 7.00 m while falling 3.00 m. Draw a diagram of the skater on the ramp to find the acceleration of the skater caused by the force of gravity.

Solution According to Newton's second law, $\sum F = mg \sin 15.0° = ma$, so $a = g \sin 15.0°$ along the surface of the ramp. Use Eq. (2-16) to relate the distance up the ramp to the speed of the skater at the end of the ramp.

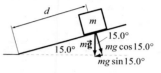

$$v_f^2 - v_i^2 = v_f^2 - 0 = 2a\Delta x = 2(g \sin 15.0°)d, \text{ so } d = v_f^2 / (2g \sin 15.0°).$$

Since the ramp is frictionless, the velocity of the skater at the end of the horizontal part of the ramp is in the x-direction with magnitude equal to v_f. So, the components of the displacement are $\Delta x = v_f \Delta t$ (1) and

$$\Delta y = v_{iy}\Delta t + \frac{1}{2}a_y(\Delta t)^2 = 0 - \frac{1}{2}g(\Delta t)^2 = -\frac{1}{2}g(\Delta t)^2 \quad (2). \text{ Solving for } \Delta t \text{ in (1) and substituting into (2) gives}$$

$$\Delta y = -\frac{1}{2}g\left(\frac{\Delta x}{v_f}\right)^2, \text{ or } v_f^2 = -\frac{g(\Delta x)^2}{2\Delta y}.$$

Substitute this result into the equation for d.

$$d = \frac{v_f^2}{2g \sin 15.0°} = \frac{-g(\Delta x)^2/(2\Delta y)}{2g \sin 15.0°} = -\frac{(\Delta x)^2}{4\Delta y \sin 15.0°} = -\frac{(7.00 \text{ m})^2}{4(-3.00 \text{ m})\sin 15.0°} = \boxed{15.8 \text{ m}}$$

65. Strategy Consider the relative motion of the two vehicles.

Solution Let north be in the $+x$-direction.
v_{JRx} = the velocity of the Jeep relative to the road = 82 km/h
v_{RFx} = the velocity of the road relative to the Ford = $-v_{FRx}$ = 48 km/h
v_{JFx} = the velocity of the Jeep relative to the (observer in the) Ford = $v_{JRx} + v_{RFx}$ = 82 km/h + 48 km/h
  = 130 km/h

So, $\vec{v}_{JFx} = \boxed{130 \text{ km/h north}}$.

69. Strategy The minimum air velocity is in the same direction as the airplane's.

Solution
210 m/s east − 160 m/s east = $\boxed{50 \text{ m/s east}}$

73. Strategy Consider the relative motion of the two vehicles. Use the component method.

Solution Let the $+y$-direction be north and the $+x$-direction be east.

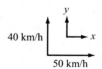

$\vec{v}_{ps}$ = the velocity of the Pierce Arrow relative to the Stanley Steamer
$\vec{v}_{pg}$ = the velocity of the Pierce Arrow relative to the ground
$\vec{v}_{sg}$ = the velocity of the Stanley Steamer relative to the ground

Compute the components of the velocity of the Pierce Arrow relative to the observer riding in the Stanley Steamer.

$v_{psx} = v_{pgx} + v_{gsx} = 50$ km/h + 0, so $\boxed{v_x = 50 \text{ km/h east}}$.

$v_{psy} = v_{pgy} + v_{gsy} = v_{pgy} - v_{sgy} = 0 - 40$ km/h and -40 km/h north = 40 km/h south, so

$\boxed{v_y = 40 \text{ km/h south}}$.

77. **(a) Strategy** Consider the relative motion of the boy and the water.

Solution

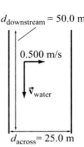

$$\Delta t = \frac{d_{across}}{v_{boy}} \text{ and } v_{water} = \frac{d_{downstream}}{\Delta t}, \text{ so}$$

$$v_{water} = \frac{d_{downstream}}{d_{across}/v_{boy}} = \frac{d_{downstream}}{d_{across}} v_{boy} = \frac{50.0 \text{ m}}{25.0 \text{ m}} (0.500 \text{ m/s}) = \boxed{1.00 \text{ m/s}}$$

(b) Strategy Use the Pythagorean theorem.

Solution Find the speed of the boy relative to the friend.

$$v_{bf} = \sqrt{(0.500 \text{ m/s})^2 + (1.00 \text{ m/s})^2} = \boxed{1.12 \text{ m/s}}$$

81. **(a) Strategy** Draw a diagram and use vector addition.

Solution Find the magnitude of the displacement.

$$|\Delta \vec{r}| = \sqrt{[600.0 \text{ km} + (300.0 \text{ km})\cos(-30.0°)]^2 + [(300.0 \text{ km})\sin(-30.0°)]^2}$$

$$= \boxed{873 \text{ km}}$$

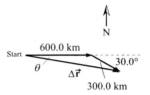

(b) Strategy Refer to the diagram in part (a). Find the angle between the initial displacement vector and $\Delta \vec{r}$.

Solution Find the direction of the displacement.

$$\theta = \tan^{-1} \frac{(300.0 \text{ km})\sin(-30.0°)}{600.0 \text{ km} + (300.0 \text{ km})\cos(-30.0°)} = \boxed{9.90° \text{ south of east}}$$

(c) Strategy The flight time is given by the quotient of the distance traveled and the speed of the jetliner.

Solution

$$\Delta t = \frac{d}{v} = \frac{600.0 \text{ km} + 300.0 \text{ km}}{400.0 \text{ km/h}} = \boxed{2.250 \text{ h}}$$

(d) Strategy The direct flight time is given by the quotient of the magnitude of the displacement and the speed of the jetliner.

Solution

$$\Delta t = \frac{|\Delta \vec{r}|}{v} = \frac{873 \text{ km}}{400.0 \text{ km/h}} = \boxed{2.18 \text{ h}}$$

85. Strategy The projectile must be displaced 75.0 m vertically in the same amount of time that it travels 350 m horizontally. The projectile may hit the headquarters on its way up, on its way down, or at its maximum height. Use $\Delta x = v_x \Delta t$ and Eq. (3-12).

Solution Solve for the initial speed, v_i.

$$\Delta x = v_x \Delta t, \text{ so } \Delta t = \frac{\Delta x}{v_x} = \frac{\Delta x}{v_i \cos\theta}.$$

$$\Delta y = v_{iy}\Delta t + \frac{1}{2}a_y(\Delta t)^2 = (v_i \sin\theta)\Delta t - \frac{1}{2}g(\Delta t)^2 = (v_i \sin\theta)\frac{\Delta x}{v_i \cos\theta} - \frac{1}{2}g\left(\frac{\Delta x}{v_i \cos\theta}\right)^2 = \Delta x \tan\theta - \frac{g(\Delta x)^2}{2v_i^2 \cos^2\theta}, \text{ so}$$

$$v_i = \sqrt{\frac{g(\Delta x)^2}{2(\Delta x \tan\theta - \Delta y)\cos^2\theta}} = \sqrt{\frac{(9.80 \text{ m/s}^2)(350 \text{ m})^2}{2[(350 \text{ m})\tan 40.0° - 75.0 \text{ m}]\cos^2 40.0°}} = \boxed{68 \text{ m/s}}.$$

89. Strategy Use the results from Problems 57 and 59.

Solution

(a) At the maximum height, $v_{fy} = 0 = v_i \sin\theta - g\Delta t$, so $\Delta t = v_i \sin\theta / g$.

Also, $\Delta x = (v_i \cos\theta)\Delta t$, so $\Delta t = \frac{\Delta x}{v_i \cos\theta}$. ($\Delta x$ is half of 0.800 m.)

Equate the expressions for Δt.

$$\frac{v_i \sin\theta}{g} = \frac{\Delta x}{v_i \cos\theta}, \text{ so } v_i^2 = \frac{g\Delta x}{\sin\theta \cos\theta}. \text{ From Problem 50, } H = \frac{v_i^2 \sin^2\theta}{2g}, \text{ so}$$

$$H = \frac{g\Delta x \sin^2\theta}{2g\sin\theta\cos\theta} = \frac{\Delta x}{2}\tan\theta = \frac{0.400 \text{ m}}{2}\tan 55.0° = \boxed{28.6 \text{ cm}}.$$

(b) Since $H \propto \tan\theta$, and since $\tan\theta$ increases if θ increases $(0 \le \theta \le 90)$, the maximum height would be $\boxed{\text{smaller}}$ $(45.0° < 55.0°)$.

(c) The range would be $\boxed{\text{larger}}$, since the range is maximized for $\theta = 45°$.

(d) Calculate v_i^2.

$$v_i^2 = \frac{(9.80 \text{ m/s}^2)(0.400 \text{ m})}{\sin 55.0° \cos 55.0°} = 8.34 \text{ m}^2/\text{s}^2$$

Calculate the maximum height and range for 45.0°.

$$H = \frac{(8.34 \text{ m}^2/\text{s}^2)\sin^2 45.0°}{2(9.80 \text{ m/s}^2)} = \boxed{21.3 \text{ cm}}$$

From Problem 48, $R = \frac{v_i^2 \sin 2\theta}{g}$, so $R = \frac{(8.34 \text{ m}^2/\text{s}^2)\sin 90.0°}{9.80 \text{ m/s}^2} = \boxed{85.1 \text{ cm}}$.

93. **Strategy** Use Eqs. (3-5), (2-12), and (3-12).

Solution Find the time of flight in terms of h and v_i.

$\Delta x = v_i \Delta t = h$, so $\Delta t = \dfrac{h}{v_i}$.

Find the time of flight in terms of v_i and g.

$h = \dfrac{1}{2} g(\Delta t)^2 = \dfrac{1}{2} g \left(\dfrac{h}{v_i} \right)^2 = \dfrac{gh^2}{2v_i^2}$, so $h = \dfrac{2v_i^2}{g}$ and $\Delta t = \dfrac{h}{v_i} = \dfrac{1}{v_i} \left(\dfrac{2v_i^2}{g} \right) = \dfrac{2v_i}{g}$.

Find the components of $\vec{v}$.

$v_x = v_i$ and $v_y = -g\Delta t = -g \left(\dfrac{2v_i}{g} \right) = -2v_i$, so $\theta = \tan^{-1} \dfrac{v_y}{v_x} = \tan^{-1} \dfrac{-2v_i}{v_i} = \tan^{-1}(-2) = -63°$,

or $\boxed{63° \text{ below the horizontal}}$.

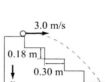

97. **Strategy** Let $+y$ be downward. $v_{ix} = v_x = v_i$ and $v_{iy} = 0$, so $\Delta x = v_i \Delta t$ and $\Delta y = \dfrac{1}{2} g(\Delta t)^2$.

Let the step number $n = \dfrac{x}{0.30 \text{ m}} = \dfrac{y}{0.18 \text{ m}}$, such that $n = 0$ to 1 represents step 1, $n = 1$ to 2 represents step 2, etc.

Solution Find the step the marble strikes first.

$n = \dfrac{v_i \Delta t}{0.30 \text{ m}} = \dfrac{g(\Delta t)^2}{2(0.18 \text{ m})}$, so $\Delta t = \dfrac{2(0.18 \text{ m})v_i}{(0.30 \text{ m})g} = 1.2 \dfrac{v_i}{g}$.

Therefore,

$n = \dfrac{v_i}{0.30 \text{ m}} \left(1.2 \dfrac{v_i}{g} \right) = (4.0 \text{ m}^{-1}) \dfrac{v_i^2}{g} = (4.0 \text{ m}^{-1}) \dfrac{(3.0 \text{ m/s})^2}{9.80 \text{ m/s}^2} = 3.7$.

The value of n is between 3 and 4, so the marble first strikes $\boxed{\text{step 4}}$.

Chapter 4

FORCE AND NEWTON'S LAWS OF MOTION

Conceptual Questions

1. In an automobile accident the force due to the collision changes the motion of the car, but the driver and passengers continue to move in accordance with Newton's first law. Seat belts supply the force necessary to change their motion and slow them down. Without seat belts people would collide with the steering wheel or windshield, for example, and stand a greater risk of injury.

5. When the handle hits the board and stops abruptly, Newton's first law says that the steel head will continue to move for a short distance, resisting changes in velocity, until the force of friction between the head and the handle has brought it to rest. It will have then moved down some to where the handle is a little wider, resulting in a tightening of the head onto the handle.

9. (a) The reading of the scale is the magnitude of the normal force pushing up on you. This equals your weight as long as the normal force and the force of gravity are the only forces acting on you, and you are at rest or moving with a constant velocity.

13. The key is that the equal and opposite forces of Newton's third law are acting on two different objects—one on the wagon and the other on you. The wagon can therefore experience a non-zero net force, which causes it to accelerate forward.

17. Newton's third law tells us that if the person on the raft walks away from the pier, the raft will in turn move toward the pier. Thus, after walking the length of the raft, it should be possible for the person with the hook on the pier to grab the raft and reel it in. Without the person on the pier to hold the raft, this technique would be of no use, as the raft would move back away from the pier on the return walk.

21. The apparent weight of the load is increased by an amount equal to the force required to accelerate it.

25. Yes, as long as the y-axis is perpendicular to the chosen x-axis. This will often simplify a problem.

Problems

1. **Strategy** Determine the forces *not* acting on the scale.

   Solution The scale is in contact with the floor, so a contact force due to the floor is exerted on the scale. The scale is in contact with the person's feet, so a contact force due to the person's feet is exerted on the scale. The scale is in the proximity of a very large mass (Earth), so the weight of the scale is a force exerted on the scale. The weight of the person is a force exerted on the person due to the very large mass, so it is not a force exerted on the scale.

3. **Strategy** There are 0.2248 pounds per newton.

   Solution Find the weight of the astronaut in newtons.
   $$175 \text{ lb} \times \frac{1 \text{ N}}{0.2248 \text{ lb}} = \boxed{778 \text{ N}}$$

Chapter 4: Force and Newton's Laws of Motion   Physics

5. **Strategy** Use graph paper to draw a diagram.

Solution Find the vector sum of the vectors.

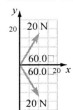

Because of symmetry, the y-components of the vectors cancel. The x-components look to be about 10 N, so the vector sum is $10\text{ N} + 10\text{ N} = 20\text{ N}$, or $\boxed{20\text{ N in the positive }x\text{-direction}}$.

9. **Strategy** Graph the vectors and their sum. Use the scale of the graph to find the magnitude of the vector sum.

Solution Use graph paper, ruler, and protractor to find the magnitude and direction of the vector sum of the two forces. The vector sum points due north. Each side of a grid square represents 10 N, so the magnitude of the net force on the sledge is about $\boxed{120\text{ N north}}$.

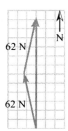

13. **Strategy** Make a scale drawing to determine which net force is greater.

Solution The scale drawing is shown.

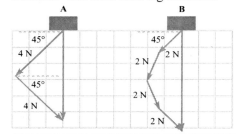

The net force magnitude on object B is greater than that on object A because two of the forces acting on B are directed at an angle greater than $45°$ with respect to the horizontal and contribute more to the downward directed net force.

15. **Strategy** Draw a free-body diagram and add the force to find the net force on the truck.

Solution The vertically directed forces balance, so the net force is due to the difference in the east-west forces.
7 kN east + 5 kN west = 7 kN east − 5 kN east = $\boxed{2\text{ kN east}}$

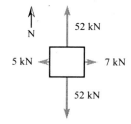

17. **Strategy** Since the hummingbird is hovering motionless, there is no net force on the hummingbird.

    Solution The weight (downward force exerted by the Earth) is equal to the upward force exerted by the air, $\boxed{0.30 \text{ N}}$.

21. **Strategy** Use Newton's second law for the vertical direction.

    Solution Draw a free-body diagram. Find the mass of the potatoes.

    $$\Sigma F_y = T - mg = ma_y, \text{ so } m = \frac{T - mg}{a_y} = \frac{46.8 \text{ N} - 39.2 \text{ N}}{1.90 \text{ m/s}^2} = \boxed{4.0 \text{ kg}}.$$

25. **Strategy** Use Newton's laws of motion.

    Solution The stone is lifted with constant velocity, so the net force on the stone in the vertical direction is zero. The force of the man's hand on the stone is equal and opposite to the force of gravity on the stone, which is mg downward. The magnitude of the total force of the man's hand on the stone is

    $$F = mg = (2.0 \text{ kg})(9.80 \text{ m/s}^2) = \boxed{20 \text{ N}}.$$

29. **Strategy** Find the net force on the airplane; and from it determine the acceleration. Then use Eq. (4-4) to find the distance traveled.

    Solution Find the net force on the airplane.

    $$\Sigma F_x = 1.800 \text{ kN} - 1.400 \text{ kN} = 0.400 \text{ kN} = ma_x, \text{ so } a_x = \frac{0.400 \text{ kN}}{m} = \frac{0.400 \text{ kN}}{1160 \text{ kg}} = 0.3448 \text{ m/s}^2.$$

    $$\Sigma F_y = 16.000 \text{ kN} - 16.000 \text{ kN} = 0 = ma_y, \text{ so } a_y = 0.$$

    Find the distance traveled.

    $$\Delta x = v_{ix}\Delta t + \frac{1}{2}a_x(\Delta t)^2 = (60.0 \text{ m/s})(60.0 \text{ s}) + \frac{1}{2}(0.3448 \text{ m/s}^2)(60.0 \text{ s})^2 = \boxed{4.22 \text{ km}}$$

33. **Strategy** Analyze the forces due to and on the three interacting objects: the woman, the chair, and the floor.

    Solution

    (a) The weight of the woman is directed downward. The forces on the woman due to the seat and armrests are directed upward and total 25 N + 25 N + 500 N = 550 N. The chair and floor must support her entire weight, so the balance of her weight to support is $600 \text{ N} - 2(25 \text{ N}) - 500 \text{ N} = 50 \text{ N}$. Thus, the floor exerts a force on the woman's feet of $\boxed{50.0 \text{ N upward}}$.

    (b) The force exerted by the floor on the chair must be equal to the weight of the chair plus the weight of the woman supported by the chair, or $600.0 \text{ N} + 100.0 \text{ N} - 50.0 \text{ N} = 650.0 \text{ N}$. Thus, the floor exerts a force on the chair of $\boxed{650.0 \text{ N upward}}$.

    (c) The two forces acting on the woman and chair system are the upward force due to the floor and the downward gravitational force due to the Earth. Let the subscripts be the following: s = woman and chair system, e = Earth, f = floor.

    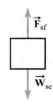

35. Strategy Consider forces acting on the fish suspended by the line.

Solution

> One force acting on the fish is an upward force on the fish by the line; its interaction partner is a downward force on the line by the fish. A second force acting on the fish is the downward gravitational force on the fish; its interaction partner is the upward gravitational force on the Earth by the fish.

37. Strategy Use Newton's first and third laws.

Solution

(a) Margie exerts a downward force on the scale equal to her weight, 543 N. According to the third law, the scale exerts an upward force on Margie equal in magnitude to the magnitude of the force exerted by Margie on it, or $\boxed{543 \text{ N}}$.

(b) Refer to part (a). The interaction partner of the force exerted on Margie by the scale is the $\boxed{\text{contact force of Margie's feet}}$ on the scale.

(c) The Earth must hold up both the scale and (indirectly) Margie, since Margie is standing on the scale. So, the Earth must push up on the scale with a force equal to the combined weight of Margie and the scale, or $543 \text{ N} + 45 \text{ N} = \boxed{588 \text{ N}}$.

(d) Refer to part (c). The interaction partner of the force exerted on the scale by the Earth is $\boxed{\text{the contact force on the Earth due to the scale}}$.

41. Strategy Find the mass using the weight of the man and the Earth's average gravitational field strength, $g = 9.80$ N/kg.

Solution Find the mass of the man.

$$m = \frac{W}{g} = \frac{0.80 \times 10^3 \text{ N}}{9.80 \text{ N/kg}} = \boxed{82 \text{ kg}}$$

45. Strategy The gravitational field strength is given by $g = GM/R^2$. Use the mass of the Earth and the gravitational field strength of the Moon and solve for R, which, in this case, is the distance from the center of the Earth. Then, subtract the radius of the Earth to find the height above the surface.

Solution Solving for R, we have

$$R = \sqrt{\frac{GM}{g}} = \sqrt{\frac{(6.674 \times 10^{-11} \text{ N} \cdot \text{m}^2/\text{kg}^2)(5.974 \times 10^{24} \text{ kg})}{1.62 \text{ m/s}^2}} = 1.57 \times 10^4 \text{ km.}$$

So, the height above the surface is $1.57 \times 10^4 \text{ km} - 6.371 \times 10^3 \text{ km} = \boxed{9.3 \times 10^3 \text{ km}}$.

49. Strategy This is the same as asking, "At what altitude is the gravitational field strength half of its value at the surface of the Earth?" $g = GM/R^2$, so let the new field strength be $g' = ng = GM/r^2$ where $n = 1/2$.

Solution Determine r in terms of R.

$$\frac{g'}{g} = \frac{ng}{g} = n = \frac{\frac{GM}{r^2}}{\frac{GM}{R^2}} = \frac{R^2}{r^2}, \text{ so } r = \frac{R}{\sqrt{n}}.$$

Find an expression for the altitude, h.

$$h = r - R = \frac{R}{\sqrt{n}} - R = R\left(\frac{1}{\sqrt{n}} - 1\right), \text{ so } h = (6.371 \times 10^3 \text{ km})\left(\frac{1}{\sqrt{1/2}} - 1\right) = \boxed{2639 \text{ km}}.$$

53. **Strategy** Consider each of the four forces and any possible relationships between them.

Solution (a) The force of the Earth pulling on the book and (d) the force of the book pulling on the Earth are an interaction pair; they are equal and opposite. (b) The force of the table pushing on the book and (c) the force of the book pushing on the table are an interaction pair; they are equal and opposite. There are two forces acting on the book: the gravitational force of Earth pulling on it and the contact force of the table pushing on it. Since the book is in equilibrium, the net force on it must be zero; therefore, the forces due to Earth and the table on the book are equal and opposite, so the pair of forces given in (a) and (b) are equal in magnitude and opposite in direction even though they are not an interaction pair.

57. **(a) Strategy** Draw a diagram and use Newton's laws of motion.

Solution According to the first law, since the skier is moving with constant velocity, the net force on the skier is zero. Calculate the force of kinetic friction.
$\Sigma F_x = f_k - mg \sin \theta = 0$, so
$f_k = mg \sin \theta = (85 \text{ kg})(9.80 \text{ N/kg}) \sin 11° = 160 \text{ N}$.
The force of kinetic friction is $\boxed{160 \text{ N up the slope}}$.

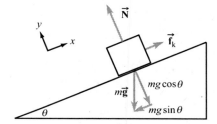

(b) Strategy Use the diagram and results from part (a).

Solution Find the normal force.
$\Sigma F_y = N - mg \cos \theta = 0$, so $N = mg \cos \theta$. Since $f_k = \mu_k N$,

$$\mu_k = \frac{f_k}{N} = \frac{mg \sin \theta}{mg \cos \theta} = \tan \theta = \tan 11° = \boxed{0.19}.$$

59. **Strategy** While the crate is sliding down the ramp, the force of kinetic friction has its maximum magnitude and is opposite in direction to the crate's motion down the incline.

Solution Find the frictional force on the crate.
$\Sigma F_y = N - mg \cos \theta = 0$, so $N = mg \cos \theta$.
$\Sigma F_x = f_k = \mu_k N = \mu_k mg \cos \theta = 0.40(18.0 \text{ kg})(9.80 \text{ N/kg}) \cos 30° = 61 \text{ N}$
The frictional force is $\boxed{61 \text{ N up the ramp}}$.

61. **Strategy** Draw free-body diagrams for each situation. Let the subscripts be the following:
b = book   t = table   e = Earth   h = hand

Solution The diagrams are shown.

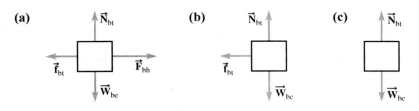

(d) **Strategy and Solution** In cases $\boxed{\text{(a) and (b)}}$, the book is accelerating; so in these cases, the net force is not zero.

(e) **Strategy and Solution** The normal force on the book is equal to its weight, $(0.50 \text{ kg})(9.80 \text{ m/s}^2) = 4.9 \text{ N}$. The net force acting on the book in part (b) is equal to the force of kinetic friction. The force of kinetic friction is opposite the direction of motion. The magnitude is $\mu_k N = 0.40(4.9 \text{ N}) = 2.0 \text{ N}$. Thus, the net force on the book is $\boxed{2.0 \text{ N opposite the direction of motion}}$.

(f) **Strategy and Solution** The free-body diagram would look $\boxed{\text{just like the diagram for part (c) and the book}}$ $\boxed{\text{would not slow down because there is no net force on the book}}$ (friction is zero).

63. (a) **Strategy** To just get the block to move, the force must be equal to the maximum force of static friction.

Solution Solve for μ_s.

$$F = f_{\max} = \mu_s N = \mu_s mg, \text{ so } \mu_s = \frac{F}{mg} = \frac{12.0 \text{ N}}{(3.0 \text{ kg})(9.80 \text{ N/kg})} = \boxed{0.41}.$$

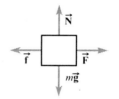

(b) **Strategy** The maximum static frictional force is now proportional to the total mass of the two blocks. The free-body diagram is the same as before, except that the mass m is now the sum of the masses of both blocks.

Solution Find the magnitude F of the force required to make the two blocks start to move.

$$F = \mu_s mg = 0.41(3.0 \text{ kg} + 7.0 \text{ kg})(9.80 \text{ N/kg}) = \boxed{40 \text{ N}}$$

65. **Strategy** Since the block moves with constant speed, there is no net force on the block. Draw the free-body diagram using this information. Let the subscripts be the following:
b = block   B = Brenda   w = wall   e = Earth

Solution Find the coefficient of kinetic friction between the wall and the block.
$\Sigma F_x = N_{bw} - F_{bB} \sin\theta = 0$, so $N_{bw} = F_{bB} \sin\theta$.
$\Sigma F_y = F_{bB} \cos\theta - F_{be} - f_{bw} = 0$, so $f_{bw} = F_{bB} \cos\theta - F_{be}$.
Since $f_{bw} = \mu_k N_{bw}$, we have

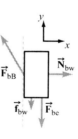

$$\mu_k = \frac{F_{bB} \cos\theta - F_{be}}{N_{bw}} = \frac{F_{bB} \cos\theta - F_{be}}{F_{bB} \sin\theta} = \cot\theta - \frac{F_{be}}{F_{bB}} \csc\theta$$
$$= \cot 30.0° - \frac{2.0 \text{ N}}{3.0 \text{ N}} \csc 30.0° = \boxed{0.4}$$

69. **Strategy** Identify each force acting on the sailboat. Draw a free-body diagram.

Solution The forces acting on the sailboat are:

1) the force of gravity
2) the vertical force of the water opposing gravity
3) the force of the wind
4) the force of the line tied to the mooring

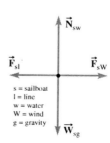

73. Strategy Use Newton's laws of motion. The lower cord supports only the lower box, whereas the upper cord supports both boxes. Draw a diagram.

Solution Find the tension in each cord.

<u>Lower cord</u>

$\Sigma F_x = T_l - m_l g \sin\theta = 0$, so

$T_l = m_l g \sin\theta = (2.0 \text{ kg})(9.80 \text{ N/kg}) \sin 25° = \boxed{8.3 \text{ N}}$.

<u>Upper cord</u>

$\Sigma F_x = T_u - m_u g \sin\theta - T_l = 0$, so

$T_u = m_u g \sin\theta + T_l = (1.0 \text{ kg})(9.80 \text{ N/kg}) \sin 25° + 8.3 \text{ N} = \boxed{12.4 \text{ N}}$.

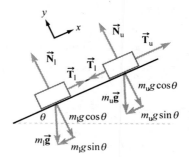

75. Strategy Recall that the tension in the rope is the same along its length.

Solution The tension is equal to the weight at the end of the rope, 120 N. Therefore, scale A reads 120 N.

There are two forces pulling downward on the pulley due to the tension of 120 N in each part of the rope. Therefore, $T_B = T_A + T_A = 2T_A = 240 \text{ N}$. Scale B reads 240 N, since it supports the pulley.

77. Strategy Use Newton's laws of motion. Draw a free-body diagram.

Solution

(a) Find the tension in the rope from which the pulley hangs.

$\Sigma F_y = T_1 \sin\theta - Mg = 0$ and $\Sigma F_x = T_1 \cos\theta - T_2 = 0$.

The tension in T_2 is due to the mass M, so $T_2 = Mg$.

Thus, $T_1 \cos\theta = Mg$ and $T_1 \sin\theta = Mg$.

According to these equations, $\cos\theta = \sin\theta$, which is true only if

$\theta = 45°$ for $0° \le \theta \le 90°$.

Therefore, $T_1 = \dfrac{Mg}{\cos 45°} = \boxed{\sqrt{2}Mg}$.

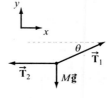

(b) As found in part (a), $\theta = \boxed{45°}$.

81. Strategy Use Newton's laws of motion. Let $+y$ be down and $+x$ to the right.

Solution Find the force $\vec{F}$ applied to the front tooth.

$\Sigma F_x = T \sin\theta - T \sin\theta = 0$ and $\Sigma F_y = T \cos\theta + T \cos\theta - F = 0$. So, we have

$F = 2T \cos\theta = 2(1.2 \text{ N}) \cos 33° = 2.0 \text{ N}$. By symmetry, the force is directed toward the back of the mouth, so

$\vec{F} = \boxed{2.0 \text{ N toward the back of the mouth}}$.

85. (a) Strategy The force required to start the block moving is that needed to overcome the maximum force of static friction. Draw a diagram.

Solution Find the applied horizontal force.

$\Sigma F_x = F - f_s = 0$, so $F = f_s = \mu_s N$.

$\Sigma F_y = N - F_g = N - mg = 0$, so $N = mg$.

So, the value of the applied force at the instant that the block

starts to slide is $F = \mu_s mg = 0.40(5.0 \text{ kg})(9.80 \text{ N/kg}) = \boxed{20 \text{ N}}$.

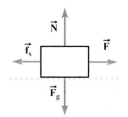

(b) Strategy The force required to keep the block moving is that needed to overcome kinetic friction. At the instant the block starts to slide, the net force on the block is the difference between the forces required to overcome static and kinetic friction.

Solution Calculate the net force.

$\Sigma F = f_s - f_k = \mu_s mg - \mu_k mg = (\mu_s - \mu_k)mg = (0.40 - 0.15)(5.0 \text{ kg})(9.80 \text{ N/kg}) = \boxed{12 \text{ N}}$

89. Strategy Use the expressions for a_y and T found in Example 4.15.

Solution

(a) $a_y = \dfrac{(m_2 - m_1)g}{m_2 + m_1} = \dfrac{(5.0 \text{ kg} - 3.0 \text{ kg})(9.80 \text{ m/s}^2)}{5.0 \text{ kg} + 3.0 \text{ kg}} = 2.5 \text{ m/s}^2$

Since $m_2 > m_1$, $\boxed{\vec{a}_1 = 2.5 \text{ m/s}^2 \text{ up and } \vec{a}_2 = 2.5 \text{ m/s}^2 \text{ down}}$.

(b) $T = \dfrac{2m_1 m_2}{m_1 + m_2} g = \dfrac{2(3.0 \text{ kg})(5.0 \text{ kg})}{3.0 \text{ kg} + 5.0 \text{ kg}}(9.80 \text{ m/s}^2) = \boxed{37 \text{ N}}$

93. Strategy Let the $+x$-direction be down the incline. Use Newton's second law and Eq. (2-16).

Solution Find the acceleration of the glider.

$\Sigma F_x = mg \sin \theta = ma_x$, so $a_x = g \sin \theta$.

Find the angle of inclination.

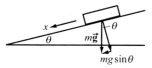

$v_{fx}^2 - v_{ix}^2 = v_{fx}^2 - 0 = 2a_x \Delta x = 2g \sin \theta \Delta x$, so

$\theta = \sin^{-1} \dfrac{v_{fx}^2}{2g\Delta x} = \sin^{-1} \dfrac{(0.250 \text{ m/s})^2}{2(9.80 \text{ m/s}^2)(0.500 \text{ m})} = \boxed{0.365°}$.

The slope is $a_x = g \sin \theta = 0.0625 \text{ m/s}^2 = 6.25 \text{ cm/s}^2$.

The positions and times are shown in the graph.

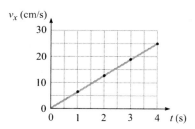

97. Strategy Draw a free-body diagram and use Newton's second law.

Solution The elevator floor pushes upward on Oliver with a force equal to the normal force.
$\Sigma F_y = N - W = N - mg = ma_y$, so $N = m(g + a_y)$.
Therefore, the magnitude of the force exerted by the floor is
$F = m(g + a_y) = (76.2 \text{ kg})(9.80 \text{ m/s}^2 - 1.37 \text{ m/s}^2) = \boxed{642 \text{ N}}$.

101. Strategy The apparent weight is given by $W' = m(g + a_y)$.

Solution

(a) Up is the positive direction. Solve for a_y.

$$W' = m(g + a_y) = mg\left(1 + \frac{a_y}{g}\right) = W\left(1 + \frac{a_y}{g}\right), \text{ so } a_y = g\left(\frac{W'}{W} - 1\right) = (9.80 \text{ m/s}^2)\left(\frac{120 \text{ lb}}{140 \text{ lb}} - 1\right) = -1.4 \text{ m/s}^2.$$

So, $\vec{a}$ is $\boxed{1.4 \text{ m/s}^2 \text{ downward}}$.

(b) With a downward acceleration, the elevator could be going up and slowing down, or going down and speeding up, so the answer is $\boxed{\text{no}}$; one cannot tell whether the elevator is speeding up or slowing down.

103. Strategy The apparent weight is given by $W' = W(1 + a_y/g)$.

Solution Up is the positive direction. Find Felipe's actual weight.

$$W' = W\left(1 + \frac{a_y}{g}\right), \text{ so } W = \frac{W'}{1 + \frac{a_y}{g}} = \frac{750 \text{ N}}{1 + \frac{2.0 \text{ m/s}^2}{9.80 \text{ m/s}^2}} = \boxed{620 \text{ N}}.$$

105. Strategy Consider the ranges and natures of the fundamental forces.

Solution Of all of the fundamental forces, $\boxed{\text{the weak force}}$ has the shortest range (about 10^{-17} m). In the Sun, the weak interaction enables thermonuclear reactions to occur, without which there would be no sunlight.

109. Strategy Consider the natures of the fundamental forces.

Solution Of the fundamental forces, $\boxed{\text{the strong force}}$ is the strongest, hence its name. It is strong, but has a very short range. But the range is just the right size (about 10^{-15} m) to be the fundamental interaction that binds quarks together to form protons, neutrons, and many exotic subatomic particles.

113. Strategy Use Newton's laws of motion.

Solution

(a) Since the airplane is cruising in a horizontal level flight (straight line) at constant velocity, it is in equilibrium and the net force is $\boxed{\text{zero}}$.

(b) The air pushes upward with a force equal to the weight of the airplane: $\boxed{2.6 \times 10^4 \text{ N}}$.

117. (a) Strategy Neglect frictional forces. Identify all of the forces acting on the car; then draw a free-body diagram.

Solution The forces are the normal force due to the road, the gravitational force due to the Earth, and the tension due to the rope. The diagram is shown including the angles.

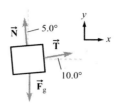

(b) Strategy Use Newton's laws of motion and the free-body diagram.

Solution Find the tension.

$$\Sigma F_x = T\cos 10.0° - N\sin 5.0° = 0, \text{ so } T = \frac{\sin 5.0°}{\cos 10.0°}N.$$

$$\Sigma F_y = N\cos 5.0° + T\sin 10.0° - F_g = 0, \text{ so } N = \frac{F_g}{\cos 5.0°} - \frac{T\sin 10.0°}{\cos 5.0°} = \frac{mg}{\cos 5.0°} - \frac{T\sin 10.0°}{\cos 5.0°}.$$

Solve for T.

$$T = \frac{\sin 5.0°}{\cos 10.0°}N = \frac{\sin 5.0°}{\cos 10.0°}\left(\frac{mg}{\cos 5.0°} - \frac{T\sin 10.0°}{\cos 5.0°}\right) = \frac{\tan 5.0°}{\cos 10.0°}mg - T\tan 5.0°\tan 10.0°, \text{ so}$$

$$T(1 + \tan 5.0°\tan 10.0°) = \frac{\tan 5.0°}{\cos 10.0°}mg \text{ and } T = \frac{\tan 5.0°(1000 \text{ kg})(9.80 \text{ N/kg})}{\cos 10.0°(1 + \tan 5.0°\tan 10.0°)} = \boxed{860 \text{ N}}.$$

121. (a) Strategy For the sum of the two forces to be in the forward ($+y$) direction, the net force in the x-direction must be zero. Draw a diagram and use Newton's laws of motion.

Solution Compute the magnitude of the force.
$$\Sigma F_x = F\sin 38° - (105 \text{ N})\sin 28° = 0, \text{ so}$$
$$F = \frac{(105 \text{ N})\sin 28°}{\sin 38°} = \boxed{80 \text{ N}}.$$

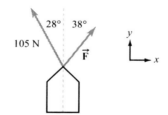

(b) Strategy Find the sum of the y-components of the two forces to find the magnitude of the net force on the barge from the two tow ropes.

Solution Find the magnitude of the force.
$$\Sigma F_y = F\cos 38° + (105 \text{ N})\cos 28° = (80 \text{ N})\cos 38° + (105 \text{ N})\cos 28° = \boxed{160 \text{ N}}$$

125. Strategy Let left be positive. Both blocks move with acceleration a (to the left). Use Newton's second law.

Solution For the two-block system:
$$F_{net} = F = (2m + m)a = 3ma$$

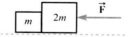

Let F_{12} be the force of the smaller block on the larger block and F_{21} be the force of the larger block on the smaller block. Also, by Newton's third law, $F_{21} = -F_{12}$.
For the smaller block: $F_{net} = F_{21} = ma$. Find F_{12}.

$$\frac{F_{21}}{F} = \frac{ma}{3ma} = \frac{1}{3}, \text{ so } \frac{F}{3} = F_{21} = -F_{12}, \text{ or } F_{12} = -\frac{F}{3}.$$

So, the force of the smaller block on the larger block is $\boxed{F/3 \text{ to the right}}$.

129. (a) Strategy Determine the maximum force of static friction and compare it to the force of the push, 5.0 N.

Solution Find the maximum force of static friction.
$$f_{s,\text{max}} = \mu_s N = \mu_s mg_{\text{Moon}} = 0.35(2.0 \text{ kg})(9.80 \text{ m/s}^2) = 6.9 \text{ N} > 5.0 \text{ N}$$
No, the puck does not move, since the maximum force of static friction is greater than the force of the push.

(b) Strategy and Solution Since 7.5 N > 6.9 N, the maximum force of static friction, yes, the puck does move.

(c) Strategy Use Newton's second law.

Solution Find the acceleration of the puck.
$$\Sigma F = F_{\text{push}} - f_k = F_{\text{push}} - \mu_k mg = ma, \text{ so}$$
$$a = \frac{F_{\text{push}}}{m} - \mu_k g = \frac{6.0 \text{ N}}{2.0 \text{ kg}} - 0.25(9.80 \text{ m/s}^2) = \boxed{0.6 \text{ m/s}^2}.$$

(d) Strategy Use Newton's second law and the fact that gravity is weaker on the Moon.

Solution The acceleration of the puck is
$$a = \frac{F_{\text{push}}}{m} - \mu_k g.$$
Since gravity is weaker on the Moon, the second term on the right side of the equation is smaller than it would be for Earth. This is the same thing as saying that the force of friction will be less on the Moon. Therefore, the acceleration of the puck is more on the Moon than on Earth.

133. Strategy Set the magnitudes of the forces on the spaceship due to the Earth and the Moon equal. (The forces are along the same line.)

Solution Find the distance from the Earth expressed as a percentage of the distance between the centers of the Earth and the Moon.
$$F_{sE} = \frac{GM_E m}{r_E^2} = F_{sM} = \frac{GM_M m}{r_M^2}, \text{ so } r_E = r_M \sqrt{\frac{M_E}{0.0123 M_E}} = 9.02 r_M.$$
Find the percentage.
$$\frac{r_E}{r_E + r_M} = \frac{9.02 r_M}{9.02 r_M + r_M} = \frac{9.02}{10.02} = 0.900$$
The distance from the Earth is 90.0% of the Earth-Moon distance.

137. (a) Strategy The tension due to the weight of the potatoes is divided evenly between the two sets of scales.

Solution Find the tension and, thus, the reading of each scale.
$$2T = mg, \text{ so } T = mg/2 = (220.0 \text{ N})/2 = \boxed{110.0 \text{ N}}.$$

(b) Strategy Scales B and D will read 110.0 N as before. Scales A and C will read an additional 5.0 N due to the weights of B and D, respectively.

Solution Find the reading of each scale.
$$T_A = 110.0 \text{ N} + 5.0 \text{ N} = \boxed{115.0 \text{ N}} = T_C \text{ and } T_B = \boxed{110.0 \text{ N}} = T_D.$$

139. **(a) Strategy** Identify the interactions between the magnet and other objects.

   Solution The interactions are:
   1) The gravitational forces between the magnet and the Earth
   2) The contact forces, normal and frictional, between the magnet and the photo
   3) The magnetic forces between the magnet and the refrigerator

   (b) Strategy Refer to part (a). Let the subscripts be the following:
   m = magnet   p = photo   e = Earth   r = refrigerator

   Solution The magnet is in equilibrium, so the horizontal pair of forces and the vertical pair of forces are equal in magnitude and opposite in direction.

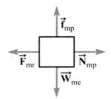

   (c) Strategy Identify the range of each force and categorize each as long-range or contact.

   Solution
   The long-range forces are gravity and magnetism. The contact forces are friction and the normal force.

   (d) Strategy Refer to part (b). W_{me} and F_{mr} are given.

   Solution
   $W_{me} = 0.14$ N, $F_{mr} = 2.10$ N, $f_{mp} = W_{me} = 0.14$ N, and $N_{mp} = F_{mr} = 2.10$ N.

141. **(a) Strategy** Scale A measures the weight of both masses. Scale B only measures the weight of the 4.0-kg mass.

   Solution Find the readings of the two scales if the masses of the scales are negligible.
   Scale A $= (10.0 \text{ kg} + 4.0 \text{ kg})(9.80 \text{ N/kg}) = \boxed{137 \text{ N}}$ and Scale B $= (4.0 \text{ kg})(9.80 \text{ N/kg}) = \boxed{39 \text{ N}}$.

   (b) Strategy Scale A measures the weight of both masses and scale B. Scale B only measures the weight of the 4.0-kg mass.

   Solution Find the readings if each scale has a mass of 1.0 kg.
   Scale A $= (10.0 \text{ kg} + 4.0 \text{ kg} + 1.0 \text{ kg})(9.80 \text{ N/kg}) = \boxed{147 \text{ N}}$ and Scale B $= \boxed{39 \text{ N}}$.

143. Strategy Let the +*x*-direction be up the incline. Use Newton's second law.

Solution

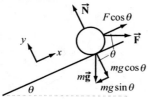

(a) Let *F* be the horizontal force.

$$\sum F_x = F \cos\theta - mg\sin\theta = 0, \text{ so } F = \boxed{mg\tan\theta}.$$

(b) To roll the crate up at constant speed, the net force is zero, so the force is that from part (a), $\boxed{mg\tan\theta}$.

(c) $\sum F_x = F\cos\theta - mg\sin\theta = ma$, so $F = \boxed{mg\tan\theta + \dfrac{ma}{\cos\theta}}$.

145. (a) Strategy Since the downward speed is decreasing at a rate of $0.10g$, the acceleration of the truck is $0.10g$ upwards. Use Newton's second law.

Solution

$\sum F_x = 0$ and $\sum F_y = T - mg = ma_y = m(0.10g)$, so $T = \boxed{1.10mg}$.

(b) **Strategy and Solution** Although the motion of the helicopter has changed, the acceleration of the truck is the same as in part (a), so the tension is the same, $\boxed{1.10mg}$.

149. (a) Strategy Set the magnitudes of the forces on the spacecraft due to the Earth and the Sun equal.

Solution Find the distance of the spacecraft from Earth.

$$F_{sS} = \frac{GM_S m}{r_S^2} = F_{sE} = \frac{GM_E m}{r_E^2}, \text{ so } \frac{r_E}{r_S} = \sqrt{\frac{M_E}{M_S}}.$$

This is the ratio of the Earth-spacecraft distance to the Sun-spacecraft distance. If this is multiplied by the Earth-Sun mean distance, the product is the distance of the spacecraft from the Earth.

$$(1.50\times10^{11} \text{ m})\sqrt{\frac{5.974\times10^{24} \text{ kg}}{1.987\times10^{30} \text{ kg}}} = \boxed{2.60\times10^8 \text{ m from Earth}}$$

(b) **Strategy** Imagine the spacecraft is a small distance *d* closer to the Earth and find out which gravitational force is stronger, the Earth's or the Sun's.

Solution At the equilibrium point the net gravitational force is zero. If the spacecraft is closer to the Earth than the equilibrium point distance from the Earth, then the force due to the Earth is greater than that due to the Sun. If the spacecraft is closer to the Sun than the equilibrium point distance from the Sun, then the force due to the Sun is greater than that due to the Earth. So, if the spacecraft is close to, but not at, the equilibrium point, the net force tends to pull it $\boxed{\text{away from}}$ the equilibrium point.

153. Strategy Choose the $+x$-axis to the right and $+y$-axis up. Use Newton's second law.

Solution

(a) For m_1:

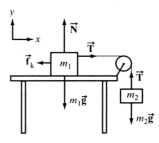

$\sum F_{1y} = N - W_1 = N - m_1 g = 0$, so $N = m_1 g$.

$\sum F_{1x} = T - f_k = T - \mu_k N = T - \mu_k m_1 g = m_1 a_{1x}$

For m_2:

$\sum F_{2x} = 0$ and $\sum F_{2y} = T - W_2 = T - m_2 g = m_2 a_{2y}$.

Now, a_{1x} and a_{2y} must be equal in magnitude, otherwise the cord will compress or expand. a_{1x} is in the $+x$-direction and a_{2y} is in the $-y$-direction.

So, let $a = a_{1x} = -a_{2y}$. Then, $T - \mu_k m_1 g = m_1 a$ and $T - m_2 g = -m_2 a$.

Subtract the second equation from the first and solve for a.

$-\mu_k m_1 g + m_2 g = m_1 a + m_2 a$

$g(m_2 - \mu_k m_1) = a(m_1 + m_2)$

$$\boxed{a = \frac{m_2 - \mu_k m_1}{m_1 + m_2} g}$$

Find T.

$T - m_2 g = -m_2 a$

$$T = m_2 g - m_2 \frac{m_2 - \mu_k m_1}{m_1 + m_2} g = \frac{m_2(m_1 + m_2) - m_2(m_2 - \mu_k m_1)}{m_1 + m_2} g = \frac{m_1 m_2 + m_2^2 - m_2^2 + m_1 m_2 \mu_k}{m_1 + m_2} g$$

$$\boxed{T = (1 + \mu_k)\frac{m_1 m_2}{m_1 + m_2} g}$$

(b) For $m_1 \ll m_2$,

$a \approx \dfrac{m_2 - \mu_k(0)}{0 + m_2} g = \boxed{g}$ and

$T \approx (1 + \mu_k)\dfrac{m_1 m_2}{0 + m_2} g = \boxed{(1 + \mu_k)m_1 g \ll m_2 g, \text{ so the tension is negligible compared to the weight of } m_2;}$. $\text{it's essentially in free fall.}$

If $m_1 \gg m_2$, the force of friction between the table and m_1 is so large that m_1 will not slide, so $a = \boxed{0}$ and

$T - m_2 g = -m_2 a = 0$, or $T = \boxed{m_2 g}$.

For $m_1 = m_2 = m$,

$a = \dfrac{m - \mu_k m}{m + m} g = \dfrac{m(1 - \mu_k)}{2m} g = \boxed{\frac{1}{2}(1 - \mu_k)g}$ and $T = (1 + \mu_k)\dfrac{m^2}{2m} g = \boxed{\frac{1}{2}(1 + \mu_k)mg}$.

(c) $\boxed{a = 0 \text{ only for } m_2 = 0; \text{ thus, there is no value at which the two masses slide with constant velocity. For } m_2 = 0, \text{ there is no tension in the cord.}}$

Chapter 5

CIRCULAR MOTION

Conceptual Questions

1. Depressing the gas pedal is not the only way to make the car accelerate. The driver can also apply the brakes or turn the car to make it accelerate.

5. In uniform circular motion, the acceleration always points toward the center of the circle. Hence it remains perpendicular to the velocity the whole time. When a projectile is launched horizontally, the acceleration is initially perpendicular to the velocity, but does not remain so.

9. When the roller coaster turns hard to the right, the inertia of a rider's upper body keeps it moving in a straight line until it runs into the wall of the car. The wall exerts a normal force on the upper body that causes it to accelerate radially with the car. Thus, no force pushes the rider to the left as they enter a turn—the rider's inertia simply carries them forward while the car moves to the right.

Problems

1. **Strategy** Find the arc length swept out by the carnival swing.

   Solution Use Eq. (5-4).
   $$s = r\theta = (8.0 \text{ m})(120°)\left(\frac{2\pi \text{ rad}}{360°}\right) = \boxed{17 \text{ m}}$$

5. **Strategy** Use Eq. (5-9) to find the angular speed of the bicycle's tires.

   Solution
   $$|\omega| = \frac{v}{r} = \frac{9.0 \text{ m/s}}{0.35 \text{ m}} = \boxed{26 \text{ rad/s}}$$

7. **(a) Strategy and Solution** There are 2π radians per revolution and 60 seconds per minute, so
   $$\left(\frac{33.3 \text{ rev}}{\text{min}}\right)\left(\frac{2\pi \text{ rad}}{\text{rev}}\right)\left(\frac{1 \text{ min}}{60 \text{ s}}\right) = \boxed{3.49 \text{ rad/s}}.$$

   (b) Strategy Use the relationship between linear speed and angular speed.

   Solution Find the speed of the doll.
   $$v = r|\omega| = (0.13 \text{ m})(3.49 \text{ rad/s}) = \boxed{0.45 \text{ m/s}}$$

9. **Strategy** Use the conversion factor between degrees and radians and $s = r\theta$, where $s = 100.0$ ft, $\theta = 1.5°$, and r is the radius of curvature.

   Solution Find the radius of curvature of a "1.5° curve".
   $$r = \frac{s}{\theta} = \frac{100.0 \text{ ft}}{1.5°}\left(\frac{360°}{2\pi \text{ rad}}\right) = \boxed{3800 \text{ ft}}$$

13. Strategy Use the relationship between angular speed and radial acceleration.

Solution The number of seconds in one day is 86,400, so the angular speed of the Earth (and baobab) is $|\omega| = 2\pi \text{ rad}/86,400 \text{ s}$. Compute the radial acceleration.

$$a_r = \omega^2 r = \omega^2 R_{\text{Earth}} = \left(\frac{2\pi \text{ rad}}{86,400 \text{ s}}\right)^2 (6.371 \times 10^6 \text{ m}) = \boxed{3.37 \text{ cm/s}^2}$$

17. (a) Strategy Use Newton's second law and the relationship between linear speed and radial acceleration.

Solution According to Newton's second law,

$$\Sigma F_r = T = ma_r = m\frac{v^2}{r} = m\frac{v^2}{L}, \text{ thus, } T = \boxed{\frac{mv^2}{L}}.$$

(b) Strategy Draw a free-body diagram for the rock. Use Newton's second law.

Solution Decompose the force into vertical (y) and radial (r) components.

$$\Sigma F_r = T_r = \frac{mv^2}{r} = \frac{mv^2}{L\cos\theta} \text{ and } \Sigma F_y = T_y - mg = 0.$$

Find the magnitude of the tension.

$$T = \sqrt{T_r^2 + T_y^2} = \sqrt{\left(\frac{mv^2}{L\cos\theta}\right)^2 + (mg)^2}$$

$$= \boxed{m\sqrt{g^2 + \left(\frac{v^2}{L\cos\theta}\right)^2}}$$

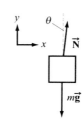

21. Strategy Let the x-axis point toward the center of curvature and the y-axis point upward. Draw a free-body diagram. Use Newton's second law and the relationship between radial acceleration and linear speed.

Solution

$$\Sigma F_y = N\cos\theta - mg = 0, \text{ so } N\cos\theta = mg, \text{ and } \Sigma F_x = N\sin\theta = ma_r = m\frac{v^2}{r}.$$

Solve for v.

$$\frac{N\sin\theta}{N\cos\theta} = \frac{m\frac{v^2}{r}}{mg}, \text{ so } v^2 = rg\tan\theta, \text{ or}$$

$$v = \sqrt{rg\tan\theta} = \sqrt{(120 \text{ m})(9.80 \text{ m/s}^2)\tan 3.0°} = \boxed{7.9 \text{ m/s}}.$$

25. Strategy Let the x-axis point toward the center of curvature and the y-axis point upward. Draw a free-body diagram. Use Newton's second law and the relationship between radial acceleration and linear speed.

Solution

(a) $\Sigma F_y = N\cos\theta - mg - f\sin\theta = 0$ and $\Sigma F_x = N\sin\theta + f\cos\theta = mv^2/r$.

Solve for N in the first equation and substitute into the second.

$N = \dfrac{f\sin\theta + mg}{\cos\theta}$, so

$\dfrac{f\sin\theta + mg}{\cos\theta}\sin\theta + f\cos\theta = m\dfrac{v^2}{r}$

$f\sin^2\theta + mg\sin\theta + f\cos^2\theta = m\dfrac{v^2}{r}\cos\theta$

$f(\sin^2\theta + \cos^2\theta) = m\dfrac{v^2}{r}\cos\theta - mg\sin\theta$

$f(1) = m\left(\dfrac{v^2}{r}\cos\theta - g\sin\theta\right)$

$f = (1400\text{ kg})\left[\dfrac{(32\text{ m/s})^2}{410\text{ m}}\cos 5.0° - (9.80\text{ m/s}^2)\sin 5.0°\right] = \boxed{2300\text{ N}}$

(b) Set the expression found for the force of friction equal to zero.

$f = 0 = m\left(\dfrac{v^2}{r}\cos\theta - g\sin\theta\right)$, so $\dfrac{v^2}{r}\cos\theta = g\sin\theta$, or

$v = \sqrt{gr\tan\theta} = \sqrt{(9.80\text{ m/s}^2)(410\text{ m})\tan 5.0°} = \boxed{19\text{ m/s}}$.

29. Strategy Use $v = r\omega$ and $\omega = 2\pi/T$, where the radius is the average Earth-Sun distance and the period is one year.

Solution

$v = r\omega = r\left(\dfrac{2\pi}{T}\right) = \dfrac{2\pi r}{T} = \dfrac{2\pi(1.50\times10^{11}\text{ m})}{1\text{ y}}\left(\dfrac{1\text{ y}}{3.156\times10^7\text{ s}}\right) = \boxed{2.99\times10^4\text{ m/s}}$

31. Strategy According to Kepler's third law, $r^3 \propto T^2$. Form a proportion.

Solution Find the orbital period of the second satellite.

$\left(\dfrac{4.0r}{r}\right)^3 = 64 = \left(\dfrac{T_{4.0}}{T}\right)^2 = \dfrac{T_{4.0}^2}{T^2}$, so $T_{4.0}^2 = 64T^2$, or $T_{4.0} = 8.0T = 8.0(16\text{ h}) = \boxed{130\text{ h}}$.

33. **Strategy** Use Eq. (5-14) with the mass of Jupiter in place of the mass of the Sun.

    Solution Solve for r.

    $$\frac{4\pi^2}{GM_J}r^3 = T^2, \text{ so } r^3 = \frac{GM_J}{4\pi^2}T^2, \text{ or } r = \sqrt[3]{\frac{GM_J}{4\pi^2}T^2}.$$

    Compute the distance from the center of Jupiter for each satellite.

    $$r_{Io} = \sqrt[3]{\frac{GM_J}{4\pi^2}T_{Io}^2} = \sqrt[3]{\frac{(6.674\times10^{-11}\ \text{N}\cdot\text{m}^2/\text{kg}^2)(1.9\times10^{27}\ \text{kg})}{4\pi^2}(1.77\ \text{d})^2\left(\frac{86,400\ \text{s}}{1\ \text{d}}\right)^2} = \boxed{420,000\ \text{km}}$$

    $$r_{Europa} = \sqrt[3]{\frac{GM_J}{4\pi^2}T_{Europa}^2} = \sqrt[3]{\frac{(6.674\times10^{-11}\ \text{N}\cdot\text{m}^2/\text{kg}^2)(1.9\times10^{27}\ \text{kg})}{4\pi^2}(3.54\ \text{d})^2\left(\frac{86,400\ \text{s}}{1\ \text{d}}\right)^2} = \boxed{670,000\ \text{km}}$$

37. **Strategy** Use Newton's second law and law of universal gravitation, as well as the relationship between radial acceleration and linear speed.

    Solution

    $$\Sigma F_r = \frac{GmM_J}{(3.0\ R_J)^2} = \frac{mv^2}{3.0R_J}$$

    Now, the gravitational field strength of Jupiter is given by $g_J = \frac{GM_J}{R_J^2}$. Find the period of the spacecraft's orbit.

    $$\frac{mv^2}{3.0R_J} = \frac{mg_J}{9.0}, \text{ so } v^2 = \frac{g_J R_J}{3.0} = \left(\frac{2\pi r}{T}\right)^2 = \frac{4\pi^2(3.0R_J)^2}{T^2}. \text{ Solving for } T, \text{ we have}$$

    $$T = 2\pi\sqrt{\frac{27R_J}{g_J}} = 2\pi\sqrt{\frac{27(71,500\times10^3\ \text{m})}{23\ \text{N/kg}}}\left(\frac{1\ \text{h}}{3600\ \text{s}}\right) = \boxed{16\ \text{h}}.$$

41. **Strategy** Use Newton's second law and the relationship between radial acceleration and linear speed.

    Solution The only forces acting on the car are gravity and the normal force of the ground pushing on the car. The radial acceleration is downward, or toward the center of the radius of curvature. Let up be positive.

    $$\Sigma F_r = N - mg = ma_r = m\left(-\frac{v^2}{r}\right)$$

    When the car is just in contact with the ground, the normal force must be zero. If the car goes any faster it will lose contact with the road. Solve for the speed.

    $$-\frac{mv^2}{r} = N - mg = 0 - mg, \text{ so } v^2 = gr, \text{ or } v = \sqrt{gr} = \sqrt{(9.80\ \text{m/s}^2)(55.0\ \text{m})} = \boxed{23.2\ \text{m/s}}.$$

43. **Strategy** Use Eq. (5-20) to find the constant angular acceleration.

    Solution Since the cyclist starts from rest, the initial angular velocity is zero.

    $$\Delta\theta = \omega_i\Delta t + \frac{1}{2}\alpha(\Delta t)^2 = (0)\Delta t + \frac{1}{2}\alpha(\Delta t)^2 = \frac{1}{2}\alpha(\Delta t)^2, \text{ so } \alpha = \frac{2\Delta\theta}{(\Delta t)^2} = \frac{2(8.0\ \text{rev})\left(\frac{2\pi\ \text{rad}}{\text{rev}}\right)}{(5.0\ \text{s})^2} = \boxed{4.0\ \text{rad/s}^2}.$$

45. Strategy Equations (5-18) and (5-19) are $\Delta\omega = \omega_f - \omega_i = \alpha\Delta t$ and $\Delta\theta = \frac{1}{2}(\omega_f + \omega_i)\Delta t$, respectively.

Solution Solve for ω_f in Eq. (5-18), then substitute the result into Eq. (5-19) and simplify.

$\Delta\omega = \omega_f - \omega_i = \alpha\Delta t$, so $\omega_f = \omega_i + \alpha\Delta t$. Substitute.

$$\Delta\theta = \frac{1}{2}(\omega_f + \omega_i)\Delta t = \frac{1}{2}(\omega_i + \alpha\Delta t + \omega_i)\Delta t = \frac{1}{2}(2\omega_i + \alpha\Delta t)\Delta t = \omega_i\Delta t + \frac{1}{2}\alpha(\Delta t)^2 = \Delta\theta, \text{ which is Eq. (5-20).}$$

47. Strategy Draw a free-body diagram for the bob. Use Newton's second law and the relationship between radial acceleration and linear speed.

Solution Refer to the figure.

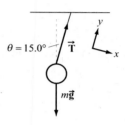

$\Sigma F_y = T - mg\cos\theta = ma_r$ and $\Sigma F_x = mg\sin\theta = ma_t$, so

$a_t = g\sin\theta = (9.80 \text{ m/s}^2)\sin 15.0° = \boxed{2.54 \text{ m/s}^2}$ and

$a_r = \dfrac{v^2}{r} = \dfrac{(1.40 \text{ m/s})^2}{0.800 \text{ m}} = \boxed{2.45 \text{ m/s}^2}$. The tension is

$T = m(a_r + g\cos\theta) = (1.00 \text{ kg})\left[2.45 \text{ m/s}^2 + (9.80 \text{ m/s}^2)\cos 15.0°\right] = \boxed{11.9 \text{ N}}$.

49. (a) Strategy Use Eq. (5-18) to find the magnitude of the constant angular acceleration.

Solution

$$|\Delta\omega| = |\alpha|\Delta t, \text{ so } |\alpha| = \frac{|\Delta\omega|}{\Delta t} = \frac{(33.3 \text{ rpm})\left(\frac{2\pi \text{ rad}}{\text{rev}}\right)\left(\frac{1 \text{ min}}{60 \text{ s}}\right)}{2.0 \text{ s}} = \boxed{1.7 \text{ rad/s}^2}.$$

(b) Strategy Use Eq. (5-20) to find the number of revolutions.

Solution

$$\Delta\theta = \omega_i\Delta t + \frac{1}{2}\alpha(\Delta t)^2 = (0)\Delta t + \frac{1}{2}\alpha(\Delta t)^2$$

$$\Delta\theta = \frac{1}{2}\alpha(\Delta t)^2$$

$$\left(\frac{1 \text{ rev}}{2\pi \text{ rad}}\right)\Delta\theta = \frac{1}{2}(1.744 \text{ rad/s}^2)(2.0 \text{ s})^2\left(\frac{1 \text{ rev}}{2\pi \text{ rad}}\right) = \boxed{0.56 \text{ rev}}$$

53. (a) Strategy Use Eq. (5-18) to find the time it takes for the rotor to come to rest.

Solution The acceleration is opposite the rotation of the rotor, so it is negative.

$$\Delta t = \frac{\omega_f - \omega_i}{\alpha} = \frac{0 - 5.0\times10^5 \text{ rad/s}}{-0.40 \text{ rad/s}^2} = \boxed{1.3\times10^6 \text{ s}}$$

(b) Strategy Use Eq. (5-21) to find the number of revolutions the rotor spun before it stopped.

Solution

$$\Delta\theta = \frac{\omega_f^2 - \omega_i^2}{2\alpha}$$

$$\left(\frac{1 \text{ rev}}{2\pi \text{ rad}}\right)\Delta\theta = \frac{0 - (5.0\times10^5 \text{ rad/s})^2}{2(-0.40 \text{ rad/s}^2)}\left(\frac{1 \text{ rev}}{2\pi \text{ rad}}\right) = \boxed{5.0\times10^{10} \text{ rev}}$$

57. Strategy Use $a_r = \omega^2 r$ to find the angular speed required.

Solution The magnitude of the radial acceleration must be the same as the magnitude of the gravitational field strength.

$$a_r = \omega^2 r = g, \text{ so } \omega = \sqrt{\frac{g}{r}} = \sqrt{\frac{9.80 \text{ m/s}^2}{0.20 \text{ m}}} = \boxed{7.0 \text{ rad/s}}.$$

61. (a) Strategy At the top, $\vec{g}$ and $\vec{a}$ are both directed downward. Draw a free-body diagram. Use Newton's second law.

Solution
$\Sigma F_r = N - mg = ma_r = m(-a_y)$, so $W' = N = mg - ma_y$.

The apparent weight is less than the true weight by ma_y. Thus, the lower weight,

$\boxed{518.5 \text{ N}}$, is measured at the top.

(b) Strategy At the bottom, $\vec{g}$ is directed downward and $\vec{a}$ is directed upward. Draw a free-body diagram.

Solution
$\Sigma F_r = N - mg = ma_r = m(a_y)$, so $W' = N = mg + ma_y$.

The apparent weight is greater than the true weight by ma_y. Thus, the higher weight,

$\boxed{521.5 \text{ N}}$, is measured at the bottom.

(c) Strategy The apparent weight at the top is given by $W'_{top} = W(1 - a_y/g)$, where $W = mg$. Use $a_r = \omega^2 r = a_y$ to find the radius.

Solution Solve for the radius r.

$$W'_{top} = W\left(1 - \frac{a_y}{g}\right) = W\left(1 - \frac{\omega^2 r}{g}\right), \text{ so } r = \frac{g}{\omega^2}\left(1 - \frac{W'_{top}}{W}\right) = \frac{9.80 \text{ m/s}^2}{(0.025 \text{ rad/s})^2}\left(1 - \frac{518.5 \text{ N}}{520.0 \text{ N}}\right) = \boxed{45 \text{ m}}.$$

65. Strategy Use the relationship between linear speed and angular speed.

Solution Compute the linear speed of the tip of the nylon cord.
$$v = \omega r = (660 \text{ rad/s})(0.23 \text{ m}) = \boxed{150 \text{ m/s}}$$

69. Strategy Use Eq. (5-20) to find the number of rotations each gear goes through in 2.0 s. Refer to Problem 68.

Solution

$$\Delta\theta_A = \omega_i \Delta t + \frac{1}{2}\alpha(\Delta t)^2$$

$$\left(\frac{1 \text{ rotation}}{2\pi \text{ rad}}\right)\Delta\theta_A = \left(\frac{1 \text{ rotation}}{2\pi \text{ rad}}\right)\left[2\pi(0.955 \text{ Hz})(2.0 \text{ s}) + \frac{1}{2}(3.0 \text{ rad/s}^2)(2.0 \text{ s})^2\right] = \boxed{2.9 \text{ rotations}}$$

From Problem 68, we know that gear B has an angular speed that is twice that of gear A. Thus, gear B rotates twice for each rotation of gear A. Therefore,

$$\left(\frac{1 \text{ rotation}}{2\pi \text{ rad}}\right)\Delta\theta_B = 2\left(\frac{1 \text{ rotation}}{2\pi \text{ rad}}\right)\left[2\pi(0.955 \text{ Hz})(2.0 \text{ s}) + \frac{1}{2}(3.0 \text{ rad/s}^2)(2.0 \text{ s})^2\right] = \boxed{5.7 \text{ rotations}}.$$

73. **(a) Strategy** Use the relationship between radial acceleration and linear speed.

   Solution

   $$a_r = \frac{v^2}{r} = \frac{(2.0\pi \text{ m/s})^2}{0.50 \text{ m}} = \boxed{8.0\pi^2 \text{ m/s}^2 = 79 \text{ m/s}^2}$$

   (b) Strategy Use Newton's second law.

   Solution

   $$\Sigma F_r = T = ma_r, \text{ so } T = (0.50 \text{ kg})(8.0\pi^2 \text{ m/s}^2) = \boxed{4.0\pi^2 \text{ N} = 39 \text{ N}}.$$

77. **Strategy** For each revolution of the flagellum, the bacterium moves the distance of the pitch.

   Solution Compute the speed of the bacterium.
   $$v = (1.0 \text{ μm/rev})(110 \text{ rev/s}) = \boxed{110 \text{ μm/s}}$$

81. **Strategy** Use the relationship between linear speed and angular speed.

   Solution

   $$v = r|\omega| = \left(\frac{0.65 \text{ m}}{2}\right)(101 \text{ rad/s})\left(\frac{1 \text{ km}}{1000 \text{ m}}\right)\left(\frac{3600 \text{ s}}{\text{h}}\right) = \boxed{120 \text{ km/h}}$$

85. **Strategy and Solution** The cutting tool moves one inch in the time $\Delta t = \dfrac{d}{v} = \dfrac{1 \text{ in}}{0.080 \text{ in/s}}$. The lathe chuck must

   complete 18 revolutions in the time Δt. Thus, the rotational speed must be $\dfrac{18 \text{ rev}}{\frac{1 \text{ in}}{0.080 \text{ in/s}}} = \boxed{1.4 \text{ rev/s}}$.

REVIEW AND SYNTHESIS: CHAPTERS 1–5

Review Exercises

1. **Strategy** Replace the quantities with their units.

    Solution Find the units of the spring constant k.

    $F = kx$, so $k = \dfrac{F}{x}$, and the units of k are $\boxed{\text{N/m}} = \dfrac{\text{kg} \cdot \text{m/s}^2}{\text{m}} = \boxed{\text{kg/s}^2}$.

5. **(a) Strategy** Find the total distance traveled and the time of travel. Then divide the distance by the time to obtain the average speed.

    Solution Find Mike's average speed.

    $\Delta x = 50.0 \text{ m} + 34.0 \text{ m} = 84.0 \text{ m}$ and $\Delta t = \dfrac{50.0 \text{ m}}{1.84 \text{ m/s}} + \dfrac{34.0 \text{ m}}{1.62 \text{ m/s}} = 48.2 \text{ s}$, so the average speed is

    $v_{av} = \dfrac{\Delta x}{\Delta t} = \dfrac{84.0 \text{ m}}{48.2 \text{ s}} = \boxed{1.74 \text{ m/s}}$.

    (b) Strategy Find Mike's total displacement and divide it by the time found in part (a) to obtain his average velocity. Let his initial direction be positive.

    Solution Mike's total displacement is $\Delta \vec{r} = 50.0 \text{ m forward} - 34.0 \text{ m back} = 16.0 \text{ m forward}$. So, his average

    velocity is $\vec{v}_{av} = \dfrac{\Delta \vec{r}}{\Delta t} = \dfrac{16.0 \text{ m forward}}{48.2 \text{ s}} = 0.332 \text{ m/s forward}$, or

    $\boxed{0.332 \text{ m/s in his original direction of motion}}$.

9. **Strategy** Let north be up. Using a ruler and a protractor, draw the force vectors to scale; then, find the sum of the force vectors graphically.

    Solution Draw the diagram and measure the length and angle of the sum of the force vectors.

    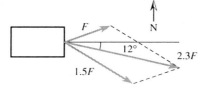

    The net force has a magnitude of about 2.3F, where F is the magnitude of the force with which Sandy pulls. The net force is at an angle of about 12° south of east. So, $\boxed{\text{the cart will go off the road toward south}}$.

48

13. **Strategy** Use Newton's second law to find the acceleration due to friction. Then use the acceleration and distance the plate must travel to determine the necessary initial speed.

    Solution According to Newton's second law, $\Sigma F_y = N - mg = 0$ and $\Sigma F_x = -f_k = ma$.

    So, $a = -\dfrac{f_k}{m} = -\dfrac{\mu_k N}{m} = -\dfrac{\mu_k mg}{m} = -\mu_k g$. Find the initial speed.

    $v_f^2 - v_i^2 = 0 - v_i^2 = 2a\Delta x = -2\mu_k g\Delta x$, so

    $v_i = \sqrt{2\mu_k g\Delta x} = \sqrt{2(0.32)(9.80 \text{ m/s}^2)(0.44 \text{ m})} = \boxed{1.7 \text{ m/s}}$.

17. **Strategy and Solution** Consider a cord attached to a wall at one end and pulled by one of the boys at the other end. The cord does not accelerate when the boy pulls it; thus, the force on the cord from the wall must be equal in magnitude to the pulling force. This situation is identical to the one in which the two boys pull from opposite ends of the cord—the tension in the cord is the same as the case when only one boy is pulling. However, if both pull from one end, the tension is doubled, so $\boxed{\text{Stefan's plan is superior and thus more likely to work}}$.

21. **(a) Strategy** Let the $+y$-direction be down and neglect air resistance.

    Solution Let the initial speed for all three rocks be v_i and the vertical distance from the cliff to the ground be h. For the first rock (thrown straight down):
    $$v_{fy}^2 - v_{iy}^2 = v_f^2 - v_i^2 = 2gh, \text{ so } v_f = \sqrt{v_i^2 + 2gh}.$$
    For the second rock (thrown straight up):
    $$v_{fy}^2 - v_{iy}^2 = v_f^2 - v_i^2 = 2gh, \text{ so } v_f = \sqrt{v_i^2 + 2gh}.$$
    For the third rock (thrown horizontally):
    $$v_{fy}^2 - v_{iy}^2 = v_{fy}^2 - 0 = v_{fy}^2 = 2gh \text{ and } v_{fx} = v_{ix} = v_i, \text{ so } v_f = \sqrt{v_{fx}^2 + v_{fy}^2} = \sqrt{v_i^2 + 2gh}.$$
    Therefore, just before the rocks hit the ground at the bottom of the cliff, $\boxed{\text{all three have the same final speed}}$.

    (b) Strategy Use the result obtained for the final speed in part (a).

    Solution Compute the final speed.
    $$v_f = \sqrt{v_i^2 + 2gh} = \sqrt{(10.0 \text{ m/s})^2 + 2(9.80 \text{ m/s}^2)(15.00 \text{ m})} = \boxed{19.8 \text{ m/s}}$$

25. **(a) Strategy** Use Kepler's third law. Ignore the mass of the cable.

    Solution Find the height H.
    $$R = \left(\frac{Gm_E T^2}{4\pi^2}\right)^{1/3} = R_E + H, \text{ so}$$
    $$H = \left(\frac{Gm_E T^2}{4\pi^2}\right)^{1/3} - R_E = \left(\frac{[6.674\times10^{-11} \text{ m}^3/(\text{kg}\cdot\text{s}^2)](5.974\times10^{24} \text{ kg})(86{,}400 \text{ s})^2}{4\pi^2}\right)^{1/3} - 6.371\times10^6 \text{ m}$$
    $$\cong \boxed{3.6\times10^7 \text{ m}}$$

(b) Strategy Use Newton's second law and law of universal gravitation.

Solution Find the tension in the cable.

$$R = R_E + \frac{H}{2} = 6.371 \times 10^6 \text{ m} + \frac{3.5874 \times 10^7 \text{ m}}{2} = 2.4308 \times 10^7 \text{ m}$$

$$\Sigma F_y = F_g - T = ma = \frac{mv^2}{R} = m\omega^2 R, \text{ so } T = \frac{Gm_E m}{R^2} - m\left(\frac{4\pi^2}{T^2}\right)R.$$

$$T = \frac{[6.674 \times 10^{-11} \text{ m}^3/(\text{kg} \cdot \text{s}^2)](5.974 \times 10^{24} \text{ kg})(100 \text{ kg})}{(2.4308 \times 10^7 \text{ m})^2} - \frac{4\pi^2(100 \text{ kg})(2.4308 \times 10^7 \text{ m})}{(86,400 \text{ s})^2} = \boxed{55 \text{ N}}$$

29. **Strategy** Use Newton's second law to determine the acceleration of each block. Then, use the accelerations to describe the motion of each block.

    Solution Let the positive x-direction be to the right.

    The accelerations of each block are $a_A = \frac{F_A}{m_A}$ and $a_B = \frac{F_B}{m_B}$. In the time Δt, the blocks are displaced by

    $$\Delta x_{A1} = \frac{1}{2}a_A(\Delta t)^2 = \frac{1}{2}\left(\frac{F_A}{m_A}\right)(\Delta t)^2 \text{ and } \Delta x_{B1} = \frac{1}{2}a_B(\Delta t)^2 = \frac{1}{2}\left(\frac{F_B}{m_B}\right)(\Delta t)^2. \text{ At the end of } \Delta t, \text{ the blocks move}$$

    with constant velocities until they meet. These velocities are $v_A = a_A\Delta t = \frac{F_A}{m_A}\Delta t$ and $v_B = a_B\Delta t = \frac{F_B}{m_B}\Delta t$. Let the

    time between Δt and the moment when the two blocks meet be Δt_2. Then, the displacements of each block

    during Δt_2 are $\Delta x_{A2} = v_A\Delta t_2 = \frac{F_A}{m_A}\Delta t\Delta t_2$ and $\Delta x_{B2} = v_B\Delta t_2 = \frac{F_B}{m_B}\Delta t\Delta t_2$. Let $d = 3.40$ m. Then, we have the

    equation $d = \frac{1}{2}\left(\frac{F_A}{m_A}\right)(\Delta t)^2 + \frac{F_A}{m_A}\Delta t\Delta t_2 - \frac{1}{2}\left(\frac{F_B}{m_B}\right)(\Delta t)^2 - \frac{F_B}{m_B}\Delta t\Delta t_2$, where the displacements of block B are

    subtracted because they are negative. Solving this equation for Δt_2 gives

    $$\Delta t_2 = \frac{d - \frac{(\Delta t)^2}{2}\left(\frac{F_A}{m_A} - \frac{F_B}{m_B}\right)}{\left(\frac{F_A}{m_A} - \frac{F_B}{m_B}\right)\Delta t} = \frac{3.40 \text{ m} - \frac{(0.100 \text{ s})^2}{2}\left(\frac{2.00 \text{ N}}{0.225 \text{ kg}} - \frac{-5.00 \text{ N}}{0.600 \text{ kg}}\right)}{\left(\frac{2.00 \text{ N}}{0.225 \text{ kg}} - \frac{-5.00 \text{ N}}{0.600 \text{ kg}}\right)(0.100 \text{ s})} = 1.92 \text{ s. So, the elapsed time from } t = 0 \text{ until}$$

    the blocks meet is $\Delta t + \Delta t_2 = 0.100 \text{ s} + 1.92 \text{ s} = \boxed{2.02 \text{ s}}$. The distance from block B's initial position to where

    the two blocks meet is negative the displacement of B.

    $$-\frac{1}{2}\left(\frac{F_B}{m_B}\right)(\Delta t)^2 - \frac{F_B}{m_B}\Delta t\Delta t_2 = -\frac{1}{2}\left(\frac{-5.00 \text{ N}}{0.600 \text{ kg}}\right)(0.100 \text{ s})^2 - \frac{-5.00 \text{ N}}{0.600 \text{ kg}}(0.100 \text{ s})(1.924 \text{ s}) = 1.65 \text{ m}$$

    The blocks meet at $\boxed{1.65 \text{ m to the left of } B\text{'s initial position}}$.

33. Strategy Use Eq. $\Delta x = (v_i \cos\theta)\Delta t$ for x and $\Delta y = (v_i \sin\theta)\Delta t - \frac{1}{2}g(\Delta t)^2$ for y.

Solution

(a) The vertical and horizontal components of the projectile are given by

$\Delta y = (v_i \sin\theta)\Delta t - \frac{1}{2}g(\Delta t)^2$ and $\Delta x = (v_i \cos\theta)\Delta t$.

When a projectile returns to its original height, $\Delta y = 0$.

$0 = (v_i \sin\theta)\Delta t - \frac{1}{2}g(\Delta t)^2 = v_i \sin\theta - \frac{1}{2}g\Delta t$, so $\Delta t = \frac{2v_i \sin\theta}{g}$.

Substitute this value for Δt into $R = \Delta x = (v_i \cos\theta)\Delta t$ to find the range.

$$R = (v_i \cos\theta)\Delta t = v_i \cos\theta \left(\frac{2v_i \sin\theta}{g}\right) = \boxed{\frac{2v_i^2 \sin\theta\cos\theta}{g}}$$

(b) Using the equation for the range found in part (a), we find that the range of the projectile if it is not intercepted by the wall is $R = \frac{2(50.0 \text{ m/s})^2 \sin 30.0°\cos 30.0°}{9.80 \text{ m/s}^2} = \boxed{221 \text{ m}}$.

(c) Find the time of flight in terms of v_i, Δx, and θ.

$\Delta x = (v_i \cos\theta)\Delta t$, so $\Delta t = \frac{\Delta x}{v_i \cos\theta}$.

Find the height at which the cannonball strikes.

$$y = y_i + (v_i \sin\theta)\Delta t - \frac{1}{2}g(\Delta t)^2 = y_i + (v_i \sin\theta)\left(\frac{\Delta x}{v_i \cos\theta}\right) - \frac{1}{2}g\left(\frac{\Delta x}{v_i \cos\theta}\right)^2 = y_i + \Delta x \tan\theta - \frac{g(\Delta x)^2}{2v_i^2 \cos^2\theta}$$

$$= 1.10 \text{ m} + (215 \text{ m})\tan 30.0° - \frac{(9.80 \text{ m/s}^2)(215 \text{ m})^2}{2(50.0 \text{ m/s})^2 \cos^2 30.0°} = \boxed{4 \text{ m}}$$

37. (a) Strategy Use Newton's law of universal gravitation. Assume the mass of the galaxy is concentrated at its center.

Solution Estimate the mass of the galaxy.

$\frac{GMm}{R^2} = \frac{mv^2}{R}$, so

$$M = \frac{v^2 R}{G} = \frac{(2.75\times10^5 \text{ m/s})^2(40{,}000 \text{ ly})}{6.674\times10^{-11} \text{ m}^3/(\text{kg}\cdot\text{s}^2)}\left(\frac{9.461\times10^{15} \text{ m}}{1 \text{ ly}}\right)$$

$= 4.3\times10^{41}$ kg or about $\boxed{216 \text{ billion solar masses}}$.

(b) **Strategy** Compute the ratio of the visible mass to the estimated mass.

Solution

$\frac{10^{11}}{2.16\times10^{11}} = \boxed{0.46}$

MCAT Review

1. **Strategy and Solution** Gravity contributes an acceleration of $-g$. Air resistance is always opposite an object's direction of motion, so the vertical component of the acceleration contributed by air resistance is negative as well. According to Newton's second law, $F = ma$, so the magnitude of the acceleration due to air resistance is $a_R = F_R/m = bv^2/m$. Since we want the vertical component of acceleration, the correct answer is $\boxed{D}$, $-g - (bvv_y)/(0.5 \text{ kg})$.

2. **Strategy** Use the result for the range derived in Problem 4.48b.

   Solution Assuming air resistance is negligible, the horizontal distance the projectile travels before returning to the elevation from which it was launched is $R = \dfrac{v_i^2 \sin 2\theta}{g} = \dfrac{(30 \text{ m/s})^2 \sin[2(40°)]}{9.80 \text{ m/s}^2} = 90$ m. Thus, the correct answer is $\boxed{C}$.

3. **Strategy and Solution** The magnitude of the horizontal component of air resistance is $F_R \cos\theta = bv^2 \cos\theta = bv(v\cos\theta) = bvv_x$. Thus, the correct answer is $\boxed{D}$.

4. **Strategy** Use Newton's second law to analyze each case. For simplicity, consider only vertical motion.

   Solution Let the positive y-direction be up.
   On the way up:
   $$\Sigma F_y = -mg - bv^2 = ma_y, \text{ so } a_y = -g - \frac{bv^2}{m}.$$
   On the way down:
   $$\Sigma F_y = -mg + bv^2 = ma_y, \text{ so } a_y = -g + \frac{bv^2}{m}.$$
   The magnitude of the acceleration is greater on the way up than on the way down. On the way up, the magnitude of the acceleration is never less than g. On the way down, it may be as small as zero. The projectile must travel the same distance in each case. So, when a projectile is rising, it begins with an initial speed which is reduced to zero relatively quickly due to the relatively large negative acceleration it experiences. When a projectile is falling, it begins with zero speed and is accelerated toward the ground by a smaller acceleration relative to when it is rising. Thus, it must take the projectile longer to reach the ground than to reach its maximum height; therefore, the correct answer is $\boxed{C}$.

5. **Strategy** Find the time it takes to cross the river. Use this time and the speed of the river to find how far downstream the raft travels while crossing. Then use the Pythagorean theorem to find the total distance traveled.

   Solution Let x be the width of the river and y be the distance traveled down the river during the crossing. The raft takes the time $\Delta t = x/v_{\text{raft}}$ to cross the river. During this time, the raft travels the distance $y = v_{\text{river}}\Delta t = v_{\text{river}}(x/v_{\text{raft}})$ down the river. Compute the distance traveled.
   $$\sqrt{x^2 + y^2} = \sqrt{x^2 + \left(\frac{v_{\text{river}}}{v_{\text{raft}}}x\right)^2} = x\sqrt{1 + \left(\frac{v_{\text{river}}}{v_{\text{raft}}}\right)^2} = (200 \text{ m})\sqrt{1 + \left(\frac{2 \text{ m/s}}{2 \text{ m/s}}\right)^2} = 283 \text{ m}$$
   The correct answer is $\boxed{C}$.

6. **Strategy** To row directly across the river, the component of the raft's velocity that is antiparallel to the current of the river must equal the speed of the current, 2 m/s.

   Solution Since the angle is relative to the shore, the antiparallel component of the raft's velocity is $(3 \text{ m/s})\cos\theta$. Set this equal to the speed of the current and solve for θ.

   $(3 \text{ m/s})\cos\theta = 2$ m/s, so $\cos\theta = \dfrac{2}{3}$, or $\theta = \cos^{-1}\dfrac{2}{3}$. The correct answer is $\boxed{D}$.

7. **Strategy** Use $\Delta y = v_{iy}\Delta t + \dfrac{1}{2}a_y(\Delta t)^2$.

   Solution Find the time it takes the rock to reach the ground.

   $$\Delta y = v_{iy}\Delta t + \frac{1}{2}a_y(\Delta t)^2 = (0)\Delta t - \frac{1}{2}g(\Delta t)^2, \text{ so } \Delta t = \sqrt{-\frac{2\Delta y}{g}} = \sqrt{-\frac{2(0-100 \text{ m})}{10 \text{ m/s}^2}} = 4.5 \text{ s}.$$

   The correct answer is $\boxed{A}$.

Chapter 6

CONSERVATION OF ENERGY

Conceptual Questions

1. Assuming the object can be treated as a point particle, the total work done on it by external forces is equal to the change in its kinetic energy. An object moving in a circle may be changing its speed as it goes around, so the total work done on it is not necessarily zero.

5. Yes, static friction can do work. As an example, imagine a book on a conveyor belt that carries it up an incline. The force of static friction on the book is directed upward along the surface of the belt and has a component that is parallel to the book's displacement. Thus, the force of static friction does positive work on the book. (At the same time, the work done on the book by gravity is negative and the total work done on the book is zero.)

9. The bicyclist requires a minimum amount of energy to climb the hill. This quantity is independent of the means that the bicyclist employs to acquire the energy and is solely a function of the height of the hill (the energy required is also affected by the work done by frictional and drag forces—the magnitude of this effect is approximately equal for any method used by the bicyclist to climb the hill and therefore doesn't affect our reasoning). After beginning the ascent, a component of the gravitational force acts in the direction opposite to the displacement thereby increasing the amount of negative work done on the rider with respect to the amount done while riding on flat land. Therefore, the rate at which the rider must do work to acquire the necessary energy is greater when pedaling uphill than when on flat land. It is thus advantageous to acquire as much energy as possible before the ascent when the amount of kinetic energy gained per amount of work done by the rider is greatest.

13. Zorba is correct. You get to a top speed sooner on the first slide, so it takes less time to get to the bottom, but the final speeds are the same from $mgh = \frac{1}{2}mv^2$.

Problems

1. **Strategy** Use Eq. (6-1).

   Solution Find the work done by Denise dragging her basket of laundry.
   $$W = F\Delta r \cos\theta = (30.0\ \text{N})(5.0\ \text{m})\cos 60.0° = \boxed{75\ \text{J}}$$

3. **Strategy and Solution** Since the book undergoes no displacement, $\boxed{\text{no work is done}}$ *on the book* by Hilda.

5. **Strategy** Use Newton's second law and Eq. (6-2).

   Solution Find the net force on the barge.
   $$\Sigma F_y = T\sin\theta - T\sin\theta = 0 \text{ and } \Sigma F_x = T\cos\theta + T\cos\theta = F_x.$$
   Find the work done on the barge.
   $$W = F_x\Delta x = (2T\cos\theta)\Delta x = 2(1.0\ \text{kN})\cos 45°(150\ \text{m}) = \boxed{210\ \text{kJ}}$$

9. **Strategy** Use Eq. (6-2). Let the x-axis point in the direction of motion.

Solution The force of friction is opposite the motion of the box, and according to Newton's second law, it is equal to $f_k = \mu_k N = \mu_k mg$. Juana's horizontal force is in the direction of motion. Solve for the displacement.
$W = F_x \Delta x$, so

$$\Delta x = \frac{W}{F_x} = \frac{W}{F - f_k} = \frac{W}{F - \mu_k mg} = \frac{74.4 \text{ J}}{124 \text{ N} - 0.120(56.8 \text{ kg})(9.80 \text{ m/s}^2)} = \boxed{1.3 \text{ m}}.$$

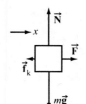

11. **Strategy** The work done on the briefcase by the executive is equal to the change in kinetic energy of the briefcase. Use Eqs. (6-6) and (6-7).

Solution Find the work done by the executive on the briefcase.

$$W = \Delta K = \frac{1}{2}m(v_f^2 - v_i^2) = \frac{1}{2}(5.00 \text{ kg})\left[(2.50 \text{ m/s})^2 - 0\right] = \boxed{15.6 \text{ J}}$$

13. **Strategy** The kinetic energy of the sack is equal to the work done on it by Sam. Use Eqs. (6-2), (6-6), and (6-7).

Solution

(a) Compute the kinetic energy of the sack.

$$\Delta K = \frac{1}{2}m(v_f^2 - v_i^2) = \frac{1}{2}mv^2 - 0 = K = W = F_x \Delta x, \text{ so } K = (2.0 \text{ N})(0.35 \text{ m}) = \boxed{0.70 \text{ J}}.$$

(b) Solve for the speed of the sack.

$$K = \frac{1}{2}mv^2, \text{ so } v = \sqrt{\frac{2K}{m}} = \sqrt{\frac{2(0.70 \text{ J})}{10.0 \text{ kg}}} = \boxed{0.37 \text{ m/s}}.$$

17. **Strategy** Use Eqs. (6-6) and (6-7).

Solution Compute the work done by the wall on the skater.

$$W_{total} = \Delta K = K_f - K_i = 0 - \frac{1}{2}mv^2 = -\frac{1}{2}(69.0 \text{ kg})(11.0 \text{ m/s})^2 = \boxed{-4.17 \text{ kJ}}$$

21. (a) **Strategy and Solution** Since the floor is level, the motion of the desk is perpendicular to the force due to gravity; therefore, the change in the desk's gravitational potential energy is $\boxed{\text{zero}}$.

(b) **Strategy** The motion of the desk is in the direction of the applied constant force. Use Eq. (6-2).

Solution Compute the work done by Justin.
$$W = F_x \Delta x = (340 \text{ N})(10.0 \text{ m}) = \boxed{3.4 \text{ kJ}}$$

(c) **Strategy and Solution** Justin did work against friction, not gravity, so the energy has been $\boxed{\text{dissipated as heat}}$ by friction between the bottom of the desk and the floor.

23. **Strategy** Use Eq. (6-9).

   Solution

   (a) Since the orange returns to its original position $(\Delta y = 0)$ and air resistance is ignored, the change in its potential energy is $\boxed{0}$.

   (b) Let the y-axis point upward and the initial position be $y = 0$.
   $$\Delta U_{grav} = mg\Delta y = (0.30 \text{ kg})(9.80 \text{ m/s}^2)(-1.0 \text{ m} - 0) = \boxed{-2.9 \text{ J}}$$

25. (a) **Strategy and Solution** Since there are two pulleys, only half the force is required to move the mass (but twice the length of rope must be pulled), so the pulley system multiplies the force exerted by a factor of $\boxed{2}$.

   (b) **Strategy** Use $\Delta U = mg\Delta h$.

   Solution Find the change in potential energy of the weight.
   $$\Delta U = mg\Delta h = (48.0 \text{ kg})(9.80 \text{ m/s}^2)(4.00 \text{ m}) = \boxed{1.88 \text{ kJ}}$$

   (c) **Strategy and Solution** By conservation of energy, the work done to lift the mass is equal to its change in potential energy, so $W = \Delta U = \boxed{1.88 \text{ kJ}}$.

   (d) **Strategy and Solution** Twice the length of rope must be pulled to do a given amount of work while applying half the force, so the length of rope pulled is $\boxed{8.00 \text{ m}}$.

27. (a) **Strategy** Use conservation of energy.

   Solution Find the speed of the cart as it passes point 3.
   $$E_1 = \frac{1}{2}mv_1^2 \text{ if } y_1 = 0 \text{ and } E_3 = \frac{1}{2}mv_3^2 + mgy_3. \quad E_f = E_3 = \frac{1}{2}mv_3^2 + mgy_3 = E_i = E_1 = \frac{1}{2}mv_1^2, \text{ so}$$
   $$v_3 = \sqrt{v_1^2 - 2gy_3} = \sqrt{(20.0 \text{ m/s})^2 - 2(9.81 \text{ m/s}^2)(10.0 \text{ m})} = \boxed{14.3 \text{ m/s}}.$$

   (b) **Strategy** Use the result of part (a), replacing 3 with 4. If the result is real—the argument of the square root is nonnegative—the cart will reach position 4.

   Solution Compute the speed of the cart at position 4.
   $$v_4 = \sqrt{v_1^2 - 2gy_4} = \sqrt{(20.0 \text{ m/s})^2 - 2(9.81 \text{ m/s}^2)(20.0 \text{ m})} = 3 \text{ m/s}$$
   The answer is $\boxed{\text{yes; the cart will reach position 4}}$.

29. **Strategy** The initial height of the rope is $l\cos\theta$ where l is the length of the rope and θ is the angle it makes with the vertical. Then $\Delta y = l\cos\theta - l = l(\cos\theta - 1)$. Use conservation of energy.

   Solution Find Bruce's speed at the bottom of the swing.
   $$\Delta K = \frac{1}{2}mv^2 - 0 = \frac{1}{2}mv^2 = -\Delta U = -mg\Delta y = mgl(1 - \cos\theta), \text{ so}$$
   $$v = \sqrt{2gl(1 - \cos\theta)} = \sqrt{2(9.80 \text{ m/s}^2)(20.0 \text{ m})(1 - \cos 35.0°)} = \boxed{8.42 \text{ m/s}}.$$

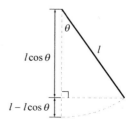

33. Strategy Use conservation of energy.

Solution

(a) Solve for the final speed of the ball.

$$\Delta K = \frac{1}{2}mv_f^2 - \frac{1}{2}mv^2 = -\Delta U = mgh, \text{ so } v_f = \boxed{\sqrt{v^2 + 2gh}}.$$

(b) By inspection of the equation found in part (a), we find that $\boxed{\text{the final speed is independent of the angle}}$.

37. Strategy Use the result for escape speed found in Example 6.8.

Solution Replace the values for Earth with those for the Moon.

$$v = \sqrt{\frac{2GM_{\text{Moon}}}{R_{\text{Moon}}}} = \frac{\sqrt{2(6.674\times10^{-11}\ \text{N}\cdot\text{m}^2/\text{kg}^2)(7.35\times10^{22}\ \text{kg})}}{1.74\times10^6\ \text{m}} = \boxed{2.37\ \text{km/s}}$$

39. Strategy Use conservation of energy and the result for escape speed found in Example 6.8.

Solution Replacing the values for Earth with those for the Zoroaster, we find that the escape speed for Zoroaster is given by $v_{\text{esc}} = \sqrt{2GM_Z/R_Z}$. Find the speed of the meteor when it hits the surface of the planet.

$$\Delta K = \frac{1}{2}mv_f^2 - \frac{1}{2}mv_i^2 = -\Delta U = \frac{GM_Z m}{R_Z}, \text{ so}$$

$$v_f = \sqrt{v_i^2 + \frac{2GM_Z}{R_Z}} = \sqrt{v_i^2 + v_{\text{esc}}^2} = \sqrt{(5.0\ \text{km/s})^2 + (12.0\ \text{km/s})^2} = \boxed{13.0\ \text{km/s}}.$$

41. Strategy Use Eq. (6-14) and form a ratio.

Solution Find the ratio of the potential energies at perigee and apogee.

$$\frac{U_{\text{perigee}}}{U_{\text{apogee}}} = \frac{-\dfrac{GmM_E}{2R_E}}{-\dfrac{GmM_E}{4R_E}} = \frac{4}{2} = \boxed{2}$$

45. Strategy Use Newton's second law and law of universal gravitation.

Solution Calculate the orbital speed.

$$\sum F_r = \frac{GM_E m}{(4.0R_E)^2} = ma_r = m\frac{v_{\text{orb}}^2}{4.0R_E}, \text{ so } v_{\text{orb}} = \sqrt{\frac{GM_E}{4.0R_E}}.$$

Calculate the escape speed.

$$K_i + U_i = \frac{1}{2}mv_{\text{esc}}^2 - \frac{GM_E m}{4.0R_E} = K_f + U_f = 0 + 0, \text{ so } v_{\text{esc}} = \sqrt{\frac{GM_E}{2.0R_E}}.$$

Find the change in speed.

$$\Delta v = v_{\text{esc}} - v_{\text{orb}} = \sqrt{\frac{GM_E}{2.0R_E}} - \sqrt{\frac{GM_E}{4.0R_E}} = \left(\frac{1}{\sqrt{2.0}} - \frac{1}{2.0}\right)\sqrt{\frac{GM_E}{R_E}}$$

$$= \left(\frac{1}{\sqrt{2.0}} - \frac{1}{2.0}\right)\sqrt{\frac{(6.674\times10^{-11}\ \text{N}\cdot\text{m}^2/\text{kg}^2)(5.974\times10^{24}\ \text{kg})}{6.371\times10^6\ \text{m}}} = \boxed{1.6\ \text{km/s}}$$

49. Strategy The work done by the hammer on the object is represented by the area between the curve and the *x*-axis.

Solution Compute the work done in driving the nail.

$W = (50 \text{ N})(0.012 \text{ m}) + (120 \text{ N})(0.050 \text{ m} - 0.012 \text{ m}) = \boxed{5.2 \text{ J}}$

51. (a) Strategy Use Hooke's law and form a proportion.

Solution Form the proportion.

$k = \dfrac{F_1}{x_1} = \dfrac{F_2}{x_2}$

Solve for x_2 to find the amount that the spring stretches.

$x_2 = \dfrac{F_2}{F_1} x_1 = \dfrac{7.0 \text{ N}}{5.0 \text{ N}} (3.5 \text{ cm}) = \boxed{4.9 \text{ cm}}$

(b) Strategy Substitute known values for F_1 and x_1 to find k.

Solution Compute the spring constant.

$k = \dfrac{F_1}{x_1} = \dfrac{5.0 \text{ N}}{3.5 \text{ cm}} = \boxed{1.4 \text{ N/cm}}$

(c) Strategy The triangular area under a forces on the spring vs. the stretch of the spring graph is equal to the work done by the forces.

Solution Compute the work done by the forces on the spring.

$W = \dfrac{1}{2} Fx = \dfrac{1}{2} (5.0 \text{ N})(0.035 \text{ m}) = \boxed{88 \text{ mJ}}$

53. (a) Strategy Set the weight of the mass equal to the force in Hooke's law.

Solution Compute the spring constant.

$W = mg = F = kx, \text{ so } k = \dfrac{mg}{x} = \dfrac{(1.4 \text{ kg})(9.80 \text{ N/kg})}{7.2 \text{ cm}} = \boxed{1.9 \text{ N/cm}}.$

(b) Strategy Use Eq. (6-24).

Solution Compute the elastic potential energy stored in the spring.

$U_{\text{elastic}} = \dfrac{1}{2} kx^2 = \dfrac{1}{2} \left[\dfrac{(1.4 \text{ kg})(9.80 \text{ N/kg})}{0.072 \text{ m}} \right] (0.072 \text{ m})^2 = \boxed{0.49 \text{ J}}$

(c) Strategy Solve for *m* in the equation for *k* found in part (a).

Solution Compute the second mass.

$m_2 = \dfrac{kx_2}{g} = \left(\dfrac{m_1 g}{x_1} \right) \dfrac{x_2}{g} = \dfrac{x_2}{x_1} m_1 = \dfrac{12.2 \text{ cm}}{7.2 \text{ cm}} (1.4 \text{ kg}) = \boxed{2.4 \text{ kg}}$

57. Strategy and Solution Let *E* be the elastic energy stored in the legs. This energy is converted into gravitational potential energy, *mgh*, when the kangaroo jumps. Since only one leg is used, and since $h \propto E$, the kangaroo can only jump half as high, or $(0.70 \text{ m})/2 = \boxed{0.35 \text{ m}}$.

61. **Strategy** Take the surface of the unstretched trampoline to be $y = 0$. Use conservation of energy and Newton's second law.

    Solution Find the spring constant from the gravitational potential energy.

    $mgh = \frac{1}{2}ky_{min}^2$, so $k = \frac{2mgh}{y_{min}^2}$.

    Use Newton's second law for the situation where the gymnast is at rest.

    $\Sigma F = ky - mg = 0$, so

    $y = \frac{mg}{k} = mg\left(\frac{y_{min}^2}{2mgh}\right) = \frac{y_{min}^2}{2h} = \frac{(-0.75 \text{ m})^2}{2(2.5 \text{ m} + 0.75 \text{ m})} = \boxed{8.7 \text{ cm}}$.

65. **Strategy** Watts are joules per second and there are 3600 seconds in 1 hour.

    Solution Show that $1 \text{ kW} \cdot \text{h} = 3.6 \text{ MJ}$.

    $1 \text{ kW} \cdot \text{h} = 10^3 \frac{\text{J}}{\text{s}} \cdot \text{h} \cdot \frac{3600 \text{ s}}{1 \text{ h}} = 3.6 \times 10^6 \text{ J} = 3.6 \text{ MJ}$

69. **Strategy** Use Eq. (6-27).

    Solution

    (a) Find the force exerted on the cyclist by the air.

    $P_a + P_c = 0$ since $\frac{\Delta U}{\Delta t} = 0$. $\vec{v}$ is antiparallel to $\vec{F}_a$.

    $P_a = F_a v \cos\theta$, so $F_a = \frac{P_a}{v\cos\theta} = \frac{-120 \text{ W}}{(6.0 \text{ m/s})\cos 180°} = \boxed{20 \text{ N}}$.

    (b) Find the speed of the cyclist.

    $v = \frac{P_c}{F_c \cos\theta} = \frac{120 \text{ W}}{(-18 \text{ N})\cos 180°} = \boxed{6.7 \text{ m/s}}$

73. **Strategy** Relate the change in gravitational potential energy of the person to the energy provided by the carbohydrate.

    Solution Find the mass of carbohydrate required for the person to climb the stairs.

    $\frac{\text{energy}}{\text{energy available per gram}} = \frac{mgh}{0.100(\text{energy per gram})} = \frac{(74 \text{ kg})(9.80 \text{ m/s}^2)(15 \text{ m})}{0.100(17.6\times 10^3 \text{ J/g})} = \boxed{6.2 \text{ g}}$

    $\boxed{\text{The other 90\% of the energy is dissipated as heat}}$.

77. **Strategy** Use conservation of energy.

    Solution Compute the required speed of the high jumper.

    $K_i + U_i = \frac{1}{2}mv^2 + 0 = \frac{1}{2}mv^2 = K_f + U_f = 0 + mgh = mgh$, so

    $v = \sqrt{2gh} = \sqrt{2(9.80 \text{ m/s}^2)(1.2 \text{ m})} = \boxed{4.8 \text{ m/s}}$.

81. **Strategy** Use Newton's second law and law of universal gravitation.

    Solution Prove that $U = -2K$ for any gravitational circular orbit.

    $$\Sigma F_r = \frac{GMm}{r^2} = ma_r = \frac{mv^2}{r}, \text{ so}$$

    $$m\frac{v^2}{r} = \frac{GMm}{r^2}$$

    $$mv^2 = \frac{GMm}{r}$$

    $$\frac{1}{2}mv^2 = \frac{1}{2}\left(\frac{GMm}{r}\right)$$

    $$K = -\frac{1}{2}U$$

    $$U = -2K$$

85. **Strategy** Use Eqs. (6-10) and (6-11) for the work and energy solution. Use Newton's second law for the force solution. Let $d = 8.0$ m, $d_1 = 5.0$ m, and $d_2 = 8.0$ m $- 5.0$ m $= 3.0$ m.

    Solution First method: Find the constant tension using work and energy.
    $W_{total} = W_c + W_{nc} = \Delta K = 0,$ since the speeds at the top and bottom of the incline

    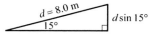

    are zero. Also, $W_c = -\Delta U$ and $W_{nc} = Td_2$.

    $W_{nc} = Td_2 = -W_c = \Delta U,$ so

    $$T = \frac{\Delta U}{d_2} = \frac{mg\Delta y}{d_2} = \frac{mg(0 - d\sin\theta)}{d_2} = -\frac{mgd\sin\theta}{d_2} = -\frac{(4.0 \text{ kg})(9.80 \text{ m/s}^2)(8.0 \text{ m})\sin 15°}{3.0 \text{ m}} = -27 \text{ N}.$$

    The sign is negative because the work done by the tension is opposite the box's motion. So, the magnitude of the tension is $\boxed{27 \text{ N}}$.

    Second method: Find the speed of the block just before the person grasps the cord
    using Newton's second law.
    $\Sigma F_y = N - mg\cos\theta = 0$ and $\Sigma F_x = mg\sin\theta = ma.$

    $$v_f^2 - v_i^2 = v_f^2 - 0 = 2g\sin\theta d_1$$

    Let $v_f = v.$
    Find the tension.

    $$\Sigma F_x = -T + mg\sin\theta = ma_x, \text{ so } a_x = -\frac{T}{m} + g\sin\theta.$$

    $$v_f^2 - v_i^2 = 0 - v^2 = -2g\sin\theta d_1 = 2a_x\Delta x = 2\left(-\frac{T}{m} + g\sin\theta\right)d_2, \text{ so}$$

    $$T = mg\sin\theta\left(\frac{d_1 + d_2}{d_2}\right) = mg\sin\theta\frac{d}{d_2} = (4.0 \text{ kg})(9.80 \text{ m/s}^2)\sin 15°\left(\frac{8.0 \text{ m}}{3.0 \text{ m}}\right) = 27 \text{ N}.$$

89. **Strategy** Use conservation of energy and Newton's second law.

Solution Find k.

$\Sigma F_y = kx_1 - mg = 0$, so $k = \dfrac{mg}{x_1}$. Find v_{max}.

$K_i + K_f = 0 + \dfrac{1}{2}mv_{max}^2 = U_i + U_f = \dfrac{1}{2}kx_{max}^2 + 0$, so

$v_{max} = x_{max}\sqrt{\dfrac{k}{m}} = x_{max}\sqrt{\dfrac{g}{x_1}} = (0.100 \text{ m} - 0.050 \text{ m})\sqrt{\dfrac{9.80 \text{ m/s}^2}{0.060 \text{ m} - 0.050 \text{ m}}} = \boxed{1.6 \text{ m/s}}$.

93. (a) **Strategy** Use the definition of average power. The change in energy is equal to the gravitational potential energy.

Solution Find the average power the motor must deliver.

$P_{av} = \dfrac{\Delta E}{\Delta t} = \dfrac{mgh}{\Delta t} = \dfrac{(1202 \text{ kg} - 801 \text{ kg})(9.80 \text{ m/s}^2)(40.0 \text{ m})}{60.0 \text{ s}} = \boxed{2.62 \text{ kW}}$

(b) **Strategy** Without the counterweight, the motor must deliver more power.

Solution Find the average power the motor must deliver.

$P_{av} = \dfrac{(1202 \text{ kg})(9.80 \text{ m/s}^2)(40.0 \text{ m})}{60.0 \text{ s}} = \boxed{7.85 \text{ kW}}$

The answer is significantly larger.

95. **Strategy** The basal metabolic rate is equal to the number of kilocalories per day required by a person resting under standard conditions.

Solution

(a) Compute Jermaine's basal metabolic rate.

$\text{BMR} = \left(\dfrac{1 \text{ kcal}}{0.010 \text{ mol}}\right)\left(\dfrac{0.015 \text{ mol}}{\text{min}}\right)\left(\dfrac{1440 \text{ min}}{\text{day}}\right) = \boxed{2200 \text{ kcal/day}}$

(b) Find the mass of fat lost.

$\dfrac{2160 \text{ kcal/day}}{9.3 \text{ kcal/g}}\left(\dfrac{2.2 \text{ lb}}{10^3 \text{ g}}\right) = 0.51 \text{ lb/day}$

Since Jermaine is not resting the entire time, he loses $\boxed{\text{more than 0.51 lb}}$.

97. **(a) Strategy** Draw a diagram and use trigonometry.

Solution Referring to the diagram, we see that when Jane is at the lowest point of her swing, $L = h + L\cos 20°$. Solving for h, we find that $h = L - L\cos 20°$.

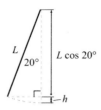

(b) Strategy We assume that no nonconservative forces act (significantly) on Jane. Thus, $\Delta E = 0$.

Solution Use conservation of energy to find Jane's speed at the lowest point of her swing.

$$\Delta E = 0 = \Delta K + \Delta U = \frac{1}{2}mv_f^2 - \frac{1}{2}mv_i^2 + mg\Delta y = \frac{1}{2}mv_f^2 - \frac{1}{2}mv_i^2 + mg(0-h), \text{ so } \frac{1}{2}v_f^2 = \frac{1}{2}v_i^2 + gh, \text{ or}$$

$$v_f = \sqrt{v_i^2 + 2gh} = \sqrt{v_i^2 + 2gL(1 - \cos 20°)} = \sqrt{(4.0 \text{ m/s})^2 + 2(9.80 \text{ m/s}^2)(7.0 \text{ m})(1 - \cos 20°)} = \boxed{4.9 \text{ m/s}}.$$

(c) Strategy When Jane's entire initial kinetic energy is converted into gravitational potential energy, she will have reached her maximum height.

Solution Use conservation of energy to find how high Jane can swing (with respect to her lowest point).

$$\Delta E = 0 = \Delta K + \Delta U = 0 - \frac{1}{2}mv_i^2 + mg\Delta y = -\frac{1}{2}mv_i^2 + mg(h_{max} - h_{min}), \text{ so}$$

$$h_{max} = \frac{v_i^2}{2g} + L(1 - \cos 20°) = \frac{(4.0 \text{ m/s})^2}{2(9.80 \text{ m/s}^2)} + (7.0 \text{ m})(1 - \cos 20°) = \boxed{1.24 \text{ m}}.$$

101. **(a) Strategy** Use Hooke's law and Newton's laws of motion.

Solution According to Hooke's law, $F_1 = k_1 x_1$ and $F_2 = k_2 x_2$.

Imagine that the springs are suspended from a ceiling such that the bottom of each is at the same height. Then a mass m is attached to the bottom of both, the springs stretch, and the system comes to equilibrium. Assume that the masses of the springs are negligible. Sum the vertical forces.

$F_1 + F_2 - W = 0$, so $W = F_1 + F_2 = k_1 x_1 + k_2 x_2$.

Assuming the springs are attached to the same point on the top of the mass, $x_1 = x_2 = x$.

$W = k_1 x_1 + k_2 x_2 = k_1 x + k_2 x = (k_1 + k_2)x = kx = W$

So, in response to a force that stretches the springs (W, in this case), the springs act like one spring with a spring constant $\boxed{k = k_1 + k_2}$.

(b) Strategy Use the result from part (a) and Eq. (6-24).

Solution Compute the potential energy stored in the spring.

$$U = \frac{1}{2}kx^2 = \frac{1}{2}(k_1 + k_2)x^2 = \frac{1}{2}(500 \text{ N/m} + 300 \text{ N/m})(0.020 \text{ m})^2 = \boxed{0.16 \text{ J}}$$

105. (a) Strategy Use Newton's second law and Eq. (6-10).

Solution Find the speed at the bottom of the incline.

$$\Delta K = \frac{1}{2}mv^2 - 0 = W_c + W_{nc} = mgh - fd \text{ so } v = \sqrt{2gh - \frac{2fd}{m}}.$$

Use Newton's second law with $+y$ perpendicular to the incline and $+x$ down the incline.

$\Sigma F_y = N - mg\cos\theta = 0$, so $N = mg\cos\theta$.

Now, $d = 0.85$ m, $h = d\sin\theta$, and $f = \mu N = \mu mg\cos\theta$.

Substitute.

$$v = \sqrt{2gd\sin\theta - \frac{2\mu mgd\cos\theta}{m}} = \sqrt{2gd(\sin\theta - \mu\cos\theta)}$$

Find the maximum compression.

$$K_f + U_f = 0 + \frac{1}{2}kx^2 = K_i + U_i = \frac{1}{2}mv^2 + 0, \text{ so}$$

$$x = v\sqrt{\frac{m}{k}} = \sqrt{\frac{2mgd}{k}(\sin\theta - \mu\cos\theta)} = \sqrt{\frac{2(0.50\text{ kg})(9.80\text{ m/s}^2)(0.85\text{ m})}{35\text{ N/m}}(\sin 30.0° - 0.25\cos 30.0°)}$$

$$= \boxed{26\text{ cm}}.$$

(b) Strategy When the block is accelerated by the spring, it attains its previous kinetic energy and speed.

Solution Find the distance along the incline, d'.

$$\Delta K = 0 - \frac{1}{2}mv^2 = W_c + W_{nc} = -mgh - fd' = -mgd'\sin\theta - \mu mgd'\cos\theta = -d'[mg(\sin\theta + \mu\cos\theta)], \text{ so}$$

$$d' = \frac{v^2}{2g(\sin\theta + \mu\cos\theta)} = \frac{2gd(\sin\theta - \mu\cos\theta)}{2g(\sin\theta + \mu\cos\theta)} = (85\text{ cm})\frac{\sin 30.0° - 0.25\cos 30.0°}{\sin 30.0° + 0.25\cos 30.0°} = \boxed{34\text{ cm}}.$$

109. Strategy Use conservation of energy.

Solution According to the graph, the potential energy in the region under consideration is 300 J. So, $U_f = 300$ J. Initially, the kinetic energy is 200 J and the potential energy is 0, so $K_i = 200$ J and $U_i = 0$. Compute the final kinetic energy; that is, the kinetic energy in the region under consideration.

$E = \Delta K + \Delta U = K_f - K_i + U_f - U_i = K_f - 200\text{ J} + 300\text{ J} - 0 = K_f + 100\text{ J} = 0$, so $K_f = -100$ J, which is impossible, since $\boxed{\text{kinetic energy cannot be negative}}$. Therefore, the answer is $\boxed{\text{no}}$, the particle cannot enter the region 3 cm $< x <$ 8 cm. Since the particle cannot enter the region, $\boxed{\text{it must remain in the region } x < 3\text{ cm}}$.

Chapter 7

LINEAR MOMENTUM

Conceptual Questions

1. The likelihood of injury resulting from jumping from a second floor window is primarily determined by the average force acting to decelerate the body.

   (a) The deceleration time interval for a person landing stiff legged on pavement is very short. The impulse-momentum theorem tells us that the average force acting on the person's feet must therefore be very large—such a person is likely to incur injuries.

   (b) Jumping into a privet hedge increases the time interval over which the body decelerates. This decreases the average force on the person's limbs and therefore decreases the likelihood of injury.

   (c) Jumping into a firefighter's net is the best option of the three. The net stretches downward, gradually bringing the person to rest. Additionally, the firefighters lower the net with their hands as the person lands to further lengthen the time interval during which the person is brought to rest.

5. The law of the conservation of linear momentum states that in the absence of external interactions, the linear momentum of a closed system is constant. Floating in free space, the astronaut and the wrench form a closed system free from interactions with other bodies. If the astronaut throws the wrench in the direction opposite the ship, conservation of momentum dictates that he must in turn move toward the ship.

9. First law: The momentum of an object is constant unless acted upon by an external force. Second law: The net force acting on an object is equal to the rate of change of the object's momentum. Third law: When two objects interact, the changes in momentum that each imparts to the other are equal in magnitude and opposite in direction.

13. An impulse must be supplied to the egg to change its momentum and bring it to rest. A good strategy is to make the time interval over which the stopping force is applied as large as possible. This will reduce the magnitude of the force required to stop the egg. One should therefore attempt to catch the egg with a swinging motion, moving the hand backwards as it is being caught, to bring it to rest as slowly and gently as possible.

17. Daryl has done his homework. If he falls when rock climbing, his rope will stretch and stop him more gradually than the rope Mary wants to buy. In a fall, the climber's momentum must go from some initial value to zero. If the time over which the momentum is decreased to zero is longer, the average force delivered by the rope is smaller.

Problems

1. **Strategy** Use the definition of linear momentum.

   Solution Find the magnitude of the total momentum of the system.
   $$\vec{p}_{total} = \vec{p}_1 + \vec{p}_2 = m\vec{v}_1 + m\vec{v}_2 = m(\vec{v}_1 + \vec{v}_2) = m[\vec{v}_1 + (-\vec{v}_1)] = 0, \text{ so the magnitude is } \boxed{0}.$$

5. **Strategy** Add the momenta of the three particles.

   Solution Find the total momentum of the system.
   $$\vec{p}_{tot} = \vec{p}_1 + \vec{p}_2 + \vec{p}_3 = m_1\vec{v}_1 + m_2\vec{v}_2 + m_3\vec{v}_3 = m_1v_1 \text{ north} + m_2v_2 \text{ south} + m_3v_3 \text{ north}$$
   $$= (m_1v_1 - m_2v_2 + m_3v_3) \text{ north} = [(3.0 \text{ kg})(3.0 \text{ m/s}) - (4.0 \text{ kg})(5.0 \text{ m/s}) + (7.0 \text{ kg})(2.0 \text{ m/s})] \text{ north}$$
   $$= \boxed{3 \text{ kg} \cdot \text{m/s north}}$$

7. Strategy The initial momentum is toward the wall and the final momentum is away from the wall.

Solution Find the change in momentum.
$$\Delta p = p_f - p_i = mv_f - mv_i = m(v_f - v_i) = (5.0 \text{ kg})(-2.0 \text{ m/s} - 2.0 \text{ m/s}) = -20 \text{ kg} \cdot \text{m/s, so}$$
$$\Delta \vec{\mathbf{p}} = \boxed{20 \text{ kg} \cdot \text{m/s in the } -x\text{-direction}}.$$

9. Strategy Use the definition of linear momentum. Let up be the positive direction.

Solution $v_f = v_{fy} = v_{iy} - g\Delta t = -g\Delta t$, since the object starts from rest.
Find Δp.
$$\Delta p = p_f - p_i = m(v_f - v_i) = m(-g\Delta t - 0) = -mg\Delta t = -(3.0 \text{ kg})(9.80 \text{ m/s}^2)(3.4 \text{ s}) = -1.0 \times 10^2 \text{ kg} \cdot \text{m/s, so}$$
$$\Delta \vec{\mathbf{p}} = \boxed{1.0 \times 10^2 \text{ kg} \cdot \text{m/s downward}}.$$

13. Strategy Use the impulse-momentum theorem. Let the positive direction be in the direction of motion.

Solution Find the average horizontal force exerted on the automobile during breaking.
$$F_{av} = \frac{\Delta p}{\Delta t} = \frac{m(v_f - v_i)}{\Delta t} = \frac{(1.0 \times 10^3 \text{ kg})(0 - 30.0 \text{ m/s})}{5.0 \text{ s}} = -6.0 \times 10^3 \text{ N}$$
So, $\vec{\mathbf{F}}_{av} = \boxed{6.0 \times 10^3 \text{ N opposite the car's direction of motion}}.$

17. (a) Strategy Use conservation of energy.

Solution Find the speed with which the pole-vaulter lands.
$$\Delta K = \frac{1}{2}mv^2 = -\Delta U = mgh, \text{ so } v = \sqrt{2gh} = \sqrt{2(9.80 \text{ m/s}^2)(6.0 \text{ m})} = \boxed{11 \text{ m/s}}.$$

(b) Strategy Use the impulse-momentum theorem.

Solution Find the average force on the pole-vaulter's body.
$\Delta p = F_{av}\Delta t$, so
$$F_{av} = \frac{\Delta p}{\Delta t} = \frac{m(v_f - v_i)}{\Delta t} = \frac{m[0 - (-v)]}{\Delta t} = \frac{mv}{\Delta t} = \frac{m}{\Delta t}\sqrt{2gh} = \frac{60.0 \text{ kg}}{0.50 \text{ s}}\sqrt{2(9.80 \text{ m/s}^2)(6.0 \text{ m})} = \boxed{1300 \text{ N}}.$$

21. Strategy Use conservation of momentum.

Solution Find the recoil speed of the thorium nucleus.
$\vec{\mathbf{p}}_i = 0 = -\vec{\mathbf{p}}_f$, so if n = nucleus and p = particle,
$$\vec{\mathbf{P}}_n + \vec{\mathbf{P}}_p = m_n \vec{\mathbf{v}}_n + m_p \vec{\mathbf{v}}_p = 0, \text{ so } |\vec{\mathbf{v}}_n| = \frac{m_p}{m_n}|-\vec{\mathbf{v}}_p| = \frac{4.0 \text{ u}}{234 \text{ u}}\left[0.050(2.998 \times 10^8 \text{ m/s})\right] = \boxed{2.6 \times 10^5 \text{ m/s}}.$$

25. Strategy Use conservation of momentum.

Solution Find the recoil speed of the railroad car.
$\vec{\mathbf{p}}_i = 0 = -\vec{\mathbf{p}}_f$, and since we are only concerned with the horizontal direction, we have:
$$m_c v_{cx} = m_s v_{sx}, \text{ so } v_{cx} = \frac{m_s}{m_c}v_{sx} = \frac{98 \text{ kg}}{5.0 \times 10^4 \text{ kg}}(105 \text{ m/s})\cos 60.0° = \boxed{0.10 \text{ m/s}}.$$

27. Strategy Use the component form of the definition of center of mass.

Solution Find the location of particle B.
Find x_{CM}.

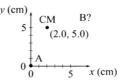

$$x_{CM} = \frac{m_A x_A + m_B x_B}{m_A + m_B} = \frac{0 + m_B x_B}{m_A + m_B}, \text{ so}$$

$$x_B = \frac{m_A + m_B}{m_B} x_{CM} = \frac{30.0 \text{ g} + 10.0 \text{ g}}{10.0 \text{ g}}(2.0 \text{ cm}) = 8.0 \text{ cm}.$$

Similarly,

$$y_B = \frac{30.0 \text{ g} + 10.0 \text{ g}}{10.0 \text{ g}}(5.0 \text{ cm}) = 20 \text{ cm}.$$

The coordinates of particle B are $(x_B, y_B) = \boxed{(8.0 \text{ cm}, 20 \text{ cm})}$.

29. Strategy Since no y-components of the positions have changed, the center of mass moves only in the x-direction. Use the component form of the definition of center of mass.

Solution Find the displacement of the center of mass of the three bodies.

$$x_i = \frac{m x_{1i} + m x_{2i} + m x_{3i}}{m + m + m} = \frac{x_{1i} + x_{2i} + x_{3i}}{3} = \frac{1 \text{ m} + 2 \text{ m} + 3 \text{ m}}{3} = 2 \text{ m}$$

$$x_f = \frac{x_{1f} + x_{2f} + x_{3f}}{3} = \frac{x_{1i} + x_{2i} + x_{3i} + 0.12 \text{ m}}{3} = \frac{6 \text{ m} + 0.12 \text{ m}}{3}$$

$$\Delta x = x_f - x_i = \frac{6 \text{ m} + 0.12 \text{ m}}{3} - 2 \text{ m} = \frac{0.12 \text{ m}}{3} = 4.0 \text{ cm}$$

The center of mass moves $\boxed{4.0 \text{ cm in the positive } x\text{-direction}}$.

33. Strategy The x-coordinate of each three-dimensional shape is midway along its horizontal dimension.

Solution Find the x-component of the center of mass of the composite object.

$$x_{CM} = \frac{m_s x_s + m_c x_c + m_r x_r}{m_s + m_c + m_r} = \frac{(200 \text{ g})(5.0 \text{ cm}) + (450 \text{ g})(10 \text{ cm} + 17/2 \text{ cm}) + (325 \text{ g})(10 \text{ cm} + 17 \text{ cm} + 16/2 \text{ cm})}{200 \text{ g} + 450 \text{ g} + 325 \text{ g}}$$

$$= \boxed{21 \text{ cm}}$$

35. Strategy The total momentum of the system is equal to the total mass of the system times the velocity of the center of mass.

Solution Find the total momentum.

$$\vec{p} = M\vec{v}_{CM} = m_A \vec{v}_A + m_B \vec{v}_B \text{ since } \vec{p} = \vec{p}_A + \vec{p}_B. \text{ Thus, } \vec{v}_{CM} = \frac{m_A \vec{v}_A + m_B \vec{v}_B}{m_A + m_B}.$$

Find the components of $\vec{v}_{CM}$.

$$v_{CMx} = \frac{m_A v_{Ax} + m_B v_{Bx}}{m_A + m_B} = \frac{(3 \text{ kg})(14 \text{ m/s}) + (4 \text{ kg})(0)}{3 \text{ kg} + 4 \text{ kg}} = 6 \text{ m/s}$$

$$v_{CMy} = \frac{(3 \text{ kg})(0) + (4 \text{ kg})(-7 \text{ m/s})}{3 \text{ kg} + 4 \text{ kg}} = -4 \text{ m/s}$$

So, the components are $(v_{CMx}, v_{CMy}) = \boxed{(6 \text{ m/s}, -4 \text{ m/s})}$.

37. **(a) Strategy** Draw a diagram and use conservation of linear momentum.

    Solution

    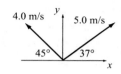

    $$\vec{p} = M\vec{v}_{CM} = \frac{M}{4}\vec{v}_1 + \frac{M}{3}\vec{v}_2 + \frac{5M}{12}\vec{v}_3 = 0 \text{ Use components.}$$

    $$\frac{M}{4}v_{1x} + \frac{M}{3}v_{2x} + \frac{5M}{12}v_{3x} = 3v_{1x} + 4v_{2x} + 5v_{3x} = 0, \text{ so}$$

    $$v_{3x} = -\frac{3v_{1x} + 4v_{2x}}{5} = -\frac{3(5.0 \text{ m/s})\cos 37° + 4(4.0 \text{ m/s})\cos 135°}{5} = -0.13 \text{ m/s}.$$

    $$\frac{M}{4}v_{1y} + \frac{M}{3}v_{2y} + \frac{5M}{12}v_{3y} = 3v_{1y} + 4v_{2y} + 5v_{3y} = 0, \text{ so}$$

    $$v_{3y} = -\frac{3v_{1y} + 4v_{2y}}{5} = -\frac{3(5.0 \text{ m/s})\sin 37° + 4(4.0 \text{ m/s})\sin 135°}{5} = -4.1 \text{ m/s}.$$

    The velocity components are $\boxed{(-0.13 \text{ m/s}, \ -4.1 \text{ m/s})}$.

    (b) Strategy and Solution Due to the law of conservation of linear momentum,
    $\boxed{\text{the center of mass of the system remains at the origin after the explosion}}$.

41. **Strategy** Use conservation of momentum.

    Solution

    (a) The collision is perfectly inelastic, so $v_{1f} = v_{2f} = v_f$. Find the speed of the two cars after the collision.

    $$m_1 v_{1i} + m_2 v_{2i} = mv_{1i} + 4.0m(0) = m_1 v_{1f} + m_2 v_{2f} = mv_f + 4.0mv_f, \text{ so } v_f = \frac{v_{1i}}{5.0} = \frac{1.0 \text{ m/s}}{5.0} = \boxed{0.20 \text{ m/s}}.$$

    (b) The cars are at rest after the collision, so $v_{1f} = v_{2f} = 0$.

    $$mv_{1i} + 4.0mv_{2i} = 0, \text{ so } v_{2i} = -\frac{v_{1i}}{4.0} = -\frac{1.0 \text{ m/s}}{4.0} = -0.25 \text{ m/s}. \text{ The initial speed was } \boxed{0.25 \text{ m/s}}.$$

43. **Strategy** Use conservation of momentum. The collision is perfectly inelastic, so $v_{1f} = v_{2f} = v_f$. Also, the block is initially at rest, so $v_{2i} = 0$.

    Solution Find the speed of the block of wood and the bullet just after the collision.
    $m_1 v_{1f} + m_2 v_{2f} = (m_1 + m_2)v_f = m_1 v_{1i} + m_2 v_{2i} = m_1 v_{1i} + m_2(0), \text{ so}$

    $$v_f = \frac{m_1}{m_1 + m_2}v_{1i} = \frac{0.050 \text{ kg}}{0.050 \text{ kg} + 0.95 \text{ kg}}(100.0 \text{ m/s}) = \boxed{5.0 \text{ m/s}}.$$

45. **Strategy** Use conservation of momentum.

    Solution Find the total momentum of the two blocks after the collision.
    $$\Delta p_2 = -\Delta p_1$$
    $$p_{2f} - p_{2i} = p_{1i} - p_{1f}$$
    $$p_{1f} + p_{2f} = p_{1i} + p_{2i}$$
    $$(m_1 + m_2)v_f = m_1 v_{1i} + m_2 v_{2i}$$
    $$p_f = (2.0 \text{ kg})(1.0 \text{ m/s}) + (1.0 \text{ kg})(0) = 2.0 \text{ kg} \cdot \text{m/s} = p_{1i}$$

    Since p_{1i} was directed to the right, and $p_f = p_{1i}$, the total momentum of the two blocks after the collision is $\boxed{2.0 \text{ kg} \cdot \text{m/s to the right}}$.

49. Strategy Use conservation of momentum. Let the positive direction be the initial direction of motion.

Solution Find the speed of the 5.0-kg body after the collision.
$m_1 v_{1f} + m_2 v_{2f} = m_1 v_{1i} + m_2 v_{2i}$, so

$$v_{2f} = \frac{m_1(v_{1i} - v_{1f}) + m_2 v_{2i}}{m_2} = \frac{(1.0 \text{ kg})\left[10.0 \text{ m/s} - (-5.0 \text{ m/s})\right] + (5.0 \text{ kg})(0)}{5.0 \text{ kg}} = \boxed{3.0 \text{ m/s}}.$$

51. Strategy The spring imparts the same (in magnitude) impulse to each block. (The same magnitude force is exerted on each block by the ends of the spring for the same amount of time.) So, each block has the same final magnitude of momentum. (The initial momentum is zero.)

Solution Find the mass of block B.
$m_B v_B = m_A v_A$, so

$$m_B = \frac{v_A}{v_B} m_A = \frac{d_A / \Delta t}{d_B / \Delta t} m_A = \frac{d_A}{d_B} m_A = \frac{1.0 \text{ m}}{3.0 \text{ m}}(0.60 \text{ kg}) = \boxed{0.20 \text{ kg}}.$$

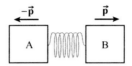

53. Strategy The collision is perfectly inelastic, so $v_{1f} = v_{2f} = v$. The block is initially at rest, so $v_{2i} = 0$ and $v_{1i} = v_i$. Use conservation of momentum.

Solution Find the speed of the bullet and block system.
$(m_{bul} + m_{blk})v = m_{bul} v_i + m_{blk}(0)$, so $v = \dfrac{m_{bul}}{m_{bul} + m_{blk}} v_i$.

Determine the time it takes the system to hit the floor.

$$\Delta y = -h = v_{iy}\Delta t - \frac{1}{2}g(\Delta t)^2 = 0 - \frac{1}{2}g(\Delta t)^2, \text{ so } \Delta t = \sqrt{\frac{2h}{g}}.$$

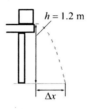

Find the horizontal distance traveled.

$$\Delta x = v_{ix}\Delta t = v\Delta t = \frac{m_{bul}}{m_{bul} + m_{blk}} v_i \sqrt{\frac{2h}{g}} = \frac{0.010 \text{ kg}}{0.010 \text{ kg} + 4.0 \text{ kg}}(400.0 \text{ m/s})\sqrt{\frac{2(1.2 \text{ m})}{9.80 \text{ m/s}^2}} = \boxed{0.49 \text{ m}}$$

57. Strategy Use conservation of momentum. Let each of the first two pieces be 45° from the positive *x*-axis (one CW, one CCW).

Solution Find the speed of the third piece.
Find v_{3x}.

$$p_{1x} + p_{2x} + p_{3x} = mv_{1x} + mv_{2x} + mv_{3x} = 0, \text{ so } v_{3x} = -v_{1x} - v_{2x} = -v\cos 45° - v\cos(-45°) = -\frac{v}{\sqrt{2}} - \frac{v}{\sqrt{2}} = -v\sqrt{2}.$$

Similarly,

$$v_{3y} = -v_{1y} - v_{2y} = -v\sin 45° - v\sin(-45°) = -\frac{v}{\sqrt{2}} + \frac{v}{\sqrt{2}} = 0, \text{ so } v_3 = |v_{3x}| = v\sqrt{2} = (120 \text{ m/s})\sqrt{2} = \boxed{170 \text{ m/s}}.$$

59. **Strategy** Use conservation of momentum. Refer to Practice Problem 7.11.

Solution

(a) Find the momentum change of the ball of mass m_1.

$$\Delta p_{1x} = -\Delta p_{2x} = m_2 v_{2ix} - m_2 v_{2fx} = m_2(0 - v_{2fx}) = -m_2 v_{2fx} = -5m_1\left[\frac{1}{4}v_i \cos(-36.9°)\right] = \boxed{-1.00m_1v_i}$$

$$\Delta p_{1y} = -\Delta p_{2y} = m_2 v_{2iy} - m_2 v_{2fy} = 5m_1(0 - v_{2fy}) = -5m_1 v_{2fy} = -5m_1\left[\frac{1}{4}v_i \sin(-36.9°)\right] = \boxed{0.751m_1v_i}$$

(b) Find the momentum change of the ball of mass m_2.

$$\Delta p_{2x} = -\Delta p_{1x} = m_1(v_{1ix} - v_{1fx}) = m_1(v_i - 0) = \boxed{m_1 v_i}$$

$$\Delta p_{2y} = -\Delta p_{1y} = m_1(v_{1iy} - v_{1fy}) = m_1(0 - v_1) = -m_1 v_1 = -m_1(0.751v_i) = \boxed{-0.751m_1v_i}$$

$$\boxed{\text{The momentum changes for each mass are equal and opposite.}}$$

61. **Strategy** Use conservation of momentum.

Solution Find v_{2f} in terms of v_{1f}.

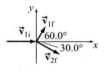

$$mv_{1fy} + mv_{2fy} = mv_{1f}\sin\theta_1 + mv_{2f}\sin\theta_2 = mv_{1iy} + mv_{2iy} = 0 + 0,\ \text{so}$$

$$v_{2f} = \frac{-\sin\theta_1}{\sin\theta_2}v_{1f} = \frac{-\sin 60.0°}{\sin(-30.0°)}v_{1f} = \boxed{1.73v_{1f}}.$$

65. **Strategy** The collision is perfectly inelastic, so the final velocities of the cars are identical. Use conservation of momentum.

Solution Let the 1700-kg car be (1) and the 1300-kg car be (2).

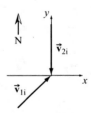

$$P_{ix} = m_1 v_{1ix} + m_2 v_{2ix} = m_1 v_{1ix} + 0 = P_{fx} = (m_1 + m_2)v_{fx},\ \text{so}\ v_{fx} = \frac{m_1}{m_1 + m_2}v_{1ix}.$$

$$P_{iy} = m_1 v_{1iy} + m_2 v_{2iy} = P_{fy} = (m_1 + m_2)v_{fy},\ \text{so}\ v_{fy} = \frac{m_1 v_{1iy} + m_2 v_{2iy}}{m_1 + m_2}.$$

Compute the final speed and the direction.

$$v_f = \sqrt{v_{fx}^2 + v_{fy}^2} = \sqrt{\left(\frac{m_1 v_{1ix}}{m_1 + m_2}\right)^2 + \left(\frac{m_1 v_{1iy} + m_2 v_{2iy}}{m_1 + m_2}\right)^2}$$

$$= \frac{\sqrt{[(1700\ \text{kg})(14\ \text{m/s})\cos 45°]^2 + [(1700\ \text{kg})(14\ \text{m/s})\sin 45° + (1300\ \text{kg})(-18\ \text{m/s})]^2}}{1700\ \text{kg} + 1300\ \text{kg}} = 6.0\ \text{m/s}$$

$$\theta = \tan^{-1}\frac{v_{fy}}{v_{fx}} = \tan^{-1}\frac{(1700\ \text{kg})(14\ \text{m/s})\sin 45° + (1300\ \text{kg})(-18\ \text{m/s})}{(1700\ \text{kg})(14\ \text{m/s})\cos 45°} = -21°$$

Thus, the final velocity of the cars is $\boxed{6.0\ \text{m/s at } 21° \text{ S of E}}$.

69. Strategy Use conservation of momentum.

Solution Let swallow 1 and its coconut be (1) and swallow 2 and its coconut be (2) (before the collision). After the collision, let swallow 1's coconut be (3), swallow 2's coconut be (4), and the tangled-up swallows be (5).

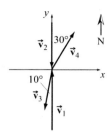

$$p_{ix} = m_1 v_{1x} + m_2 v_{2x} = 0 + 0 = p_{fx} = m_3 v_{3x} + m_4 v_{4x} + m_5 v_{5x}, \text{ so } v_{5x} = -\frac{m_3 v_{3x} + m_4 v_{4x}}{m_5}.$$

$$p_{iy} = m_1 v_{1y} + m_2 v_{2y} = m_1 v_1 + m_2 v_2 = p_{fy} = m_3 v_{3y} + m_4 v_{4y} + m_5 v_{5y}, \text{ so }$$

$$v_{5y} = \frac{m_1 v_1 + m_2 v_2 - m_3 v_{3y} - m_4 v_{4y}}{m_5}.$$

Compute the final speed of the tangled swallows, v_5.

$$v_5 = \sqrt{v_{5x}^2 + v_{5y}^2} = \frac{1}{m_5}\sqrt{[-(m_3 v_{3x} + m_4 v_{4x})]^2 + (m_1 v_1 + m_2 v_2 - m_3 v_{3y} - m_4 v_{4y})^2}$$

$$= \frac{\sqrt{\begin{array}{l}[(0.80 \text{ kg})(13 \text{ m/s})\cos 260° + (0.70 \text{ kg})(14 \text{ m/s})\cos 60°]^2 \\ +[(1.07 \text{ kg})(20 \text{ m/s}) + (0.92 \text{ kg})(-15 \text{ m/s}) - (0.80 \text{ kg})(13 \text{ m/s})\sin 260° - (0.70 \text{ kg})(14 \text{ m/s})\sin 60°]^2\end{array}}}{0.270 \text{ kg} + 0.220 \text{ kg}}$$

$$= 20 \text{ m/s}$$

Compute the direction.

$$\theta = \tan^{-1}\frac{v_{5y}}{v_{5x}}$$

$$= \tan^{-1}\frac{(1.07 \text{ kg})(20 \text{ m/s}) + (0.92 \text{ kg})(-15 \text{ m/s}) - (0.80 \text{ kg})(13 \text{ m/s})\sin 260° - (0.70 \text{ kg})(14 \text{ m/s})\sin 60°}{-(0.80 \text{ kg})(13 \text{ m/s})\cos 260° - (0.70 \text{ kg})(14 \text{ m/s})\cos 60°}$$

$$= -72°$$

Since $v_{5x} < 0$ and $v_{5y} > 0$, the velocity vector is located in the second quadrant, so the angle is $180° - 72° = 108°$ from the positive x-axis or $18°$ west of north. Thus, the velocity of the birds immediately after the collision is $\boxed{20 \text{ m/s at } 18° \text{ W of N}}$.

71. Strategy Use conservation of momentum. The collision is perfectly inelastic, so $v_{1f} = v_{2f} = v_f$.

Solution Find the speed of the cars just after the collision.

$$m_1 v_{1f} + m_2 v_{2f} = (m_1 + m_2)v_f = m_1 v_{1i} + m_2 v_{2i} = m_1 v_i + 0, \text{ so}$$

$$v_f = \frac{m_1 g}{(m_1 + m_2)g} v_i = \frac{13.6 \text{ kN}}{13.6 \text{ kN} + 9.0 \text{ kN}}(17.0 \text{ m/s}) = \boxed{10.2 \text{ m/s}}.$$

73. Strategy Use the definition of linear momentum.

Solution Find the magnitude of the total momentum of the ship and the crew.

$$p_{tot} = m_{tot}v = (2.0 \times 10^3 \text{ kg} + 4.8 \times 10^4 \text{ kg})(1.0 \times 10^5 \text{ m/s}) = \boxed{5.0 \times 10^9 \text{ kg} \cdot \text{m/s}}$$

75. **Strategy** Use the impulse-momentum theorem.

Solution Find the average force exerted by the ground on the ball.

$$F_{av} = \frac{\Delta p}{\Delta t}$$

$$= \frac{m\sqrt{(\Delta v_x)^2 + (\Delta v_y)^2}}{\Delta t}$$

$$= \frac{0.060 \text{ kg}}{0.065 \text{ s}} \sqrt{\left[(53 \text{ m/s})\cos 18° - (54 \text{ m/s})\cos(-22°)\right]^2 + \left[(53 \text{ m/s})\sin 18° - (54 \text{ m/s})\sin(-22°)\right]^2} = \boxed{34 \text{ N}}$$

77. **Strategy** The center of mass of each block is its center. Add up the individual center of mass components to find the components of the center of mass of the block structure.

Solution

$$x_{CM} = \frac{mx_1 + 2mx_2 + 5mx_3 + 4mx_4}{12m} = \frac{x_1 + 2x_2 + 5x_3 + 4x_4}{12} = \frac{0 + 2(1.0 \text{ in}) + 5(2.0 \text{ in}) + 4(3.0 \text{ in})}{12} = 2.0 \text{ in}$$

$$y_{CM} = \frac{6my_1 + 4my_2 + my_3 + my_4}{12m} = \frac{6y_1 + 4y_2 + y_3 + y_4}{12} = \frac{6(0) + 4(1.0 \text{ in}) + 2.0 \text{ in} + 3.0 \text{ in}}{12} = 0.75 \text{ in}$$

$$z_{CM} = \frac{9mz_1 + 3mz_2}{12m} = \frac{9z_1 + 3z_2}{12} = \frac{9(0) + 3(1.0 \text{ in})}{12} = 0.25 \text{ in}$$

The center of mass of the block structure is located at $\boxed{(2.0 \text{ in}, 0.75 \text{ in}, 0.25 \text{ in})}$.

81. **(a) Strategy** The initial momentum of the baseball is $p_i = mv_i$. The final momentum is zero.

Solution Compute the change in momentum.
$$\Delta p = p_f - p_i = 0 - mv_i = -mv_i = -(0.15 \text{ kg})(35 \text{ m/s}) = -5.3 \text{ kg} \cdot \text{m/s}$$

Thus, the change in momentum was $\boxed{5.3 \text{ kg} \cdot \text{m/s opposite the ball's direction of motion}}$.

(b) Strategy and Solution According to the impulse-momentum theorem, the impulse applied to the ball is equal to the change in the momentum of the ball, or $\boxed{5.3 \text{ kg} \cdot \text{m/s opposite the ball's direction of motion}}$.

(c) Strategy Use the impulse momentum theorem.

Solution Since the acceleration is assumed constant, the time it takes for the ball to come to a complete stop is $\Delta t = \Delta x / v_{av}$. Compute the average force applied to the ball by the catcher's glove.

$$\vec{F}_{av} = \frac{\Delta \vec{p}}{\Delta t} = v_{av} \frac{\Delta \vec{p}}{\Delta x} = \left(\frac{35 \text{ m/s}}{2}\right) \frac{5.25 \text{ kg} \cdot \text{m/s opposite the direction of motion}}{0.050 \text{ m}}$$

$$= \boxed{1.8 \text{ kN opposite the ball's direction of motion}}$$

85. **Strategy** We must determine the initial speeds of the two cars. The collision is perfectly inelastic, so the final velocities of the cars are identical. Use conservation of momentum and the work-kinetic energy theorem.

Solution Let the 1100-kg car be (1) and the 1300-kg car be (2). Use the work-kinetic energy theorem to determine the kinetic energy and, thus, the initial speed of the wrecked cars, which is the final speed of the collision.

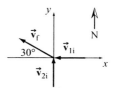

$$W = F\Delta r = f_k \Delta r = -\mu_k mg \Delta r = \Delta K = 0 - \frac{1}{2}mv_i^2, \text{ so } v_i = \sqrt{2\mu_k g \Delta r}.$$

Thus, the final speed of the collision is $v_f = \sqrt{2\mu_k g \Delta r}$.

Find the initial speeds.

$$p_{ix} = m_1 v_{1ix} + m_2 v_{2ix} = m_1 v_{1i} + 0 = p_{fx} = (m_1 + m_2)v_{fx}, \text{ so}$$

$$v_{1i} = \frac{m_1 + m_2}{m_1}v_{fx} = \frac{m_1 + m_2}{m_1}\sqrt{2\mu_k g \Delta r}\cos 150° = \frac{2400 \text{ kg}}{1100 \text{ kg}}\sqrt{2(0.80)(9.80 \text{ m/s}^2)(17 \text{ m})}\cos 150° \left(\frac{1 \text{ km/h}}{0.2778 \text{ m/s}}\right)$$
$$= -110 \text{ km/h}.$$

$$p_{iy} = m_1 v_{1iy} + m_2 v_{2iy} = 0 + m_2 v_{2i} = p_{fy} = (m_1 + m_2)v_{fy}, \text{ so}$$

$$v_{2i} = \frac{m_1 + m_2}{m_2}v_{fy} = \frac{m_1 + m_2}{m_2}\sqrt{2\mu_k g \Delta r}\sin 150° = \frac{2400 \text{ kg}}{1300 \text{ kg}}\sqrt{2(0.80)(9.80 \text{ m/s}^2)(17 \text{ m})}\sin 150° \left(\frac{1 \text{ km/h}}{0.2778 \text{ m/s}}\right)$$
$$= 54 \text{ km/h}.$$

Since $110 > 70$, $\boxed{\text{the lighter car was speeding}}$.

89. **Strategy** Use conservation of energy and momentum. Let $2m = m_B = 2m_A$.

Solution Find the maximum kinetic energy of A alone and, thus, its speed just before it strikes B.
$$\Delta K = \frac{1}{2}mv_1^2 - 0 = -\Delta U = mgh - 0, \text{ so } v_1 = \sqrt{2gh}.$$
Use conservation of momentum to find the speed of the combined bobs just after impact. The collision is perfectly inelastic, so $v_{Af} = v_{Bf} = v_2$.

$$m_A v_{Af} + m_B v_{Bf} = (m + 2m)v_2 = m_A v_{Ai} + m_B v_{Bi} = mv_1 + 0, \text{ so } v_2 = \frac{1}{3}v_1.$$

Find the maximum height.
$$\Delta K = 0 - \frac{1}{2}mv_2^2 = -\frac{1}{2}m\left(\frac{1}{3}\sqrt{2gh}\right)^2 = -\Delta U = 0 - mgh_2, \text{ so } h_2 = \boxed{\frac{1}{9}h}.$$

93. Strategy Use conservation of momentum and Eq. (6-6) for the kinetic energies. Since the radium nucleus is at rest, $\vec{p}_i = \vec{p}_{Ra} = 0$.

Solution

(a) Find the ratio of the speed of the alpha particle to the speed of the radon nucleus.

$p_f = m_{Rn}v_{Rn} + m_\alpha v_\alpha = p_i = 0$, so $m_\alpha v_\alpha = -m_{Rn}v_{Rn}$. Therefore,

$$\frac{v_\alpha}{v_{Rn}} = \frac{m_{Rn}}{m_\alpha} = \frac{222\text{ u}}{4\text{ u}} = \frac{222}{4} = \boxed{\frac{111}{2}}, \text{ where the negative was dropped because speed is nonnegative.}$$

(b) Since the initial momentum is zero, $\vec{p}_{Rn} = -\vec{p}_\alpha$; therefore, $\dfrac{|\vec{p}_\alpha|}{|\vec{p}_{Rn}|} = \dfrac{p_\alpha}{p_{Rn}} = \boxed{1}$.

(c) Find the ratio of the kinetic energies.

$$\frac{K_\alpha}{K_{Rn}} = \frac{\frac{1}{2}m_\alpha v_\alpha^2}{\frac{1}{2}m_{Rn}v_{Rn}^2} = \frac{m_\alpha}{m_{Rn}}\left(\frac{v_\alpha}{v_{Rn}}\right)^2 = \frac{4\text{ u}}{222\text{ u}}\left(\frac{111}{2}\right)^2 = \boxed{\frac{111}{2}}$$

Chapter 8

TORQUE AND ANGULAR MOMENTUM

Conceptual Questions

1. To maximize the torque, locate it as far as possible from the rotation axis: along the lower edge.

5. For a body to be in equilibrium, both the net force and the net torque acting on it must equal zero. To satisfy the first requirement, the two forces must be equal in magnitude and opposite in direction. To satisfy the second requirement, the two forces must act along the same line—a net torque would otherwise act to rotate the object.

9. An object's moment of inertia depends on how its mass is distributed with respect to the axis of rotation. The farther the mass is from the axis, the greater the object's moment of inertia. When animals have leg muscles that are concentrated close to the hip joint, their legs have relatively small moments of inertia. This makes it easier for them to rotate their legs, allowing them to run faster.

13. The vertical component of the angular momentum of the system (merry-go-round and child) is conserved throughout this process, since there are no external torques about the vertical axis of the merry-go-round. When the child moves out to the rim, the rotational inertia of the system increases, because the child is located farther from the axis. To conserve angular momentum, the angular velocity must therefore decrease. Noting that the rotational kinetic energy can be written as $L^2/(2I)$ and that L remains constant while I increases, we see that the rotational kinetic energy of the system decreases.

17. The astronaut and satellite constitute an isolated system. The initial angular momentum of the system is zero. When the astronaut tries to remove the bolt, both he and the satellite will rotate. They will rotate in opposite directions so that the total angular momentum of the system remains zero. To put it another way, when the astronaut applies a torque to a part of the satellite, the satellite applies an equal and opposite torque to him. The astronaut must anchor the satellite and himself somehow before trying to remove the bolt.

21. The melting of Earth's polar ice caps would distribute some of its mass from locations near its rotation axis to locations that are on average farther from its rotation axis. The rotational inertia of a sphere is greater if its mass is distributed farther from its axis of rotation—the Earth's moment of inertia would therefore increase. Angular momentum conservation requires that the product of the Earth's rotational inertia and its angular velocity be constant. A larger moment of inertia must be accompanied by a smaller angular velocity—the melting of the caps would therefore increase the length of the day.

Problems

1. **Strategy and Solution** I has units $\text{kg} \cdot \text{m}^2$. ω^2 has units $(\text{rad/s})^2$. So, $\frac{1}{2}I\omega^2$ has units
   $\text{kg} \cdot \text{m}^2 \cdot \text{rad}^2/\text{s}^2 = \text{kg} \cdot \text{m}^2/\text{s}^2 = \text{J}$, which is a unit of energy.

3. **Strategy** $I = \frac{2}{5}MR^2$ for a solid sphere and mass density is $\rho = M/V$.

   Solution

   (a) $M = \rho V = \rho \frac{4}{3}\pi R^3$ for a solid sphere. Form a proportion.

   $$\frac{M_{\text{child}}}{M_{\text{adult}}} = \left(\frac{R_{\text{child}}}{R_{\text{adult}}}\right)^3 = \left(\frac{1}{2}\right)^3 = \frac{1}{8}, \text{ so the mass is } \boxed{\text{reduced by a factor of 8}}.$$

   (b) Form a proportion.

   $$\frac{I_{\text{child}}}{I_{\text{adult}}} = \frac{1}{8}\left(\frac{R_{\text{child}}}{R_{\text{adult}}}\right)^2 = \frac{1}{8}\left(\frac{1}{2}\right)^2 = \frac{1}{32}$$

   The rotational inertia is $\boxed{\text{reduced by a factor of 32}}$.

5. **Strategy** Find the rotational inertia in each case by using Eq. (8-2).

   Solution

   (a) $I = m(r^2 + 0^2 + 0^2 + r^2) = 2mr^2 = 2(3.0 \text{ kg})(0.50 \text{ m})^2 = \boxed{1.5 \text{ kg} \cdot \text{m}^2}$

   (b) $I = m(0^2 + r^2 + 0^2 + r^2) = 2mr^2 = 2(3.0 \text{ kg})(0.50 \text{ m}/\sqrt{2})^2 = \boxed{0.75 \text{ kg} \cdot \text{m}^2}$

   (c) $I = m(r^2 + r^2 + r^2 + r^2) = 4mr^2 = 4(3.0 \text{ kg})(0.50 \text{ m}/\sqrt{2})^2 = \boxed{1.5 \text{ kg} \cdot \text{m}^2}$

9. (a) **Strategy and Solution** Since a significant fraction of the wheel's kinetic energy is rotational, to model it as
       if it were sliding without friction would be unjustified. So, the answer is $\boxed{\text{no}}$.

   (b) **Strategy** Use Eq. (8-1) and form a proportion.

   Solution Find the fraction of the total kinetic energy that is rotational.

   $$\frac{K_{\text{rot}}}{K_{\text{total}}} = \frac{4\left(\frac{1}{2}I\omega^2\right)}{\frac{1}{2}Mv^2 + 4\left(\frac{1}{2}I\omega^2\right)} = \frac{1}{\frac{Mv^2}{4I\omega^2} + 1} = \frac{1}{1 + \frac{Mv^2}{4I(v^2/R^2)}} = \frac{1}{1 + \frac{MR^2}{4I}} = \frac{1}{1 + \frac{(1300 \text{ kg})(0.35 \text{ m})^2}{4(0.705 \text{ kg} \cdot \text{m}^2)}} = \boxed{0.017}$$

13. **Strategy** Use Eq. (8-3).

    Solution Find the magnitude of the torque.
    $$|\tau| = F_\perp r = mgr = (40.0 \text{ kg})(9.80 \text{ N/kg})(2.0 \text{ m}) = \boxed{780 \text{ N} \cdot \text{m}}$$

17. Strategy Use Eq. (8-4).

Solution Find the net torque in each case.

(a) $\Sigma\tau = F(r_{2\perp} - r_{1\perp}) = Fx_2 - Fx_1 = F(x_2 - x_1) = Fd$, since $d = x_2 - x_1$.

(b) $\Sigma\tau = F(r_{2\perp} - r_{1\perp}) = F(r_2 \sin\theta_2) - F(r_1 \sin\theta_1) = Fx_2 - Fx_1 = F(x_2 - x_1) = Fd$

21. Strategy The center of gravity is located at the center of mass. Let the origin be at the center of the door.

Solution Due to symmetry, $y_{CM} = 0$.

$$x_{CM} = \frac{m_1 x_1 + m_2 x_2}{M} = \frac{m_1(0) + m_2(x)}{M} = \frac{W_2 x}{W} = \frac{(5.0\ N)(-0.75\ m)}{5.0\ N + 300.0\ N} = -0.012\ m$$

The center of gravity is located $\boxed{\text{1.2 cm toward the doorknob as measured from the center of the door.}}$

25. (a) Strategy Use the work-kinetic energy theorem.

Solution Find the work done spinning up the wheel.

$\omega_f = 120$ rpm

0.62 m

$$W = \Delta K = \frac{1}{2} I \omega_f^2 = \frac{1}{2}(MR^2)\omega_f^2$$
$$= \frac{1}{2}(182\ kg)(0.62\ m)^2[(120\ rev/min)(1/60\ min/s)(2\pi\ rad/rev)]^2 = \boxed{5.5\ kJ}$$

(b) Strategy Use the equations for rotational motion with constant acceleration and the relationship between work, torque, and angular displacement.

Solution Find the torque.

$$W = \tau\Delta\theta = \tau(\omega_{av}\Delta t),\ so\ \tau = \frac{W}{\omega_{av}\Delta t} = \frac{5.5\times10^3\ J}{(120\ rev/min)(1/60\ min/s)(2\pi\ rad/rev)(30.0\ s)/2} = \boxed{29\ N\cdot m}.$$

27. Strategy Choose the axis of rotation at the fulcrum. Use Eqs. (8-8).

Solution Find the force required to lift the load.
$\Sigma\tau = 0 = -F_A \cos\theta(2.4\ m) + F_{load} \cos\theta(1.2\ m)$, so

$$F_A = \frac{1.2\ m}{2.4\ m} F_{load} = 0.50mg = 0.50(20.0\ kg)(9.80\ N/kg) = \boxed{98\ N}.$$

29. Strategy Choose the rotation axis at the edge of the base of the sculpture that is in contact with the floor as it is tipped. The angle that the base makes with the floor is the same angle that the force due to gravity makes with the vertical axis of the sculpture.

Solution Set the net torque equal to zero at the equilibrium point to find the maximum angle.
$\Sigma\tau = 0 = -mgb\sin\theta + mga\cos\theta$, where $b = 1.80\ m$ and $a = (1.10\ m)/2 = 0.550\ m$.
Solve for the angle.

$$b\sin\theta = a\cos\theta,\ so\ \theta = \tan^{-1}\frac{a}{b} = \tan^{-1}\frac{0.550\ m}{1.80\ m} = \boxed{17.0°}.$$

33. Strategy Use Eqs. (8-8).

Solution Choose the axis of rotation at the point of contact between the driveway and the ladder.
$\Sigma F_x = 0 = f - N_w$, so $f = N_w$.

$\Sigma \tau = 0 = N_w (4.7 \text{ m}) - W_1 (2.5 \text{ m}) \cos \theta - W_p \left(\dfrac{3.0 \text{ m}}{4.7 \text{ m}} \right)(5.0 \text{ m}) \cos \theta$, so $N_w = \dfrac{\cos \theta}{4.7 \text{ m}} \left[W_1 (2.5 \text{ m}) + W_p \left(\dfrac{15 \text{ m}}{4.7} \right) \right]$.

Find θ.

$4.7 \text{ m} = (5.0 \text{ m}) \sin \theta$, so $\theta = \sin^{-1} \dfrac{4.7}{5.0}$.

Calculate f.

$$f = N_w = \frac{\cos \sin^{-1} \frac{4.7}{5.0}}{4.7 \text{ m}} \left[(120 \text{ N})(2.5 \text{ m}) + (680 \text{ N}) \left(\frac{15 \text{ m}}{4.7} \right) \right] = 180 \text{ N}$$

So, the force of friction is $\boxed{180 \text{ N toward the wall}}$.

35. Strategy Use Eqs. (8-8).

Solution Choose the axis of rotation at the hinge.
$\Sigma \tau = 0 = T(2.38 \text{ m}) \sin 35° - (80.0 \text{ N})(1.50 \text{ m}) - (120.0 \text{ N})(3.00 \text{ m})$, so
$$T = \frac{(80.0 \text{ N})(1.50 \text{ m}) + (120.0 \text{ N})(3.00 \text{ m})}{(2.38 \text{ m}) \sin 35°} = \boxed{350 \text{ N}}.$$
Find F_x and F_y.
$\Sigma F_x = 0 = -T \cos 35° + F_x$ and $\Sigma F_y = 0 = F_y + T \sin 35° - 80.0 \text{ N} - 120.0 \text{ N}$, so
$F_x = T \cos 35° = (350 \text{ N}) \cos 35° = \boxed{290 \text{ N}}$ and $F_y = -(351.6 \text{ N}) \sin 35° + 80.0 \text{ N} + 120.0 \text{ N} = \boxed{-2 \text{ N}}$.

$\boxed{\text{The magnitude of } F_y \text{ is small}}$ compared to that of F_x and T.

37. Strategy Use Eqs. (8-8). Choose the axis of rotation at the point where the beam meets the store.

Solution The tension in the cable cannot exceed 417 N. Sum the torques.
$\Sigma \tau = 0 = T \sin \theta (1.50 \text{ m}) - (50.0 \text{ N})(0.75 \text{ m}) - (200.0 \text{ N})(1.00 \text{ m})$
Solve for θ and substitute 417 N (the breaking strength) for T.

$$\theta = \sin^{-1} \frac{(50.0 \text{ N})(0.75 \text{ m}) + (200.0 \text{ N})(1.00 \text{ m})}{(417 \text{ N})(1.50 \text{ m})} = 22.3°$$

The minimum angle is $\boxed{22.3°}$.

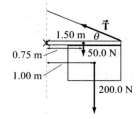

41. Strategy Use Eqs. (8-8). Choose the axis of rotation at the point of contact of the normal force.

Solution Find the tension in the Achilles tendon, F_A.
$\Sigma \tau = 0 = -F_A (4.60 \text{ cm} + 12.8 \text{ cm}) + F_T (12.8 \text{ cm})$ and $\Sigma F_y = 0 = N + F_A - F_T$, or $F_T = N + F_A$, so

$$-F_A (17.4 \text{ cm}) + (N + F_A)(12.8 \text{ cm}) = 0, \text{ or } F_A = \frac{(12.8 \text{ cm})(750 \text{ N})}{4.6 \text{ cm}} = 2100 \text{ N}.$$

Find the force that the tibia exerts on the ankle joint, F_T.

$$F_T = N + F_A = 750 \text{ N} + \frac{12.8}{4.6}(750 \text{ N}) = 2800 \text{ N}$$

The forces are: $\boxed{\text{tendon, 2100 N upward and tibia, 2800 N downward}}$.

43. **Strategy** Use Eqs. (8-8). Choose the axis of rotation at the elbow.

    Solution Find the force exerted by the biceps muscle.
    $\Sigma \tau = 0 = -W_m(35.0 \text{ cm}) - W_a(16.5 \text{ cm}) + F_b(5.00 \text{ cm})\sin\theta$, so

    $$F_b = \frac{W_m(35.0 \text{ cm}) + W_a(16.5 \text{ cm})}{(5.00 \text{ cm})\sin\theta} = \frac{(9.9 \text{ N})(35.0 \text{ cm}) + (18.0 \text{ N})(16.5 \text{ cm})}{(5.00 \text{ cm})\frac{30.0 \text{ cm}}{\sqrt{(30.0 \text{ cm})^2 + (5.00 \text{ cm})^2}}} = \boxed{130 \text{ N}}.$$

45. **Strategy** Refer to Figure 8.32. First find the magnitude of the force exerted by the back F_b by analyzing the torques about an axis at the sacrum; then, find the horizontal component of the extreme force on the sacrum F_s. Use Eqs. (8-8).

    Solution Sum the torques to find F_b.

    $\Sigma \tau = 0 = F_b(44 \text{ cm})\sin 12° - (10 \text{ kg})(9.80 \text{ m/s}^2)(76 \text{ cm}) - (55 \text{ kg})(9.80 \text{ m/s}^2)(38 \text{ cm})$, so

    $$F_b = \frac{(10 \text{ kg})(9.80 \text{ m/s}^2)(76 \text{ cm}) + (55 \text{ kg})(9.80 \text{ m/s}^2)(38 \text{ cm})}{(44 \text{ cm})\sin 12°} = 3053 \text{ N}.$$

    The only forces with components in the horizontal direction are those due to the back and the sacrum. Find the horizontal component of the extreme force, F_{sx}.

    $\Sigma F_x = 0 = F_{sx} - F_b\cos 12°$, so $F_{sx} = F_b\cos 12° = (3053 \text{ N})\cos 12° = \boxed{3.0 \text{ kN}}$.

    $\dfrac{(3053 \text{ N})\cos 12°}{540 \text{ N}} = 5.5$, so the force is $\boxed{\text{about 5.5 times larger}}$ than that from his torso alone!

49. **Strategy** Use the rotational form of Newton's second law and Eq. (5-21).

    Solution Find the torque that the motor must deliver.

    $I = \frac{1}{2}MR^2$ for a uniform disk, so

    $$\Sigma\tau = I\alpha = \frac{1}{2}MR^2\left(\frac{\omega_f^2 - \omega_i^2}{2\Delta\theta}\right) = \frac{MR^2\omega_f^2}{4\Delta\theta} = \frac{(0.22 \text{ kg})\left(\frac{0.305 \text{ m}}{2}\right)^2(3.49 \text{ rad/s})^2}{4(2.0 \text{ rev})(2\pi \text{ rad/rev})} = \boxed{0.0012 \text{ N·m}}.$$

53. **Strategy** The rotational inertia of the wheel is $I = MR^2$. Use the rotational form of Newton's second law.

    Solution Find the magnitude of the average torque.

    $$\left|\Sigma\tau_{av}\right| = I\alpha = MR^2\left|\frac{\Delta\omega}{\Delta t}\right| = (2 \text{ kg})(0.30 \text{ m})^2\left(\frac{4.00 \text{ rev/s}}{50 \text{ s}}\right)\left(\frac{2\pi \text{ rad}}{\text{rev}}\right) = \boxed{0.09 \text{ N·m}}$$

55. **Strategy** The rotational inertia is $I = \frac{1}{2}MR^2$. Use the rotational form of Newton's second law and Eq. (5-18).

    Solution

    (a) $\alpha = \dfrac{\Sigma\tau}{I} = \dfrac{FR + FR}{\frac{1}{2}MR^2} = \dfrac{4F}{MR} = \dfrac{4(10.0 \text{ N})}{(180 \text{ kg})(2.0 \text{ m})} = \boxed{0.11 \text{ rad/s}^2}$

    (b) $\omega_f = \omega_i + \alpha\Delta t = 0 + (0.11 \text{ rad/s}^2)(4.0 \text{ s}) = \boxed{0.44 \text{ rad/s}}$

57. Strategy Follow the steps to derive the rotational from of Newton's second law.

Solution

(a) According to Newton's second law, $F_i = m_i a_i$, so $a_i = F_i / m_i$.

(b) The torque is the product of the perpendicular component of the force and the shortest distance between the rotation axis and the point of application of the force, so $\tau_i = F_i r_i = m_i a_i r_i$.

(c) The tangential acceleration is related to the angular acceleration by $a_i = r_i \alpha$, so $\tau_i = m_i (r_i \alpha) r_i = m_i r_i^2 \alpha$.

(d) Summing the torques and using the definition of rotational inertia, we have
$$\sum_{i=1}^{N} \tau_i = \sum_{i=1}^{N} m_i r_i^2 \alpha = \left(\sum_{i=1}^{N} m_i r_i^2 \right) \alpha = I\alpha.$$

61. Strategy The sphere is rolling on a horizontal surface, so its total energy is equal to its total kinetic energy. Use conservation of energy.

Solution Compute the total energy.
$$E_{\text{total}} = K_{\text{tr}} + K_{\text{rot}} = \frac{1}{2} mv^2 + \frac{1}{2} I\omega^2 = \frac{1}{2} mv^2 + \frac{1}{2}\left(\frac{2}{5} mr^2\right)\left(\frac{v}{r}\right)^2 = \frac{1}{2} mv^2 + \frac{1}{5} mv^2 = \frac{7}{10} mv^2$$
$$= \frac{7}{10}(0.600 \text{ kg})(5.00 \text{ m/s})^2 = 10.5 \text{ J}$$

Find the height achieved by the sphere.
$$\Delta U = mgh = -\Delta K = K, \text{ so } h = \frac{K}{mg} = \frac{10.5 \text{ J}}{(0.600 \text{ kg})(9.80 \text{ N/kg})} = \boxed{1.79 \text{ m}}.$$

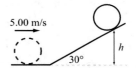

63. Strategy Let $h = 17.0$ m, m be the mass of the bucket, and M be the mass of the cylinder. The tangential speed of the cylinder is the same as the linear speed of the bucket, since they are attached by a rope. Use conservation of energy. The rotational inertia of a uniform solid cylinder is $\frac{1}{2} MR^2$.

Solution Find the speed of the bucket when it reaches the bottom of the well.
$$\Delta K = \frac{1}{2} mv^2 + \frac{1}{2} I\omega^2 = \frac{1}{2} mv^2 + \frac{1}{2}\left(\frac{1}{2} MR^2\right)\left(\frac{v}{R}\right)^2 = \frac{1}{2} mv^2 + \frac{1}{4} Mv^2 = -\Delta U = mgh, \text{ so } v = \sqrt{\frac{4mgh}{2m+M}}.$$

Compute how long it will take for the bucket to fall to the bottom of the well.
$$\Delta y = h = \frac{1}{2}(v_{\text{fy}} + v_{\text{iy}})\Delta t, \text{ so}$$
$$\Delta t = \frac{2h}{v_{\text{fy}} + v_{\text{iy}}} = \frac{2h}{\sqrt{\frac{4mgh}{2m+M}} + 0} = \sqrt{\frac{h(2m+M)}{mg}} = \sqrt{\frac{(17.0 \text{ m})[2(1.10 \text{ kg}) + 2.60 \text{ kg}]}{(1.10 \text{ kg})(9.80 \text{ m/s}^2)}} = \boxed{2.75 \text{ s}}.$$

65. **(a) Strategy** Use conservation of energy and the relationship between speed and radial acceleration.

   Solution At the top of the loop, the sphere's speed must be at least the speed that results in a radial acceleration of g.

   $$\frac{v^2}{r} = g, \text{ so } v^2 = gr.$$

   The sphere's kinetic energy is $\frac{1}{2}mv^2 = \frac{1}{2}mgr$, and it must equal the potential energy difference

   $mgh - mg(2r)$. Thus, $\frac{1}{2}r = h - 2r$ or $h = \boxed{\dfrac{5}{2}r}$.

   (b) Strategy The rotational inertia of a uniform solid sphere is $\frac{2}{5}mr^2$. Use conservation of energy.

   Solution Find the kinetic energy of the sphere.

   $$K = \frac{1}{2}mv^2 + \frac{1}{2}\left(\frac{2}{5}mr^2\right)\left(\frac{v^2}{r^2}\right) = \frac{7}{10}mv^2 = \frac{7}{10}mgr$$

   Find h.

   $$\Delta K = \frac{7}{10}mgr = -\Delta U = mgh - mg(2r), \text{ so } h = \boxed{\dfrac{27}{10}r}.$$

69. **Strategy** The rotational inertia of a uniform disk is $I = \frac{1}{2}MR^2$. Use Eq. (8-14).

   Solution Find the magnitude of the angular momentum of the turntable.
   $$L = I\omega = \frac{1}{2}MR^2\omega = \frac{1}{2}(5.00 \text{ kg})(0.100 \text{ m})^2(0.550 \text{ rev/s})(2\pi \text{ rad/rev}) = \boxed{0.0864 \text{ kg} \cdot \text{m}^2/\text{s}}$$

73. **Strategy** Since the torque is constant, it is equal to the change in angular momentum divided by the time interval.

   Solution Find the time to stop the spinning wheel
   $$\tau = \frac{\Delta L}{\Delta t}, \text{ so } \Delta t = \frac{\Delta L}{\tau} = \frac{-6.40 \text{ kg} \cdot \text{m}^2/\text{s}}{-4.00 \text{ N} \cdot \text{m}} = \boxed{1.60 \text{ s}}.$$

75. **Strategy** Use conservation of angular momentum and Eq. (8-14).

   Solution Find the skater's final angular velocity.
   $$L_i = I_i\omega_i = L_f = I_f\omega_f, \text{ so } \omega_f = \frac{I_i}{I_f}\omega_i = \frac{2.50}{1.60}(10.0 \text{ rad/s}) = \boxed{15.6 \text{ rad/s}}.$$

77. **Strategy** The rotational inertias of the wheel and guinea pig are $I_w = MR^2$ and $I_g = mR^2$, respectively, where M is the mass of the wheel, m is the mass of the guinea pig, and R is the radius of the wheel. Use conservation of angular momentum and $v = r\omega$.

   Solution Find the angular velocity of the wheel.
   $$L_w = L_g, \text{ so } I_w\omega_w = MR^2\omega_w = I_g\omega_g = mR^2\omega_g = mRv_g.$$
   Thus, $\omega_w = \dfrac{mv_g}{MR} = \dfrac{(0.500 \text{ kg})(0.200 \text{ m/s})}{(2.00 \text{ kg})(0.400 \text{ m})} = \boxed{0.125 \text{ rad/s}}.$

81. **Strategy** The average torque is equal to the magnitude of the change in angular momentum divided by the time interval.

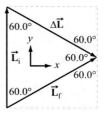

Solution Let $\vec{L}_i = L$ in the $+y$-direction. Then $\Delta\vec{L}$ has components $\Delta L_x = L\sin\theta$ and $\Delta L_y = L\cos\theta - L = L(\cos\theta - 1)$. So,

$$|\Delta\vec{L}| = \sqrt{(L\sin\theta)^2 + [L(\cos\theta - 1)]^2} = L\sqrt{\sin^2 60.0° + (\cos 60.0° - 1)^2} = 1.00L.$$

Compute the magnitude of the required torque.

$$\tau = \left|\frac{\Delta\vec{L}}{\Delta t}\right| = \frac{1.00L}{\Delta t} = \frac{1.00I\omega}{\Delta t} = \frac{\frac{1}{2}mr^2\omega}{\Delta t}$$

$$= \frac{(1.00\times10^5 \text{ kg})(2.00 \text{ m})^2(300.0 \text{ rpm})}{2(3.00 \text{ s})}\left(\frac{2\pi \text{ rad}}{\text{rev}}\right)\left(\frac{1 \text{ min}}{60 \text{ s}}\right) = \boxed{2.10\times10^6 \text{ N}\cdot\text{m}}$$

85. **Strategy** Use Eq. (8-3). The force due to the weight is mg.

Solution Find the torque.

$$\tau = F_\perp r = mgr = (10.0 \text{ kg})(9.80 \text{ N/kg})(1.0 \text{ m}) = \boxed{98 \text{ N}\cdot\text{m}}$$

87. **Strategy** The rotational inertia of the gymnast is $\frac{1}{3}m(2r)^2$, where $r = 1.0$ m. Use conservation of energy.

Solution Find the angular speed at the bottom of the swing.

$$\Delta K = \frac{1}{2}I\omega^2 = \frac{1}{2}\left[\frac{1}{3}m(2r)^2\right]\omega^2 = \frac{2}{3}mr^2\omega^2 = -\Delta U = mg(2r) = 2rmg, \text{ so}$$

$$\omega = \sqrt{\frac{3g}{r}} = \sqrt{\frac{3(9.80 \text{ m/s}^2)}{1.0 \text{ m}}} = \boxed{5.4 \text{ rad/s}}.$$

89. **Strategy** Use conservation of energy.

Solution Find the final speeds of each object.
Solid sphere:

$$\Delta K = K_{tr} + K_{rot} = \frac{1}{2}mv^2 + \frac{1}{2}I\omega^2 = \frac{1}{2}mv^2 + \frac{1}{2}\left(\frac{2}{5}mr^2\right)\left(\frac{v}{r}\right)^2 = \frac{1}{2}mv^2 + \frac{1}{5}mv^2 = \frac{7}{10}mv^2 = -\Delta U = U_i - U_f = mgh,$$

so $v = \sqrt{\frac{10gh}{7}}$.

Hollow sphere:

$$\frac{1}{2}mv^2 + \frac{1}{2}\left(\frac{2}{3}mr^2\right)\left(\frac{v}{r}\right)^2 = \frac{1}{2}mv^2 + \frac{1}{3}mv^2 = \frac{5}{6}mv^2 = mgh, \text{ so } v = \sqrt{\frac{6gh}{5}}.$$

Solid cylinder:

$$\frac{1}{2}mv^2 + \frac{1}{2}\left(\frac{1}{2}mr^2\right)\left(\frac{v}{r}\right)^2 = \frac{1}{2}mv^2 + \frac{1}{4}mv^2 = \frac{3}{4}mv^2 = mgh, \text{ so } v = \sqrt{\frac{4gh}{3}}.$$

Hollow cylinder:

$$\frac{1}{2}mv^2 + \frac{1}{2}(mr^2)\left(\frac{v}{r}\right)^2 = \frac{1}{2}mv^2 + \frac{1}{2}mv^2 = mv^2 = mgh, \text{ so } v = \sqrt{gh}.$$

Cube:

$\frac{1}{2}mv^2 = mgh$, so $v = \sqrt{2gh}$.

So, $v_{\text{cube}} > v_{\text{solid sphere}} > v_{\text{solid cylinder}} > v_{\text{hollow sphere}} > v_{\text{hollow cylinder}}$.

The objects reach the bottom in the following order from first to last: cube, solid sphere, solid cylinder, hollow sphere, and hollow cylinder.

93. **(a) Strategy** The rotational inertia of a uniform solid disk is $I = \frac{1}{2}MR^2$.

Solution Compute the rotational inertia.

$$I = \frac{1}{2}MR^2 = \frac{1}{2}(200.0\ \text{kg})(0.40\ \text{m})^2 = \boxed{16\ \text{kg} \cdot \text{m}^2}$$

(b) Strategy Use Eq. (8-1).

Solution Compute the initial rotational kinetic energy.

$$K_{\text{rot}} = \frac{1}{2}I\omega^2 = \frac{1}{2}(16\ \text{kg}\cdot\text{m}^2)(3160\ \text{rad/s})^2 = \boxed{8.0\times10^7\ \text{J}}$$

(c) Strategy and Solution The ratio of the rotational to the translational kinetic energies is

$$\frac{K_{\text{rot}}}{K_{\text{tr}}} = \frac{K_{\text{rot}}}{\frac{1}{2}mv^2} = \frac{2(8.0\times10^7\ \text{J})}{(1000.0\ \text{kg})(22.4\ \text{m/s})^2} = \boxed{320}.$$

(d) Strategy Set the work done by air resistance equal to the stored energy in the flywheel.

Solution Find the distance d the car can travel.

$$Fd = K_{\text{rot}}, \text{ so } d = \frac{K_{\text{rot}}}{F} = \frac{8.0\times10^7\ \text{J}}{670.0\ \text{N}} = \boxed{120\ \text{km}}.$$

97. **Strategy** The system is in equilibrium. Use Eqs. (8-8).

Solution Find h.

$$\frac{h}{(1.26\ \text{m})/2} = \tan 75°, \text{ so } h = [(1.26\ \text{m})/2]\tan 75° = (0.630\ \text{m})\tan 75°.$$

At the top of the ladder, each leg exerts a horizontal force on the other. These forces are equal in magnitude and opposite in direction, since the system is in equilibrium. Let the magnitude of this force be F. The tension T in the rope is directed to the left at the connection point on the right leg, so for the right leg, we have $\Sigma F_x = F - T = 0$ or $T = F$.

Calculate the torque about the contact point of the right leg of the ladder and the ground.

$$\Sigma\tau = (0.630\ \text{m})mg - Fh = 0, \text{ so } T = F = \frac{(0.630\ \text{m})mg}{h} = \frac{(0.630\ \text{m})mg}{(0.630\ \text{m})\tan 75°} = \frac{mg}{\tan 75°}.$$

The tension in the rope is the same along its length, so

$$T_{\text{rope}} = \frac{mg}{\tan 75°} = \frac{(42\ \text{kg})(9.80\ \text{N/kg})}{\tan 75°} = \boxed{110\ \text{N}}.$$

101. Strategy The rotational inertial of a uniform disk is $I = \frac{1}{2}MR^2$. Use Eq. (8-14).

Solution Find the magnitude of the angular momentum of the disk.
$$L = I\omega = \frac{1}{2}MR^2\omega = \frac{1}{2}(2.0 \text{ kg})(0.100 \text{ m})^2(3.0 \text{ rev/s})(2\pi \text{ rad/rev}) = \boxed{0.19 \text{ kg}\cdot\text{m}^2/\text{s}}$$

105. Strategy The system is in equilibrium. Choose the axis of rotation at the ankle.

Solution Find the force that each calf muscle needs to exert while the woman is standing.
$\Sigma\tau = 0 = 2F(4.4 \text{ cm})\sin 81° - mg(3.0 \text{ cm})$, so
$$F = \frac{mg(3.0 \text{ cm})}{2(4.4 \text{ cm})\sin 81°} = \frac{(68 \text{ kg})(9.80 \text{ N/kg})(3.0 \text{ cm})}{2(4.4 \text{ cm})\sin 81°} = \boxed{230 \text{ N}}.$$

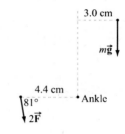

109. Strategy Since the bike travels with constant velocity, the acceleration is zero and $\Sigma\tau = 0$.

Solution Find the magnitude of the force with which the chain pulls.
$$\Sigma\tau = 0 = fr_2 - F_C r_1, \text{ so } F_C = \frac{r_2}{r_1}f = 6.0(3.8 \text{ N}) = \boxed{23 \text{ N}}.$$

113. Strategy Choose the axis of rotation at the contact point between the horizontal surface and the tip of the left leg.

Solution Find the maximum wind speeds in which the blowfly and dog can stand.

(a) $\tau_{\text{net}} = 0 = F_{\text{wind}}r\sin\theta - mgr\cos\theta$, so
$$\frac{mg}{\tan\theta} = F_{\text{wind}} = cAv^2 \text{ or } v = \sqrt{\frac{mg}{cA\tan\theta}} = \sqrt{\frac{(0.070\times10^{-3} \text{ kg})(9.80 \text{ m/s}^2)}{(1.3 \text{ N}\cdot\text{s}^2/\text{m}^4)(0.10\times10^{-4} \text{ m}^2)\tan 30.0°}} = \boxed{9.6 \text{ m/s}}.$$

(b) $$v = \sqrt{\frac{(0.070\times10^{-3} \text{ kg})(9.80 \text{ m/s}^2)}{(1.3 \text{ N}\cdot\text{s}^2/\text{m}^4)(0.10\times10^{-4} \text{ m}^2)\tan 80.0°}} = \boxed{3.1 \text{ m/s}}$$

(c) $$v = \sqrt{\frac{(10.0 \text{ kg})(9.80 \text{ m/s}^2)}{(1.3 \text{ N}\cdot\text{s}^2/\text{m}^4)(0.030 \text{ m}^2)\tan 80.0°}} = \boxed{21 \text{ m/s}}$$

REVIEW AND SYNTHESIS: CHAPTERS 6–8

Review Exercises

1. **(a) Strategy** Multiply the extension per mass by the mass to find the maximum extension required.

   Solution
   $$\left(\frac{1.0 \text{ mm}}{25 \text{ g}}\right)(5.0 \text{ kg})\left(\frac{1000 \text{ g}}{1 \text{ kg}}\right)\left(\frac{1 \text{ m}}{1000 \text{ mm}}\right) = \boxed{0.20 \text{ m}}$$

   (b) Strategy Set the weight of the mass equal to the magnitude of the force due to the spring scale. Use Hooke's law.

   Solution
   $$\text{Weight} = mg = kx, \text{ so } k = \frac{mg}{x} = \frac{(5.0 \text{ kg})(9.80 \text{ N/kg})}{0.20 \text{ m}} = \boxed{250 \text{ N/m}}.$$

5. **(a) Strategy** Use the conservation of energy.

   Solution Find the work done by friction.

   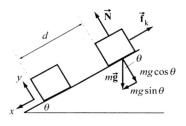

   $$W_{\text{total}} = W_{\text{friction}} + W_{\text{grav}} = W_{\text{friction}} + mgd \sin\theta = \Delta K = 0 - \frac{1}{2}mv_i^2, \text{ so}$$

   $$W_{\text{friction}} = -\frac{1}{2}mv_i^2 - mgd \sin\theta = -m\left(\frac{1}{2}v_i^2 + gd \sin\theta\right)$$

   $$= -(100 \text{ kg})\left[\frac{1}{2}(2.00 \text{ m/s})^2 + (9.80 \text{ m/s}^2)(1.50 \text{ m})\sin 30.0°\right]$$

   $$= -940 \text{ J}.$$

   Thus, the energy dissipated by friction was $\boxed{940 \text{ J}}$.

   (b) Strategy Use Newton's second law.

   Solution Find the normal force on the crate.
   $$\Sigma F_y = N - mg\cos\theta = 0, \text{ so } N = mg\cos\theta.$$

   Since $v_{fx}^2 - v_{ix}^2 = 0 - v_i^2 = 2a_x\Delta x = 2a_x d$, the acceleration of the crate is $-v_i^2/(2d)$.
   Find the force of sliding friction.

   $$\Sigma F_x = -f_k + mg\sin\theta = -\mu_k mg\cos\theta + mg\sin\theta = ma_x = -m\frac{v_i^2}{2d}, \text{ so}$$

   $$\mu_k = \tan\theta + \frac{v_i^2}{2dg\cos\theta} = \tan 30.0° + \frac{(2.00 \text{ m/s})^2}{2(1.50 \text{ m})(9.80 \text{ m/s}^2)\cos 30.0°} = \boxed{0.734}.$$

9. **Strategy** The collision is inelastic. Use conservation of momentum and energy.

Solution Write equations using conservation of momentum and energy.

momentum: $mv_i = (m + M)v_f$

energy: $\frac{1}{2}(m + M)v_f^2 = (m + M)g\Delta y$

Find the initial speed of the putty.

$$\frac{1}{2}(m + M)\left(\frac{mv_i}{m + M}\right)^2 = (m + M)g\Delta y$$

$$\left(\frac{m}{m + M}\right)^2 v_i^2 = 2g\Delta y$$

$$v_i = \sqrt{2g\Delta y \left(\frac{m + M}{m}\right)^2} = \sqrt{2(9.8 \text{ m/s}^2)(1.50 \text{ m})\left(\frac{0.50 \text{ kg} + 2.30 \text{ kg}}{0.50 \text{ kg}}\right)^2} = \boxed{30 \text{ m/s}}$$

13. **Strategy** Use conservation of energy. Let $d = 2.05$ m. Then, the ramp rises $h = d\sin 5.00°$. The rotational inertia of a uniform sphere is $\frac{2}{5}mr^2$.

Solution Find the speed of the ball when it reaches the top of the ramp.

$$0 = \Delta K + \Delta U = \frac{1}{2}mv_f^2 + \frac{1}{2}I\omega_f^2 - \frac{1}{2}mv_i^2 - \frac{1}{2}I\omega_i^2 + mgh$$

$$= \frac{1}{2}mv_f^2 + \frac{1}{2}\left(\frac{2}{5}mr^2\right)\left(\frac{v_f}{r}\right)^2 - \frac{1}{2}mv_i^2 - \frac{1}{2}\left(\frac{2}{5}mr^2\right)\left(\frac{v_i}{r}\right)^2 + mgh$$

$$= \frac{7}{10}mv_f^2 - \frac{7}{10}mv_i^2 + mgh, \text{ so}$$

$$v_f = \sqrt{v_i^2 - \frac{10}{7}gh} = \sqrt{(2.20 \text{ m/s})^2 - \frac{10}{7}(9.80 \text{ m/s}^2)(2.05 \text{ m})\sin 5.00°} = \boxed{1.53 \text{ m/s}}.$$

17. **Strategy** Use conservation of energy. The energy delivered to the fluid in the beaker plus the kinetic energies of the pulley, spool, axle, paddles, and the block are equal to the work done by gravity on the block, which is negative the change in the block's gravitational potential energy. The rotational inertia of the pulley (uniform solid disk) is $\frac{1}{2}m_pr^2$.

Solution Let the energy delivered to the fluid be E, the distance the block falls be h, and the rotational inertia of the spool, axle, and paddles be $I_s = 0.00140 \text{ kg} \cdot \text{m}^2$. Since the radii of the pulley and the spool are the same (r), their tangential speeds are the same, so let $v_p = v_s = v$.

$$m_bgh = \frac{1}{2}m_bv_b^2 + \frac{1}{2}I_p\omega_p^2 + \frac{1}{2}I_s\omega_s^2 + E = \frac{1}{2}m_bv_b^2 + \frac{1}{2}\left(\frac{1}{2}m_pr^2\right)\left(\frac{v}{r}\right)^2 + \frac{1}{2}I_s\left(\frac{v}{r}\right)^2 + E$$

The tangential speeds of the pulley and spool are equal to the speed of the block.

$$m_bgh = \frac{1}{2}m_bv_b^2 + \frac{1}{4}m_pv^2 + \frac{1}{2}I_s\frac{v^2}{r^2} + E = \frac{1}{2}m_bv^2 + \frac{1}{4}m_pv^2 + \frac{1}{2}I_s\frac{v^2}{r^2} + E, \text{ so}$$

$$E = m_bgh - \frac{v^2(2m_b + m_p + 2I_s/r^2)}{4}$$

$$= (0.870 \text{ kg})(9.80 \text{ m/s}^2)(2.50 \text{ m}) - \frac{(3.00 \text{ m/s})^2[2(0.870 \text{ kg}) + 0.0600 \text{ kg} + 2(0.00140 \text{ kg} \cdot \text{m}^2)/(0.0300 \text{ m})^2]}{4}$$

$$= \boxed{10.3 \text{ J}}.$$

21. **Strategy** Use energy conservation to find the speed of Jones just before he grabs Smith. Then, use momentum conservation to find the speed of both just after. Finally, again use energy conservation to find the final height.

    Solution Find Jones's speed, v_J.

    $$\frac{1}{2}m_J v_J^2 = m_J g h_J, \text{ so } v_J = \sqrt{2gh_J}.$$

    Find the speed of both, v.

    $$p_i = m_J v_J = p_f = (m_J + m_S)v, \text{ so } v = \frac{m_J v_J}{m_J + m_S} = \frac{m_J \sqrt{2gh_J}}{m_J + m_S}.$$

    Find the final height, h.

    $$(m_J + m_S)gh = \frac{1}{2}(m_J + m_S)\left(\frac{m_J\sqrt{2gh_J}}{m_J + m_S}\right)^2, \text{ so } h = \frac{m_J^2 h_J}{(m_J + m_S)^2} = \frac{(78.0\text{ kg})^2(3.70\text{ m})}{(78.0\text{ kg} + 55.0\text{ kg})^2} = \boxed{1.27\text{ m}}.$$

25. **Strategy** Use conservation of linear momentum.

    Solution

    $$p_{ix} = m_b v_b = p_{fx} = (m_b + m_c)v_{fx}, \text{ so } v_{fx} = \frac{m_b v_b}{m_b + m_c}.$$

    $$p_{iy} = m_c v_c = p_{fy} = (m_b + m_c)v_{fy}, \text{ so } v_{fy} = \frac{m_c v_c}{m_b + m_c}.$$

    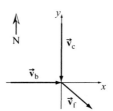

    Compute the magnitude of the final velocity.

    $$v = \sqrt{v_{fx}^2 + v_{fy}^2} = \sqrt{\left(\frac{m_b v_b}{m_b + m_c}\right)^2 + \left(\frac{m_c v_c}{m_b + m_c}\right)^2} = \frac{\sqrt{(m_b v_b)^2 + (m_c v_c)^2}}{m_b + m_c}$$

    $$= \frac{\sqrt{[(2.00\text{ kg})(2.70\text{ m/s})]^2 + [(1.50\text{ kg})(-3.20\text{ m/s})]^2}}{2.00\text{ kg} + 1.50\text{ kg}} = 2.06\text{ m/s}$$

    Compute the angle.

    $$\theta = \tan^{-1}\frac{v_{fy}}{v_{fx}} = \tan^{-1}\frac{\frac{m_c v_c}{m_b + m_c}}{\frac{m_b v_b}{m_b + m_c}} = \tan^{-1}\frac{m_c v_c}{m_b v_b} = \tan^{-1}\frac{(1.50\text{ kg})(-3.20\text{ m/s})}{(2.00\text{ kg})(2.70\text{ m/s})} = -41.6°$$

    The velocity of the block and the clay after the collision is $\boxed{2.06\text{ m/s at }41.6° \text{ S of E}}$.

29. (a) Strategy Consider the work-kinetic energy theorem and the impulse momentum theorem.

Solution Since the Romulan ship is twice as massive as the Vulcan ship, the Romulan ship will not travel as far as the Vulcan ship for the same engine force, since $\Delta x = (1/2)a(\Delta t)^2 = (1/2)(F/m)(\Delta t)^2$. Since $W = F\Delta x = \Delta K$, $\boxed{\text{the Vulcan ship will have the greater kinetic energy}}$. Since $\Delta p = F\Delta t$, $\boxed{\text{the ships will have the same momentum}}$.

(b) Strategy Consider the work-kinetic energy theorem and the impulse momentum theorem.

Solution Since the distances and the forces are the same, and since $W = F\Delta x = \Delta K$, $\boxed{\text{the ships will have the same kinetic energy}}$. Since $\Delta x = (1/2)a(\Delta t)^2 = (1/2)(F/m)(\Delta t)^2$, the more massive Romulan ship will have to fire its engines longer than the Vulcan ship to travel the same distance. Since $\Delta p = F\Delta t$ and the forces are the same, $\boxed{\text{the Romulan ship will have the greater momentum}}$.

(c) Strategy Refer to parts (a) and (b).

Solution For part (a), we have the following:
Vulcan:

$$\Delta K = W = F\Delta x = F\left[\frac{F}{2m}(\Delta t)^2\right] = \frac{(9.5\times10^6 \text{ N})^2(100 \text{ s})^2}{2(65,000 \text{ kg})} = 6.9\times10^{12} \text{ J}$$

$$\Delta p = F\Delta t = (9.5\times10^6 \text{ N})(100 \text{ s}) = 9.5\times10^8 \text{ kg}\cdot\text{m/s}$$

Romulan:

$$\Delta K = W = F\Delta x = F\left[\frac{F}{2m}(\Delta t)^2\right] = \frac{(9.5\times10^6 \text{ N})^2(100 \text{ s})^2}{2(130,000 \text{ kg})} = 3.5\times10^{12} \text{ J}$$

$$\Delta p = F\Delta t = (9.5\times10^6 \text{ N})(100 \text{ s}) = 9.5\times10^8 \text{ kg}\cdot\text{m/s}$$

$\boxed{\text{In part (a), the momenta are the same, } 9.5\times10^8 \text{ kg}\cdot\text{m/s, but the kinetic energies differ: Vulcan at } 6.9\times10^{12} \text{ J and Romulan at } 3.5\times10^{12} \text{ J.}}$

For part (b), we have the following:
Vulcan:

$$\Delta K = W = F\Delta x = (9.5\times10^6 \text{ N})(100 \text{ m}) = 9.5\times10^8 \text{ J}$$

Since $K = \frac{1}{2}mv^2 = \frac{p^2}{2m}$, $p = \sqrt{2mK} = \sqrt{2(65,000 \text{ kg})(9.5\times10^8 \text{ J})} = 1.1\times10^7 \text{ kg}\cdot\text{m/s}$.

Romulan:

$$\Delta K = W = F\Delta x = (9.5\times10^6 \text{ N})(100 \text{ m}) = 9.5\times10^8 \text{ J}$$

$$p = \sqrt{2mK} = \sqrt{2(2\times65,000 \text{ kg})(9.5\times10^8 \text{ J})} = 1.6\times10^7 \text{ kg}\cdot\text{m/s}.$$

$\boxed{\text{In part (b), the kinetic energies are the same, } 9.5\times10^8 \text{ J, but the momenta differ: Vulcan at } 1.1\times10^7 \text{ kg}\cdot\text{m/s and Romulan at } 1.6\times10^7 \text{ kg}\cdot\text{m/s.}}$

33. **Strategy** Use conservation of energy. m is the mass of one wheel. M is the total mass of the system. v is the speed of the center of mass of the system (which is the same as the speed of a point on either wheel).

    Solution

    (a) $K_{rot} = \dfrac{1}{2}I\omega^2 = \dfrac{1}{2}mv^2 = K_{trans}$ for one wheel. $K_{rot,total} = 2 \cdot \dfrac{1}{2}mv^2 = mv^2$ and $K_{trans,total} = \dfrac{1}{2}Mv^2$.

    $$K_{total} = U_i$$
    $$mv^2 + \dfrac{1}{2}Mv^2 = MgH$$
    $$v^2(2m + M) = 2MgH$$
    $$v = \sqrt{\dfrac{2MgH}{2m + M}} = \sqrt{\dfrac{2(80.0 \text{ kg})(9.8 \text{ m/s}^2)(20.0 \text{ m})}{2(1.5 \text{ kg}) + 80.0 \text{ kg}}} = \boxed{19 \text{ m/s}}$$

    (b) Since the speed depends upon the combined total mass of the system, the speed at the bottom would not be the same for a less massive rider. The answer is $\boxed{\text{no}}$.

MCAT Review

1. **Strategy** Use conservation of momentum.

   Solution
   $$p_i = mv_i = p_f = mv_f + p_{wall}, \text{ so } p_{wall} = m(v_i - v_f) = (0.2 \text{ kg})[2.0 \text{ m/s} - (-1.0 \text{ m/s})] = 0.6 \text{ kg} \cdot \text{m/s}.$$
   The correct answer is $\boxed{D}$.

2. **Strategy** Use Hooke's law.

   Solution Let up be the positive direction. The gravitational force on the mass is
   $F = mg = (0.10 \text{ kg})(-9.80 \text{ m/s}^2) = -0.98 \text{ N}$. Solving for the spring constant in Hooke's law, we have
   $k = -\dfrac{F}{x} = -\dfrac{-0.98 \text{ N}}{0.15 \text{ m}} = 6.5 \text{ N/m}$. Thus, the correct answer is $\boxed{D}$.

3. **Strategy** The net torque is zero.

   Solution
   $\Sigma\tau = 0 = F(0.60 \text{ m}) - (1.0 \times 10^{-7} \text{ kg})(9.80 \text{ m/s}^2)(0.40 \text{ m})$, so
   $F = \dfrac{(1.0 \times 10^{-7} \text{ kg})(9.80 \text{ m/s}^2)(0.40 \text{ m})}{0.60 \text{ m}} = 6.5 \times 10^{-7} \text{ N}.$
   The correct answer is $\boxed{B}$.

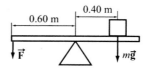

4. **Strategy** Determine the speed of the first ball just before in collides with the second. The collision is completely inelastic; that is, the balls stick together. Use conservation of momentum to find the speed of the balls after the collision.

   Solution Find the speed of the first ball just before the collision.
   $v_{fx} - v_{ix} = v_1 - 0 = a_x \Delta t$, so $v_1 = (10 \text{ m/s}^2)(2.0 \text{ s}) = 20 \text{ m/s}$.
   Find the speed v of the balls just after the collision.
   $$p_i = m_1 v_1 = p_f = (m_1 + m_2)v, \text{ so } v = \frac{m_1 v_1}{m_1 + m_2} = \frac{(0.50 \text{ kg})(20 \text{ m/s})}{0.50 \text{ kg} + 1.0 \text{ kg}} = 6.7 \text{ m/s.}$$
   The correct answer is $\boxed{B}$.

5. **Strategy** Use Newton's second law and Eq. (6-27).

   Solution The gravitational force working against the motion of the car as it climbs the hill is $mg \sin 10°$, so the additional power required is

   $$P_{car} = -P_{grav} = -Fv\cos 180° = (mg \sin 10°)v = (1000 \text{ kg})(10 \text{ m/s}^2)\sin 10°(15 \text{ m/s})$$
   $$= 1.5 \times 10^5 \times \sin 10° \text{ W}.$$
   The correct answer is $\boxed{D}$.

6. **Strategy** Find the vertical distance the patient would have climbed had the treadmill been stationary (and very long). Then, find the work done by the patient on the treadmill.

   Solution The "distance" walked along the incline is $(2 \text{ m/s})(600 \text{ s}) = 1200 \text{ m}$.
   Thus, the vertical distance climbed is $(1200 \text{ m})\sin 30° = 600 \text{ m}$. The work done is

   $$W = Fd = mgd = (90 \text{ kg})(10 \text{ m/s}^2)(600 \text{ m}) = 0.54 \text{ MJ}.$$
   The correct answer is $\boxed{C}$.

7. **Strategy** Find the angle between the force exerted by the patient and the patient's velocity. Use Eq. (6-27).

   Solution The force due to gravity is down, so the force exerted by the patient is up. The velocity is directed at the angle of the incline, or 30° above the horizontal, so the angle between the force and the velocity is 60°. Compute the mechanical power output of the patient.
   $$P = Fv \cos\theta = mgv \cos\theta = (100 \text{ kg})(10 \text{ m/s}^2)(3 \text{ m/s})\cos 60° = 1500 \text{ W}$$
   The correct answer is $\boxed{B}$.

8. **Strategy and Solution** The force pushing each friction pad is normal to the wheel; that is, it is the normal force in $f_k = \mu_k N$. Solve for the normal force.
   $$N = \frac{f_k}{\mu_k} = \frac{20 \text{ N}}{0.4} = 50 \text{ N}$$

   This is the total force. The force pushing each friction pad is half this, or 25 N. The correct answer is $\boxed{B}$.

9. **Strategy** Find the average tangential speed at the friction pads. Then, use the relationship between tangential speed and radial acceleration.

Solution

The average tangential speed is $v = \dfrac{4800 \text{ m}}{20 \text{ min}} \times \dfrac{1 \text{ min}}{60 \text{ s}} = 4.0$ m/s. The radial acceleration is

$a_r = \dfrac{v^2}{r} = \dfrac{(4.0 \text{ m/s})^2}{0.3 \text{ m}} = 50$ m/s^2. The correct answer is $\boxed{\text{D}}$.

10. **Strategy** Use the work-kinetic energy theorem.

Solution The work done by friction on the wheel is $W = -f_k d$, where d is the linear distance the wheel passes between the pads before it stops. Relate d to the kinetic energy of the wheel.

$W_{\text{total}} = -f_k d = \Delta K = 0 - K_i$, so $d = \dfrac{K_i}{f_k}$.

Divide d by the circumference of a circle with radius 0.3 m to find the number of rotations.

$\dfrac{d}{2\pi r} = \dfrac{K_i}{2\pi r f_k} = \dfrac{30 \text{ J}}{2\pi (0.3 \text{ m})(20 \text{ N})} = 0.8$ rotations

Since $0.8 < 1$, the correct answer is $\boxed{\text{A}}$.

11. **Strategy** Compute the average mechanical power output of the cyclist and compare it to the power consumed by the wheel at the friction pads.

Solution The metabolic power available for work is $535 \text{ W} - 85 \text{ W} = 450$ W. Since the efficiency is 20%, the average mechanical power output of the cyclist is $0.20 \times 450 \text{ W} = 90$ W. The average tangential speed of the

wheel is $v = \dfrac{4800 \text{ m}}{20 \text{ min}} \times \dfrac{1 \text{ min}}{60 \text{ s}} = 4.0$ m/s. Therefore, the power consumed by the friction pads is

$P = f_k v = (20 \text{ N})(4.0 \text{ m/s}) = 80$ W. Thus, the difference between the average mechanical power output of the cyclist and the power consumed by the wheel at the friction pads is $90 \text{ W} - 80 \text{ W} = 10$ W.

The correct answer is $\boxed{\text{B}}$.

12. **Strategy and Solution** Increasing the force on the friction pads would increase the power consumed by the wheel at the friction pads (because $P = Fv$). So, if the cyclist is pedaling at the same rate and the power consumed by the friction pads increases, the difference between the two decreases and the fraction of mechanical power output of the cyclist consumed by the wheel at the friction pad increases. Thus, the correct answer is $\boxed{\text{D}}$.

13. **Strategy** Relate the cyclist's average metabolic rate to the energy released per volume of oxygen consumed, the time on the bike, and volume of oxygen consumed.

Solution The cyclist's average metabolic rate while riding is 535 W. The total energy used during 20 minutes is

$(535 \text{ W})(20 \text{ min}) \dfrac{60 \text{ s}}{1 \text{ min}} = 642{,}000$ J. The total energy released by the consumption of oxygen is $(20{,}000 \text{ J/L})V$, where V is the volume of oxygen consumed. Equating these two expressions and solving for V gives the number of liters of oxygen the cyclist consumes.

$(20{,}000 \text{ J/L})V = 642{,}000$ J, so $V = \dfrac{642{,}000 \text{ J}}{20{,}000 \text{ J/L}} = 32 \text{ L} \approx 30$ L. The correct answer is $\boxed{\text{B}}$.

14. **Strategy and Solution** Since the force has been reduced by 50% and the distance has been doubled, the cyclist does the same amount of work $[W = 0.50F(2\Delta x) = F\Delta x]$. So, the energy transmitted in the second workout is equal to the energy transmitted in the first. The correct answer is $\boxed{C}$.

15. **Strategy** The circumference of a circle is $C = 2\pi r$. A wheel moves a distance equal to its circumference during each rotation. The wheel rotates twice during each rotation of the pedals.

    Solution The circumference of a circle with a radius of 0.15 m is $2\pi(0.15$ m$)$. The circumference of a circle with a radius of 0.3 m is $2\pi(0.3$ m$)$. During each rotation of the pedals, a point on the wheel at a radius of 0.3 m moves a distance $2[2\pi(0.3$ m$)]$. The ratio of the distance moved by a pedal to the distance moved by a point on

    the wheel located at a radius of 0.3 m in the same amount of time is $\dfrac{2\pi(0.15 \text{ m})}{2[2\pi(0.3 \text{ m})]} = 0.25$.

    The correct answer is $\boxed{A}$.

16. **Strategy** Use the definition of power.

    Solution
    $$P = \frac{\Delta E}{\Delta t}, \text{ so } \Delta t = \frac{\Delta E}{P} = \left(\frac{300 \text{ kcal}}{500 \text{ W}}\right)\left(\frac{4186 \text{ J}}{1 \text{ kcal}}\right)\left(\frac{1 \text{ min}}{60 \text{ s}}\right) = 41.9 \text{ min. The correct answer is } \boxed{D}.$$

17. **Strategy** Consider the distance a point on the wheel travels for each situation.

    Solution The circumference of a circle with a radius of 0.3 m is $2\pi(0.3$ m$)$. The circumference of a circle with a radius of 0.4 m is $2\pi(0.4$ m$)$. During each rotation, a point on a wheel travels a distance equal to the circumference. The force on the wheel is the same in each case, but the distance traveled by a point on the wheel is greater for a greater radius. In this case, the distance is 0.4 m/(0.3 m) $= 1.33$ times farther or 33%. Since work is equal to the product of force times distance, the work done on the wheel per revolution is 33% more. Thus, the correct answer is $\boxed{C}$.

Chapter 9

FLUIDS

Conceptual Questions

1. A manometer (with one side open) measures gauge pressure. A barometer measures absolute pressure. A tire pressure gauge and a sphygmomanometer both measure gauge pressure.

5. There is practically no atmosphere on the Moon and hence practically zero pressure. Drinking from a glass with a straw would be impossible, since a pressure difference is required to push the liquid up the straw. With a sealed juice box, the astronaut could supply the necessary pressure by squeezing the box.

9. The pressure of the atmosphere decreases with altitude. Therefore, the balloon gradually expands as it rises.

13. The rate at which air moves up a chimney is determined by the size of the pressure change along the chimney's length. Bernoulli's equation tells us that on a windless day, the pressure difference between the two ends of the chimney is proportional to the height of the chimney. On a windy day, the velocity difference between the two ends causes an additional pressure difference—resulting in an increased draft.

17. To get optimal use from hydraulic systems, the fluid used in mediating the operation of the system must be very nearly incompressible. Liquids meet this requirement, but gases like air are compressible. Thus, proper operation of a hydraulic device requires that gases be "bled" from the system.

Problems

1. Strategy Use the definition of average pressure.

Solution Compute the average pressure.

$$P_{av} = \frac{F}{A} = \frac{500 \text{ N}}{1.0 \text{ cm}^2}\left(\frac{100 \text{ cm}}{1 \text{ m}}\right)^2\left(\frac{1 \text{ atm}}{101.3\times10^3 \text{ Pa}}\right) = \boxed{49 \text{ atm}}$$

5. Strategy The average pressure is the force applied to the floor divided by the contact area.

Solution

The baby applies a pressure of $P_b = \dfrac{F}{A} = \dfrac{m_b g}{3\left(\frac{1}{4}\pi d_s^2\right)} = \dfrac{4m_b g}{3\pi d_s^2}$.

The adult applies a pressure of $P_a = \dfrac{F}{A} = \dfrac{m_a g}{4\left(\frac{1}{4}\pi d_c^2\right)} = \dfrac{m_a g}{\pi d_c^2}$.

The ratio of these two pressures is $\dfrac{P_b}{P_a} = \dfrac{4m_b g}{3\pi d_s^2}\left(\dfrac{m_a g}{\pi d_c^2}\right)^{-1} = \dfrac{4m_b d_c^2}{3m_a d_s^2} = \dfrac{4(10 \text{ kg})(0.060 \text{ m})^2}{3(60 \text{ kg})(0.020 \text{ m})^2} = 2.0.$

$\boxed{\text{The baby applies 2.0 times as much pressure as the adult.}}$

9. **Strategy** The work done by the small piston must equal that done on the car.

   Solution Find the distance that the small piston must be pushed downward to raise the car 1.0 cm.
   $F_c/F_p = A/a$ since $\Delta P = F_c/A = F_p/a$.

   $$W_p = F_p d_p = W_c = F_c d_c, \text{ so } d_p = \frac{F_c}{F_p} d_c = \frac{A}{a} d_c = 100.0(0.010 \text{ m}) = \boxed{1.0 \text{ m}}.$$

13. **Strategy** Use Eq. (9-4).

    Solution Compute the pressure on the fish.

    $$P = P_{atm} + \rho g d = 1.0 \text{ atm} + (1025 \text{ kg/m}^3)(9.80 \text{ m/s}^2)(10 \text{ m})\left(\frac{1 \text{ atm}}{1.013 \times 10^5 \text{ Pa}}\right) = \boxed{2.0 \text{ atm}}$$

17. **Strategy** Use the definition of average pressure and Eq. (9-4).

    Solution Find the magnitude of the force exerted by the water on the bottom of the container.

    $$P = P_{atm} + \rho g d = \frac{F}{A}, \text{ so}$$

    $$F = (P_{atm} + \rho g d)A = \left[1.013 \times 10^5 \text{ Pa} + (1.0 \times 10^3 \text{ kg/m}^3)(9.80 \text{ m/s}^2)(11.0 \text{ m})\right](5.00 \text{ m}^2) = \boxed{1.0 \text{ MN}}.$$

19. **(a) Strategy** Use Eq. (9-4).

    Solution Compute the pressure increase.
    $$\Delta P = \rho_{water} g d = (1.00 \times 10^3 \text{ kg/m}^3)(9.80 \text{ m/s}^2)(35.0 \text{ m}) = \boxed{343 \text{ kPa}}$$

    (b) Strategy Use Eq. (9-3).

    Solution Compute the pressure decrease.
    $$\Delta P = -\rho_{air} g h = -(1.20 \text{ kg/m}^3)(9.80 \text{ m/s}^2)(35 \text{ m}) = -410 \text{ Pa}$$
    The pressure decreases by $\boxed{410 \text{ Pa}}$.

21. **Strategy** Use the appropriate conversion factors to convert the gauge pressure into the various pressure units.

    Solution

    (a) $(32 \text{ lb/in}^2)\left(\dfrac{1.013 \times 10^5 \text{ Pa}}{14.7 \text{ lb/in}^2}\right) = \boxed{2.2 \times 10^5 \text{ Pa}}$

    (b) $(32 \text{ lb/in}^2)\left(\dfrac{760.0 \text{ torr}}{14.7 \text{ lb/in}^2}\right) = \boxed{1700 \text{ torr}}$

    (c) $(32 \text{ lb/in}^2)\left(\dfrac{1 \text{ atm}}{14.7 \text{ lb/in}^2}\right) = \boxed{2.2 \text{ atm}}$

25. Strategy The amount the fluid rises is one-half the difference of levels, or $\Delta h_{oil}/2$. Use Eq. (9-5).

Solution

(a) Find the amount that the fluid level rises.

$$\Delta P = \rho_{Hg}g\Delta h_{Hg} = \rho_{oil}g\Delta h_{oil}, \text{ so } \frac{\Delta h_{oil}}{2} = \frac{\rho_{Hg}}{2\rho_{oil}}\Delta h_{Hg} = \frac{13.6 \text{ g/cm}^3}{2(0.90 \text{ g/cm}^3)}(0.74 \text{ cm Hg}) = \boxed{5.6 \text{ cm}}.$$

(b) $\frac{\Delta h_{oil}}{2}\left(\frac{\rho_{oil}}{\rho_{Hg}}\right) = (5.6 \text{ cm})\frac{0.90 \text{ g/cm}^3}{13.6 \text{ g/cm}^3} = \boxed{0.37 \text{ cm}}$

27. Strategy and Solution Since the goose has 25% of its volume submerged, its density is 25% of water's, or about $\boxed{250 \text{ kg/m}^3}$.

29. (a) Strategy The relationship between the fraction of a floating object's volume that is submerged to the ratio of the object's density to the fluid in which it floats is $V_f/V_o = \rho_o/\rho_f$. Since the water contains ice, use the density of water at $0°C$.

Solution Find the percent of the volume of ice that is submerged when it floats in water.

$$\frac{V_{submerged}}{V_{ice}} = \frac{\rho_{ice}}{\rho_{water}} = \frac{917 \text{ kg/m}^3}{999.87 \text{ kg/m}^3} = 0.917, \text{ or } \boxed{91.7\%}$$

(b) Strategy and Solution The specific gravity and the fraction of the object submerged in water are the same for objects that float, so the specific gravity of ice is $\boxed{0.917}$.

33. Strategy Compare the densities of a block of ebony and ethanol.

Solution The density of a block of ebony is between 1000 and 1300 kg/m^3. The density of ethanol is 790 kg/m^3. Since the density of ebony is more than that of ethanol, the block will sink; therefore, $\boxed{100\%}$ of the volume of the block of ebony is submerged.

37. Strategy Find the volume of the coin using the density of water and the mass of the displaced water. Then find its density using its mass as measured in air and compare this to the density of gold.

Solution The mass of water displaced is $49.7 \text{ g} - 47.1 \text{ g} = 2.6 \text{ g}$. Since the density of water is about 1.00 g/cm^3, the volume of the coin is $2.6 \text{ cm}^3 = 2.6\times10^{-6} \text{ m}^3$. So, the density of the coin is

$$\rho = \frac{m}{V} = \frac{49.7\times10^{-3} \text{ kg}}{2.6\times10^{-6} \text{ m}^3} = 19,000 \text{ kg/m}^3.$$ The density of gold is 19,300 kg/m^3, so $\boxed{\text{yes}}$, you should get excited, since the coin may be genuine.

39. **Strategy** Let the $+y$-direction be upward. Use Newton's second law and Eq. (9-7).

    Solution

    (a) $\Sigma F_y = F_B - mg = ma$, so

    $$a = \frac{F_B}{m} - g = \frac{\rho_w g V}{m} - g = g\left(\frac{\rho_w V}{\rho V} - 1\right) = (9.80 \text{ m/s}^2)\left(\frac{1.00 \text{ g/cm}^3}{0.50 \text{ g/cm}^3} - 1\right) = 9.8 \text{ m/s}^2.$$

    Thus, $\vec{a} = \boxed{9.8 \text{ m/s}^2 \text{ upward}}$.

    (b) $a = (9.80 \text{ m/s}^2)\left(\dfrac{1.00 \text{ g/cm}^3}{0.750 \text{ g/cm}^3} - 1\right) = 3.3 \text{ m/s}^2$, so $\vec{a} = \boxed{3.3 \text{ m/s}^2 \text{ upward}}$.

    (c) $a = (9.80 \text{ m/s}^2)\left(\dfrac{1.00 \text{ g/cm}^3}{0.125 \text{ g/cm}^3} - 1\right) = 68.6 \text{ m/s}^2$, so $\vec{a} = \boxed{68.6 \text{ m/s}^2 \text{ upward}}$.

41. **Strategy** Use Eq. (9-13).

    Solution Find the speed of the water as it passes through the nozzle.

    $$A_2 v_2 = A_1 v_1, \text{ so } v_2 = \frac{A_1}{A_2} v_1 = \frac{\pi r_1^2}{\pi r_2^2} v_1 = \left(\frac{1.0 \text{ cm}}{0.20 \text{ cm}}\right)^2 (2.0 \text{ m/s}) = \boxed{50 \text{ m/s}}.$$

43. (a) **Strategy** Use Eq. (9-13).

    Solution Find the speed of the water in the hose.

    $$v_2 = \frac{A_1}{A_2} v_1 = \frac{\pi r_1^2}{\pi r_2^2} v_1 = \left(\frac{r_1}{r_2}\right)^2 v_1 = \left(\frac{1.00 \text{ mm}}{8.00 \text{ mm}}\right)^2 (25.0 \text{ m/s}) = \boxed{39.1 \text{ cm/s}}$$

    (b) **Strategy** Use Eq. (9-12).

    Solution Compute the volume flow rate.

    $$\frac{\Delta V}{\Delta t} = A_1 v_1 = \pi(1.00 \times 10^{-3} \text{ m})^2 (25.0 \text{ m/s}) = \boxed{78.5 \text{ cm}^3/\text{s}}$$

    (c) **Strategy** Use Eq. (9-11).

    Solution Compute the mass flow rate.

    $$\frac{\Delta m}{\Delta t} = \rho A_1 v_1 = (1.00 \text{ g/cm}^3)(78.5 \text{ cm}^3/\text{s}) = \boxed{78.5 \text{ g/s}}$$

45. Strategy Use Eqs. (9-13) and (9-14). Since the pipe is horizontal, $y_1 = y_2$.

Solution Let the larger end be labeled 2. Find the speed of the water at the narrow end in terms of the speed at the larger end.

$A_1 v_1 = A_2 v_2$, so $v_1 = \dfrac{A_2}{A_1} v_2$.

Find the pressure at the narrow end of the segment of pipe.

$$P_1 + \frac{1}{2}\rho v_1^2 = P_2 + \frac{1}{2}\rho v_2^2, \text{ so } P_1 = P_2 + \frac{1}{2}\rho\left[v_2^2 - \left(\frac{A_2}{A_1}v_2\right)^2\right]$$

$$= 1.20\times10^5 \text{ Pa} + \frac{1}{2}(1.00\times10^3 \text{ kg/m}^3)(0.040 \text{ m/s})^2\left[1 - \left(\frac{50.0 \text{ cm}^2}{0.500 \text{ cm}^2}\right)^2\right] = \boxed{1.12\times10^5 \text{ Pa}}.$$

47. Strategy Use Eq. (9-14).

Solution The potential energy difference is relatively small, so Bernoulli's equation becomes

$$P_1 + \frac{1}{2}\rho v_1^2 = P_2 + \frac{1}{2}\rho v_2^2, \text{ or } P_1 - P_2 = \frac{1}{2}\rho v_2^2 - \frac{1}{2}\rho v_1^2.$$

Estimate the force.

$$F = \Delta P A = (P_1 - P_2)A = \left(\frac{1}{2}\rho v_2^2 - \frac{1}{2}\rho v_1^2\right)A = \frac{1}{2}A\rho(v_2^2 - v_1^2)$$

$$= \frac{1}{2}(28 \text{ m}^2)(1.3 \text{ kg/m}^3)[(190 \text{ m/s})^2 - (160 \text{ m/s})^2] = \boxed{1.9\times10^5 \text{ N}}.$$

49. Strategy Use Eqs. (9-13) and (9-14).

Solution $y_1 = y_2$ and $v_1 \gg v_2$ since $A_2 \gg A_1$ ($d_2 \gg d_1$), so Bernoulli's equation gives

$$\Delta P = \frac{1}{2}\rho v_1^2 = \frac{1}{2}(1.0\times10^3 \text{ kg/m}^3)(25 \text{ m/s})^2 = \boxed{310 \text{ kPa}}.$$

53. Strategy Use Eq. (9-15).

Solution Show that viscosity has SI units of pascal-seconds.
Solve for η.

$$\frac{\Delta V}{\Delta t} = \frac{\pi}{8}\frac{\frac{\Delta P}{L}}{\eta}r^4, \text{ so } \eta = \frac{\pi}{8}\frac{\Delta P}{L}\frac{\Delta t}{\Delta V}r^4. \text{ Thus, the units of } \eta \text{ are } \frac{\text{Pa}}{\text{m}}\cdot\frac{\text{s}}{\text{m}^3}\cdot\text{m}^4 = \text{Pa}\cdot\text{s}.$$

57. Strategy Use Eq. (9-15). Form a ratio of the volume flow rates.

Solution Find the total flow rate in system C.

$$\frac{\frac{\Delta V}{\Delta t}\,C}{\frac{\Delta V}{\Delta t}\,B} = \frac{4\left[\dfrac{\pi\Delta P_C r^4}{8\eta\left(\frac{L}{2}\right)}\right]}{2\left(\dfrac{\pi\Delta P_B r^4}{8\eta L}\right)} = \frac{4\Delta P_C}{\Delta P_B}, \text{ so } \frac{\Delta V}{\Delta t}\,C = \frac{4\Delta P_C}{\Delta P_B}\left(\frac{\Delta V}{\Delta t}\,B\right) = 4\left(\frac{2.0\times10^5 \text{ Pa}}{4.0\times10^5 \text{ Pa}}\right)(0.020 \text{ m}^3/\text{s}) = \boxed{0.040 \text{ m}^3/\text{s}}.$$

61. Strategy Use Eq. (9-16) and Newton's second law.

Solution Find the viscosity of the second liquid.

$$\Sigma F_y = F_D + F_B - m_s g = 6\pi\eta r v + m_l g - m_s g = 6\pi\eta r v + (m_l - m_s)g = 6\pi\eta r v - \frac{4}{3}\pi r^3(\rho_s - \rho_l)g = 0,$$

so $\eta = \dfrac{\frac{4}{3}\pi r^3(\rho_s - \rho_l)g}{6\pi r v} = \dfrac{2r^2(\rho_s - \rho_l)g}{9v}.$

Find the viscosity of the second liquid by forming a proportion.

$$\frac{\eta_2}{\eta_1} = \frac{\frac{2r^2(\rho_s - \rho_l)g}{9v_2}}{\frac{2r^2(\rho_s - \rho_l)g}{9v_1}} = \frac{v_1}{v_2}, \text{ so } \eta_2 = \frac{v_1}{1.2v_1}\eta_1 = \frac{\eta_1}{1.2} = \frac{0.5\ \text{Pa}\cdot\text{s}}{1.2} = \boxed{0.4\ \text{Pa}\cdot\text{s}}.$$

65. Strategy For a viscous drag force, $v_t \propto m$. For a turbulent drag force, $v_t \propto \sqrt{m}$. For the data in the table, compute m/v_t and m/v_t^2.

Solution The data, m/v_t, and m/v_t^2 are organized in the table.

m (g)	8	12	16	20	24	28
v_t (cm/s)	1.0	1.5	2.0	2.5	3.0	3.5
$\dfrac{m}{v_t}$ (g$\cdot$s/cm)	8	8.0	8.0	8.0	8.0	8.0
$\dfrac{m}{v_t^2}$ (g$\cdot$s^2/cm^2)	8	5.3	4.0	3.2	2.7	2.3

Since m/v_t is constant, the drag force is primarily viscous.

69. Strategy Use Eq. (9-17).

Solution Form a proportion to find the pressure inside the air bubble.

$$\frac{\Delta P_2}{\Delta P_1} = \frac{\frac{2\gamma}{r_2}}{\frac{2\gamma}{r_1}}, \text{ so } \Delta P_2 = \Delta P_1 \frac{r_1}{r_2} = (10\ \text{Pa})\frac{r_1}{2r_1} = \boxed{5\ \text{Pa}}.$$

73. **(a) Strategy** The mass of the water is equal to its density times its volume.

Solution The weight of the water in the straw is
$$mg = (\rho V)g = \rho(\pi r^2 h)g = (1.00\times10^3 \text{ kg}/\text{m}^3)\pi(0.00250 \text{ m})^2(8.00 \text{ m})(9.80 \text{ m/s}^2) = \boxed{1.54 \text{ N}}.$$

(b) Strategy Equate the pressures and solve for the force on the top of the barrel. Use the definition of pressure.

Solution Find the force with which the water in the barrel pushes up on the top of the barrel.
$$F_b = \frac{A_b}{A_s}F_s = \frac{r_b^2}{r_s^2}F_s = \frac{(25.0 \text{ cm})^2}{(0.250 \text{ cm})^2}(1.54 \text{ N}) = \boxed{1.54\times10^4 \text{ N}}.$$

(c) Strategy Consider the nature of pressure in a column of fluid.

Solution For a given depth, the pressure is the same everywhere, so the very tall, narrow column of water is as effective as having a whole barrel of water filled to the same height and pushing upward on the barrel top.

77. **(a) Strategy** Assume that the change in the height of the water level in the vat is negligible and that the pressures at the top of the vat and at the outlet are the same. Use Bernoulli's equation.

Solution Let the top of the vat be labeled 1. With the above assumptions, Bernoulli's equation becomes
$$\rho g y_1 = \rho g y_2 + \frac{1}{2}\rho v_2^2, \text{ so } v_2 = \sqrt{2g(y_1 - y_2)} = \sqrt{2(9.80 \text{ m/s}^2)(1.80 \text{ m})} = \boxed{5.94 \text{ m/s}}.$$

(b) Strategy and Solution The density "falls out" of Bernoulli's equation in our calculation of the speed, so as long as we can assume Bernoulli's equation applies, it doesn't matter what fluid is in the vat.

(c) Strategy and Solution Since the speed is directly proportional to the square root of the gravitational field strength, $\sqrt{1.6/9.80} = 0.40$, the speed would be reduced by a factor of 0.40.

81. **Strategy** Use Eq. (9-14) with $v_1 = v_2 = 0$.

Solution Find the height between the basement and the seventh floor.
$$P_1 + \rho g y_1 = P_2 + \rho g y_2, \text{ so } y_1 - y_2 = \frac{P_2 - P_1}{\rho g} = \frac{4.10\times10^5 \text{ Pa} - 1.85\times10^5 \text{ Pa}}{(1.00\times10^3 \text{ kg}/\text{m}^3)(9.80 \text{ m/s}^2)} = \boxed{23.0 \text{ m}}.$$

85. **Strategy** Use the continuity equation. The speed of an object that has fallen a distance h from rest is $v = \sqrt{2gh}$.

Solution Find the diameter of the water flow after the water has fallen 30 cm.
$$A_1 v_1 = \frac{1}{4}\pi d_1^2 v_1 = A_2 v_2 = \frac{1}{4}\pi d_2^2 v_2, \text{ so } d_2 = d_1\sqrt{\frac{v_1}{\sqrt{2gh}}} = (2.2 \text{ cm})\sqrt{\frac{0.62 \text{ m/s}}{\sqrt{2(9.80 \text{ m/s}^2)(0.30 \text{ m})}}} = \boxed{1.1 \text{ cm}}.$$

89. **(a) Strategy** Use Eqs. (9-7), (9-16), and Newton's second law.

Solution Find the terminal velocity of the bubbles.

$$\Sigma F_y = F_B - F_D - m_a g = \rho_w gV - 6\pi\eta r v_t - \rho_a gV = -6\pi\eta r v_t + (\rho_w - \rho_a)\tfrac{4}{3}\pi r^3 g = 0, \text{ so}$$

$$v_t = \frac{(\rho_w - \rho_a)\tfrac{4}{3}\pi r^3 g}{6\pi\eta r} = \frac{2r^2 g(\rho_w - \rho_a)}{9\eta}$$

$$= \frac{2(1.0\times10^{-3}\ \text{m})^2(9.80\ \text{m/s}^2)(1.00\times10^3\ \text{kg/m}^3 - 1.20\ \text{kg/m}^3)}{9(1.0\times10^{-3}\ \text{Pa}\cdot\text{s})} = 2.2\ \text{m/s}.$$

Thus, $\vec{v}_t = \boxed{2.2\ \text{m/s up}}$.

(b) Strategy Divide the change in pressure by the change in time and use the result from part (a).

Solution

$$\Delta P = \rho g\Delta y, \text{ so } \frac{\Delta P}{\Delta t} = \rho g\frac{\Delta y}{\Delta t} = \rho g v_t = (1.00\times10^3\ \text{kg/m}^3)(9.80\ \text{m/s}^2)(2.175\ \text{m/s}) = \boxed{21\ \text{kPa/s}}.$$

93. **Strategy** Use Eq. (9-7) and Newton's second law.

Solution Find an expression for d.

$$\Sigma F_y = F_B - mg = \rho gV - mg = \rho gAd - mg = \rho g(\pi r^2)d - mg = 0, \text{ so } d = \frac{m}{\pi\rho r^2}.$$

$$\boxed{d \text{ is not a linear function of } \rho: d = \frac{m}{\pi\rho r^2}.}$$

97. **Strategy** Use the relationships between pressure, density, force, area, and height.

Solution Find the density of the liquid.

$$\Delta P = \frac{W_1}{A} = \frac{\rho_1 gV_1}{A} = \rho_w g\Delta y_w, \text{ so}$$

$$\rho_1 = \frac{\rho_w\Delta y_w A}{V_1} = \frac{\rho_w\Delta y_w \pi r^2}{\pi r^2 h} = \frac{\rho_w\Delta y_w}{h} = \frac{(1.0\ \text{g/cm}^3)[0.45\ \text{m} - (0.50\ \text{m} - 0.30\ \text{m})]}{0.30\ \text{m}} = \boxed{0.83\ \text{g/cm}^3}.$$

101. **(a) Strategy** Use Eq. (9-15).

Solution Find the percentage that the blood pressure difference between the ends of the artery increased.

$$\frac{\Delta P_a - \Delta P_u}{\Delta P_u}\times100\% = \frac{\left(\frac{8\eta L}{\pi r_a^4}\right)\frac{\Delta V}{\Delta t} - \left(\frac{8\eta L}{\pi r_u^4}\right)\frac{\Delta V}{\Delta t}}{\left(\frac{8\eta L}{\pi r_u^4}\right)\frac{\Delta V}{\Delta t}}\times100\% = \frac{\frac{1}{r_a^4} - \frac{1}{r_u^4}}{\frac{1}{r_u^4}}\times100\% = \left[\left(\frac{r_u}{r_a}\right)^4 - 1\right]\times100\%$$

$$= \left[\left(\frac{1}{0.75}\right)^4 - 1\right]\times100\% = \boxed{220\%}$$

(b) Strategy Divide the absolute value of the change in flow rate by the original flow rate.

Solution Compute the factor of blood flow decrease.

$$\left|\frac{\text{change in flow rate}}{\text{original flow rate}}\right| = \frac{\frac{\Delta V}{\Delta t}_u - \frac{\Delta V}{\Delta t}_a}{\frac{\Delta V}{\Delta t}_u} = 1 - \frac{\frac{\Delta V}{\Delta t}_a}{\frac{\Delta V}{\Delta t}_u} = 1 - \frac{\frac{\pi\Delta P r_a^4}{8\eta L}}{\frac{\pi\Delta P r_u^4}{8\eta L}} = 1 - \left(\frac{r_a}{r_u}\right)^4 = 1 - (0.75)^4 = \boxed{0.68}$$

Chapter 10

ELASTICITY AND OSCILLATIONS

Conceptual Questions

1. Young's modulus does not tell us which is stronger. Instead, it tells us which is more resistant to deformation for a given stress. The ultimate strength would tell us which is stronger—i.e., which can withstand the greatest stress.

5. The compressive force experienced by the columns is greater at the bottom than at the top, because the bottom must support the weight of the column itself in addition to whatever the column is holding up. By increasing the cross-sectional area of the bottom of the column, the stress it experiences is reduced. Tapering columns so that they are thicker at the base prevents the stress at the bottom from being too large.

9. The tension in the bungee cord at the lowest point would be greater than the person's weight, because there is an upward acceleration. In fact, the tension would have its maximum value at the bottom, because that is where the upward acceleration is the greatest.

13. To produce the same strain, the ratio of the force to the cross-sectional area must remain unchanged. The total cross-sectional area of the two wires together is twice the original area. Thus, the force applied to the two wires must be doubled as well. Modeling a thick wire as a bundle of thin wires, the preceding argument explains why the force to produce a given strain must be proportional to the cross-sectional area—and thus why the strain depends on the stress.

17. In the mass-spring system, the restoring force supplied by the spring is independent of the object's mass. Thus, the larger inertia of a more massive object produces a longer period. The restoring force for small amplitude oscillations of the pendulum is the horizontal component of the tension in the string. In this case, the magnitude of the tension is approximately equal to the weight of the bob. Although a more massive bob has more inertia, it also has a proportionally larger restoring force. Thus, the period of oscillation of the pendulum is independent of the mass.

Problems

1. **Strategy** The stress is proportional to the strain. Use Eq. (10-4).

    Solution Find the vertical compression of the beam.
    $$Y\frac{\Delta L}{L} = \frac{F}{A}, \text{ so } \Delta L = \frac{FL}{YA} = \frac{(5.8\times10^4 \text{ N})(2.5 \text{ m})}{(200\times10^9 \text{ Pa})(7.5\times10^{-3} \text{ m}^2)} = \boxed{0.097 \text{ mm}}.$$

3. **Strategy** The stress is proportional to the strain. Use Eq. (10-4).

    Solution Find how much the wire stretches.
    $$\Delta L = \frac{FL}{YA} = \frac{(5.0\times10^3 \text{ N})(2.0 \text{ m})}{(9.2\times10^{10} \text{ Pa})(5.0 \text{ mm}^2)(10^{-3} \text{ m/mm})^2} = \boxed{2.2 \text{ cm}}$$

5. **Strategy** Form a proportion of the elongations of the left and right wires. Use Eq. (10-4).

   Solution Find how far the midpoint moves.

   $$\frac{\Delta L_L}{\Delta L_R} = \frac{\frac{FL}{YA_L}}{\frac{FL}{YA_R}} = \frac{A_R}{A_L} = \frac{\pi(2r)^2}{\pi r^2} = 4, \text{ so } \Delta L_L = 4\Delta L_R \text{ and } \Delta L = \Delta L_L + \Delta L_R \text{ is the total elongation, 1.0 mm.}$$

   $$\Delta L = \Delta L_L + \Delta L_R = \Delta L_L + \frac{1}{4}\Delta L_L = \frac{5}{4}\Delta L_L, \text{ so } \Delta L_L = \frac{4}{5}\Delta L = \frac{4}{5}(1.0 \text{ mm}) = \boxed{0.80 \text{ mm}}.$$

9. **Strategy** Refer to Fig. 10.4c. The stress is proportional to the strain.

   Solution Calculate Young's moduli for tension and compression of bone.
   Tension:
   For tensile stress and strain, the graph is far from being linear, but for relatively small values of stress and strain, it is approximately linear. So, for small values of tensile stress and strain, Young's Modulus is

   $$Y = \frac{\text{stress}}{\text{strain}} = \frac{5.0\times10^7 \text{ N}/\text{m}^2}{0.0033} = \boxed{1.5\times10^{10} \text{ N}/\text{m}^2}.$$

   Compression:

   Similarly, for small values of compressive stress and strain, $Y = \frac{-4.5\times10^7 \text{ N}/\text{m}^2}{-0.0050} = \boxed{9.0\times10^9 \text{ N}/\text{m}^2}.$

11. **Strategy** Set the stress equal to the tensile strength of the hair to find the diameter of the hair.

    Solution Find the diameter of the hair.

    $$\text{tensile strength} = \frac{F}{A} = \frac{F}{\frac{1}{4}\pi d^2}, \text{ so } d = \sqrt{\frac{4F}{\pi(\text{tensile strength})}} = \sqrt{\frac{4(1.2 \text{ N})}{\pi(2.0\times10^8 \text{ Pa})}} = \boxed{8.7\times10^{-5} \text{ m}}.$$

13. **Strategy** The stress on the copper wire must be less than its elastic limit.

    Solution Find the maximum load that can be suspended from the copper wire.

    $$\frac{F}{A} < \text{elastic limit, so } F < \pi r^2(\text{elastic limit}) = \pi(0.0010 \text{ m})^2(2.0\times10^8 \text{ Pa}) = \boxed{630 \text{ N}}.$$

17. **Strategy** Use Eqs. (10-1), (10-2), and (10-4).

    Solution

    (a) $\text{stress} = \dfrac{F}{A} = \dfrac{7.0\times10^4 \text{ N}}{25\times10^{-4} \text{ m}^2} = \boxed{2.8\times10^7 \text{ Pa}}$

    (b) $\text{strain} = \dfrac{\Delta L}{L} = \dfrac{F}{YA} = \dfrac{7.0\times10^4 \text{ N}}{(6.0\times10^{10} \text{ Pa})(25\times10^{-4} \text{ m}^2)} = \boxed{4.7\times10^{-4}}$

    (c) $\Delta L = \dfrac{FL}{YA} = \dfrac{(7.0\times10^4 \text{ N})(2.0 \text{ m})}{(6.0\times10^{10} \text{ Pa})(25\times10^{-4} \text{ m}^2)} = \boxed{9.3\times10^{-4} \text{ m}}$

    (d) Set the compressive strength equal to the stress to find the maximum weight the column can support.
    $$\frac{F}{A} = \text{compressive strength, so } F = (\text{compressive strength})A = (2.0\times10^8 \text{ Pa})(25\times10^{-4} \text{ m}^2) = \boxed{5.0\times10^5 \text{ N}}.$$

21. Strategy Use Hooke's law for volume deformations.

Solution Find the change in volume of the sphere.

$$\Delta P = -B \frac{\Delta V}{V}, \text{ so } \Delta V = -\frac{V \Delta P}{B} = -\frac{(1.00 \text{ cm}^3)(9.12 \times 10^6 \text{ Pa})}{160 \times 10^9 \text{ Pa}} = -57 \times 10^{-6} \text{ cm}^3.$$

The volume of the steel sphere would decrease by 57×10^{-6} cm^3.

25. Strategy Use Hooke's law for shear deformations.

Solution Find the magnitude of the tangential force.

$$\frac{F}{A} = \frac{F}{L^2} = S \frac{\Delta x}{L}, \text{ so } F = S \Delta x L = (940 \text{ Pa})(0.64 \times 10^{-2} \text{ m})(0.050 \text{ m}) = \boxed{0.30 \text{ N}}.$$

29. Strategy The amplitude is half the maximum distance. Use Eqs. (10-21) and (10-22).

Solution Find the maximum velocity and maximum acceleration of the prong.

$$v_m = \omega A = 2\pi f A = 2\pi (440.0 \text{ Hz}) \left(\frac{2.24}{2} \times 10^{-3} \text{ m} \right) = \boxed{3.10 \text{ m/s}}$$

$$a_m = \omega^2 A = 4\pi^2 f^2 A = 4\pi^2 (440.0 \text{ Hz})^2 \left(\frac{2.24}{2} \times 10^{-3} \text{ m} \right) = \boxed{8560 \text{ m/s}^2}$$

33. Strategy Use Eqs. (10-21) and (10-22).

Solution

(a) Find v_m and a_m in terms of f. Then compare high- and low-frequency sounds.

$v_m = \omega A = 2\pi f A \propto f$ and $a_m = \omega^2 A = 4\pi^2 f^2 A \propto f^2$, so v_m and a_m are greatest for $\boxed{\text{high frequency}}$.

(b) $v_m = 2\pi (20.0 \text{ Hz})(1.0 \times 10^{-8} \text{ m}) = \boxed{1.3 \times 10^{-6} \text{ m/s}}$

$a_m = 4\pi^2 (20.0 \text{ Hz})^2 (1.0 \times 10^{-8} \text{ m}) = \boxed{1.6 \times 10^{-4} \text{ m/s}^2}$

(c) $v_m = 2\pi (20.0 \times 10^3 \text{ Hz})(1.0 \times 10^{-8} \text{ m}) = \boxed{0.0013 \text{ m/s}}$

$a_m = 4\pi^2 (20.0 \times 10^3 \text{ Hz})^2 (1.0 \times 10^{-8} \text{ m}) = \boxed{160 \text{ m/s}^2}$

35. Strategy The angular frequency of oscillation is inversely proportional to the square root of the mass. Form a proportion.

Solution Find the new value of ω.

$$\omega \propto \sqrt{\frac{1}{m}}, \text{ so } \frac{\omega_f}{\omega_i} = \frac{\sqrt{1/m_f}}{\sqrt{1/m_i}} = \sqrt{\frac{m_i}{m_f}} = \sqrt{\frac{1}{4.0}} = \frac{1}{2.0}. \text{ Therefore, } \omega_f = \frac{\omega_i}{2.0} = \frac{10.0 \text{ rad/s}}{2.0} = \boxed{5.0 \text{ rad/s}}.$$

37. **Strategy** Use Eqs. (10-20a) and (10-20b) and Newton's second law.

Solution Find the spring constant.

$$\Sigma F_y = kx - m_{child}g = 0, \text{ so } k = \frac{m_{child}g}{x} = \frac{(24 \text{ kg})(9.80 \text{ m/s}^2)}{0.28 \text{ m}} = 840 \text{ N/m}.$$

Find the mass of the wooden horse.

$$\omega_{child} = \sqrt{\frac{k}{m_{child} + m_{horse}}}, \text{ so } m_{horse} = \frac{k}{\omega_{child}^2} - m_{child}.$$

Find the oscillation frequency of the spring when no one is sitting on the horse.

$$f_{horse} = \frac{\omega_{horse}}{2\pi} = \frac{1}{2\pi}\sqrt{\frac{k}{m_{horse}}} = \frac{1}{2\pi}\sqrt{\frac{k}{\frac{k}{\omega_{child}^2} - m_{child}}} = \frac{1}{2\pi}\sqrt{\frac{840 \text{ N/m}}{\frac{840 \text{ N/m}}{(0.88 \text{ Hz})^2(2\pi \text{ rad/cycle})^2} - 24 \text{ kg}}} = \boxed{2.5 \text{ Hz}}$$

39. **Strategy** Use Eqs. (10-21) and (10-22) and Newton's second law.

Solution Find the radio's maximum displacement and maximum speed, and the maximum net force exerted on it.

(a) $a_m = \omega^2 A$, so $A = \dfrac{a_m}{\omega^2} = \dfrac{98 \text{ m/s}^2}{4\pi^2(120 \text{ Hz})^2} = \boxed{1.7\times10^{-4} \text{ m}}$.

(b) $v_m = \omega A = \omega\dfrac{a_m}{\omega^2} = \dfrac{a_m}{\omega} = \dfrac{98 \text{ m/s}^2}{2\pi(120 \text{ Hz})} = \boxed{0.13 \text{ m/s}}$

(c) According to Newton's second law, $F_m = ma_m = (5.24 \text{ kg})(98 \text{ m/s}^2) = \boxed{510 \text{ N}}$.

41. (a) **Strategy** Use Newton's second law and Eq. (10-22).

Solution Find the maximum force acting on the diaphragm.
$$F_m = ma_m = m\omega^2 A = 4\pi^2(0.0500 \text{ kg})(2.0\times10^3 \text{ Hz})^2(1.8\times10^{-4} \text{ m}) = \boxed{1.4 \text{ kN}}.$$

(b) **Strategy** The maximum elastic potential energy of the diaphragm is equal to the total mechanical energy.

Solution Find the mechanical energy of the diaphragm.
$$E = U = \frac{1}{2}m\omega^2 A^2 = 2\pi^2(0.0500 \text{ kg})(2.0\times10^3 \text{ Hz})^2(1.8\times10^{-4} \text{ m})^2 = \boxed{0.13 \text{ J}}.$$

43. (a) **Strategy** The speed is maximum when the spring and mass system is at its equilibrium point. Use Newton's second law.

Solution Find the extension of the spring.
$$\Sigma F_y = kx - mg = 0, \text{ so } x = \frac{mg}{k} = \frac{(0.60 \text{ kg})(9.80 \text{ N/kg})}{15 \text{ N/m}} = \boxed{0.39 \text{ m}}.$$

(b) **Strategy** Use Eqs. (10-20a) and (10-21).

Solution Find the maximum speed of the body.
$$v_m = \omega A = \sqrt{\frac{k}{m}}x = \sqrt{\frac{15 \text{ N/m}}{0.60 \text{ kg}}}(0.39 \text{ m}) = \boxed{2.0 \text{ m/s}}$$

45. Strategy Use Hooke's law, Newton's second law, and Eq. (10-20c).

Solution Find the "spring constant" of the boat. At equilibrium,

$\Sigma F_y = kx - m_{person}g = 0$, so $k = \dfrac{m_{person}g}{x}$.

Compute the period of oscillation.

$$T = 2\pi\sqrt{\frac{m_{total}}{k}} = 2\pi\sqrt{\frac{m_{total}}{m_{person}g/x}} = 2\pi\sqrt{\frac{m_{total}x}{m_{person}g}} = 2\pi\sqrt{\frac{(47 \text{ kg} + 92 \text{ kg})(0.080 \text{ m})}{(92 \text{ kg})(9.80 \text{ m/s}^2)}} = \boxed{0.70 \text{ s}}$$

49. (a) Strategy The object will oscillate up and down with an amplitude determined by the spring constant and the mass of the spring.

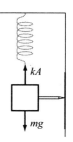

Solution Find the amplitude of the motion. At the equilibrium point, the net force on the object is zero.

$\Sigma F_y = kA - mg = 0$, so $A = mg/k = (0.306 \text{ kg})(9.80 \text{ m/s}^2)/(25 \text{ N/m}) = 12$ cm.

The object will move up and down a total vertical distance of $2A = 24$ cm. Thus, the pattern traced on the paper by the pen is $\boxed{\text{a vertical straight line of length 24 cm}}$.

(b) Strategy and Solution As the paper moves to the left at constant speed while the pen oscillates vertically in SHM, the pen traces a pattern of

$\boxed{\text{a positive cosine plot of amplitude 12 cm}}$.

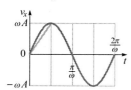

51. Strategy Use the definition of average speed and Eq. (10-21). In (d), graph v_x on the vertical axis and t on the horizontal axis.

Solution

(a) The average speed is the total distance traveled divided by the time of travel.

$$v_{av} = \frac{\Delta x}{\Delta t} = \frac{4A}{T} = \frac{4A}{2\pi/\omega} = \boxed{\frac{2}{\pi}\omega A}$$

(b) The maximum speed for SHM is $v_m = \boxed{\omega A}$.

(c) $\dfrac{v_{av}}{v_m} = \dfrac{\frac{2}{\pi}\omega A}{\omega A} = \boxed{\dfrac{2}{\pi}}$

(d) Graph $v_x(t)$ and a line from the origin to v_m.

If the acceleration were constant so that the speed varied linearly, the average speed would be 1/2 of the maximum velocity. Since the actual speed is always larger than what it would be for constant acceleration, the average speed must be larger.

53. (a) Strategy Use Eq. (10-20a) to find the spring constant. Then, find the elastic potential energy using $\frac{1}{2}kx^2$.

Solution Find the spring constant.

$\omega = \sqrt{\dfrac{k}{m}}$, so $k = \omega^2 m = (2.00 \text{ Hz})^2 (2\pi \text{ rad/cycle})^2 (0.2300 \text{ kg}) = 36.3 \text{ N/m}$.

The equation for the elastic potential energy is

$U(t) = \dfrac{1}{2}(36.3 \text{ N/m})(0.0800 \text{ m})^2 \sin^2\left[(2.00 \text{ Hz})(2\pi \text{ rad/cycle})t\right] = (116 \text{ mJ})\sin^2\left[(4.00\pi \text{ s}^{-1})t\right]$.

Since the sine function is squared, the period of $U(t)$ is half that of a sine function or

$T = \dfrac{\pi}{\omega} = \dfrac{\pi}{4.00\pi \text{ s}^{-1}} = 250 \text{ ms}$. Graph $U(t)$.

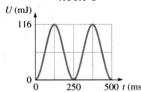

(b) Strategy Find the kinetic energy using $\frac{1}{2}mv_x^2$.

Solution The equation for the kinetic energy is

$K(t) = \dfrac{1}{2}(0.2300 \text{ kg})(2.00 \text{ Hz})^2 \left(\dfrac{2\pi \text{ rad}}{\text{cycle}}\right)^2 (0.0800 \text{ m})^2 \cos^2\left[(2.00 \text{ Hz})\left(\dfrac{2\pi \text{ rad}}{\text{cycle}}\right)t\right]$

$= (116 \text{ mJ})\cos^2\left[(4.00\pi \text{ s}^{-1})t\right]$.

Since the cosine function is squared, the period of $K(t)$ is half that of a cosine function or

$T = \pi/\omega = \pi/(4.00\pi \text{ s}^{-1}) = 250 \text{ ms}$, which is the same as $U(t)$. Graph $K(t)$.

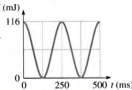

(c) Strategy Add $U(t)$ and $K(t)$ and graph the result.

Solution

$E(t) = U(t) + K(t) = (116 \text{ mJ})\sin^2\left[(4.00\pi \text{ s}^{-1})t\right] + (116 \text{ mJ})\cos^2\left[(4.00\pi \text{ s}^{-1})t\right]$

$= (116 \text{ mJ})\left\{\sin^2\left[(4.00\pi \text{ s}^{-1})t\right] + \cos^2\left[(4.00\pi \text{ s}^{-1})t\right]\right\} = (116 \text{ mJ})(1) = 116 \text{ mJ}$

Graph $E(t) = U(t) + K(t)$.

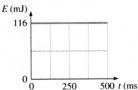

(d) Strategy and Solution Friction does nonconservative work on the object, thus,

$\boxed{U, K, \text{ and } E \text{ would gradually be reduced to zero}}$.

57. **Strategy and Solution** According to Eq. (10-26b), $T = 2\pi\sqrt{\dfrac{L}{g}}$, which does not depend upon the mass.

   Therefore, $T = \boxed{1.5 \text{ s}}$.

61. **Strategy** Use Eq. (10-26b) to find the length of the pendulum. Then, form a ratio of the lengths.

   Solution Solve for L.
   $$T = 2\pi\sqrt{\frac{L}{g}}, \text{ so } L = \frac{gT^2}{4\pi^2}.$$
   Form a proportion.
   $$\frac{L_2}{L_1} = \frac{T_2^2}{T_1^2} = \left(\frac{1.00 \text{ s}}{0.950 \text{ s}}\right)^2 = \boxed{1.11}$$

65. **Strategy** The total mechanical energy of a pendulum is $E = \frac{1}{2}m\omega^2 A^2$. Form a proportion.

   Solution Find the mechanical energy of the pendulum.
   $$\frac{E_2}{E_1} = \frac{A_2^2}{A_1^2}, \text{ so } E_2 = \left(\frac{A_2}{A_1}\right)^2 E_1 = \left(\frac{3.0 \text{ cm}}{2.0 \text{ cm}}\right)^2 (5.0 \text{ mJ}) = \boxed{11 \text{ mJ}}.$$

69. **Strategy** $E = \frac{1}{2}m\omega^2 A^2 \propto A^2$ for a pendulum.

   Solution Find the percent decrease of the oscillator's energy in ten cycles.
   $$\frac{\Delta E}{E} \times 100\% = \frac{\Delta A^2}{A^2} \times 100\% = \frac{(1 - 0.0500)^2 - 1^2}{1^2} \times 100\% = \frac{0.9500^2 - 1^2}{1^2} \times 100\% = \boxed{-9.75\%}$$

73. **Strategy** Use Eq. (10-20c).

   Solution

   (a) The period is directly proportional to the square root of the mass, and the period for the fish is longer than that for the weight, so the fish weighs $\boxed{\text{more}}$ than the weight.

   (b) Form a proportion. Let f = fish and w = weight.
   $$T = 2\pi\sqrt{\frac{m}{k}}, \text{ so } \frac{T_f}{T_w} = \sqrt{\frac{m_f}{m_w}} = \sqrt{\frac{W_f}{W_w}}. \text{ Thus, } W_f = \left(\frac{T_f}{T_w}\right)^2 W_w = \left(\frac{220}{65}\right)^2 (4.90 \text{ N}) = \boxed{56 \text{ N}}.$$

77. Strategy Graph x on the vertical axis and t on the horizontal axis. Analyze the slope of the graph (the magnitude of which is the speed) in terms of the distance between the dots to determine the fastest and slowest speeds of the mass.

Solution Graph $x(t) = -(10 \text{ cm})\cos[(1.57 \text{ s}^{-1})t]$.

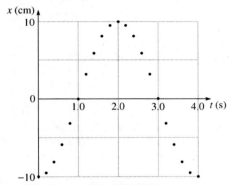

The distance between adjacent dots should be the least at the endpoints and greatest at the center, so its speed is lowest at the endpoints and fastest at its equilibrium position.

81. Strategy Since the body begins with its maximum amplitude at $t = 0$, the body oscillates according to a cosine function $(\cos 0 = 1)$. Use Newton's second law, Hooke's law, and Eq. (10-20a).

Solution Find the amplitude A.

$$\Sigma F_y = kA - mg = 0, \text{ so } A = \frac{mg}{k} = \frac{4.0 \text{ N}}{250 \text{ N/m}} = 1.6 \text{ cm}.$$

Find the angular frequency ω.

$$\omega = \sqrt{\frac{k}{m}} = \sqrt{\frac{kg}{mg}} = \sqrt{\frac{(250 \text{ N/m})(9.80 \text{ m/s}^2)}{4.0 \text{ N}}} = 25 \text{ rad/s}$$

Thus, the equation describing the motion of the body is $\boxed{y = (1.6 \text{ cm})\cos\left[(25 \text{ rad/s})t\right]}$.

85. Strategy Use Eq. (10-2) to find the tensile stress. Then compare the tensile stress to the elastic limit of steel piano wire.

Solution Find the tensile stress in the piano wire in Problem 90.

$$\text{tensile stress} = \frac{F}{A} = \frac{T}{\frac{1}{4}\pi d^2} = \frac{4T}{\pi d^2} = \frac{4(402 \text{ N})}{\pi(0.80\times10^{-3} \text{ m})^2} = 8.0\times10^8 \text{ Pa} < 8.26\times10^8 \text{ Pa}$$

The tensile stress is $\boxed{8.0\times10^8 \text{ Pa}}$; it is just under the elastic limit.

89. **Strategy** The force on the column is its weight. Use Eq. (10-2) and the relationship between density, mass, and volume.

   Solution

   (a) Calculate the compressive stress at the bottom of the column.

   $$\text{compressive stress} = \frac{F}{A} = \frac{mg}{A} = \frac{\rho V g}{A} = \frac{\rho h A g}{A} = \boxed{\rho g h}$$

   (b) Find the absolute limit to the height of a cylindrical column, regardless of how wide it is.

   $$h_m = \frac{\text{compressive strength}}{\rho g} = \frac{2.0 \times 10^8 \text{ Pa}}{(2.7 \times 10^3 \text{ kg/m}^3)(9.80 \text{ m/s}^2)} = \boxed{7.6 \text{ km}}$$

   (c) It is unlikely that someone would want to build a marble column taller than 7.6 km. So, the answer is $\boxed{\text{no}}$; this limit is of little practical concern. No beanstalk could ever reach a height of 7.6 km; its height is limited by other means.

93. (a) **Strategy** Use conservation of energy. Do *not* assume SHM.

   Solution Find the speed of the pendulum bob at the bottom of its swing.

   $$\Delta K = \frac{1}{2} mv^2 = -\Delta U = mgL, \text{ so } v = \boxed{\sqrt{2gL}}.$$

   (b) **Strategy** Assume (incorrectly, for such a large amplitude) that the motion *is* SHM. Use Eqs. (10-21) and (10-26a).

   Solution Find the speed of the pendulum bob at the bottom of its swing.

   The amplitude A is a quarter of the circumference of a circle with radius L, or $\dfrac{2\pi L}{4} = \dfrac{\pi}{2} L.$

   Assuming SHM, $v_m = \omega A = \sqrt{\dfrac{g}{L}} \left(\dfrac{\pi}{2} L \right) = \boxed{\dfrac{\pi}{2} \sqrt{gL}}$. Since $v \propto \omega \propto \dfrac{1}{T}$, a smaller speed implies a larger

   period. Since $\dfrac{v_m}{v} = \dfrac{\frac{\pi}{2}\sqrt{gL}}{\sqrt{2gL}} = \dfrac{\pi}{2\sqrt{2}} > 1$, the period of a pendulum for large amplitudes is $\boxed{\text{larger}}$ than that

   given by Eq. (10-26b).

Chapter 11

WAVES

Conceptual Questions

1. Wrapping a thick coil of copper wire around a piano string increases the string's mass density and therefore decreases the speed of waves traveling along it. The fundamental wavelength is fixed by the length of the string—the decreased wave velocity must therefore be accompanied by a decrease in the frequency at which the string vibrates.

5. Words spoken by two people at the same time are comprehensible because sound waves travel through each other—interfering while superimposed, but returning to their original waveform as they again separate.

9. Since transverse waves do not travel through the core while longitudinal waves do, some part of the core is a molten, viscous liquid that cannot support the transmission of a transverse wave. A longitudinal wave can create compressions and rarefactions in the liquid and travel on through.

Problems

1. **Strategy** Form a proportion with the intensities, treating the Sun as an isotropic source. Use Eq. (11-1).

   Solution Find the intensity of the sunlight that reaches Jupiter.
   $$\frac{I_J}{I_E} = \frac{\frac{P}{4\pi r_J^2}}{\frac{P}{4\pi r_E^2}} = \frac{r_E^2}{r_J^2}, \text{ so } I_J = \left(\frac{r_E}{r_J}\right)^2 I_E = \left(\frac{1}{5.2}\right)^2 (1400 \text{ W/m}^2) = \boxed{52 \text{ W/m}^2}.$$

3. **Strategy** Form a proportion with the intensities, treating the jet airplane as an isotropic source. Use Eq. (11-1).

   Solution Find the intensity of the sound waves at the ears of the person.
   $$\frac{I_2}{I_1} = \frac{P/(4\pi r_2^2)}{P/(4\pi r_1^2)} = \left(\frac{r_1}{r_2}\right)^2, \text{ so } I_2 = \left(\frac{r_1}{r_2}\right)^2 I_1 = \left(\frac{5.0 \text{ m}}{120 \text{ m}}\right)^2 (1.0\times10^2 \text{ W/m}^2) = \boxed{170 \text{ mW/m}^2}.$$

5. **Strategy** The power equals the intensity times the area.

   Solution Find the rate at which the Sun emits electromagnetic waves.
   $$P = IA = I(4\pi R_E^2) = 4\pi(1.4\times10^3 \text{ W/m}^2)(1.50\times10^{11} \text{ m})^2 = \boxed{4.0\times10^{26} \text{ W}}$$

7. **Strategy** Refer to the figure. Use the definition of average speed.

   Solution

   (a) Find the speed.
   $$v_x = \frac{\Delta x}{\Delta t} = \frac{1.80 \text{ m} - 1.50 \text{ m}}{0.20 \text{ s}} = 1.5 \text{ m/s}$$
   Find the position.
   $$x_f = x_i + v\Delta t = 1.80 \text{ m} + (1.5 \text{ m/s})(3.00 \text{ s} - 0.20 \text{ s}) = \boxed{6.0 \text{ m}}$$

(b) $t_f = \dfrac{x_f - x_i}{v_x} + t_i = \dfrac{4.00 \text{ m} - 1.80 \text{ m}}{1.5 \text{ m/s}} + 0.20 \text{ s} = \boxed{1.7 \text{ s}}$

9. **Strategy** Use Eq. (11-4).

   Solution Find the speed of the transverse waves on the string.

   $v = \sqrt{\dfrac{F}{\mu}} = \sqrt{\dfrac{90.0 \text{ N}}{3.20 \times 10^{-3} \text{ kg/m}}} = \boxed{168 \text{ m/s}}$

13. **Strategy** Use Eq. (11-5).

    Solution Find the wavelength.
    $\lambda = vT = (75.0 \text{ m/s})(5.00 \times 10^{-3} \text{ s}) = \boxed{0.375 \text{ m}}$

15. **Strategy** Use Eq. (11-6).

    Solution Find the frequencies.

    (a) $f = \dfrac{v}{\lambda} = \dfrac{340 \text{ m/s}}{1.0 \text{ m}} = \boxed{340 \text{ Hz}}$

    (b) $f = \dfrac{v}{\lambda} = \dfrac{3.0 \times 10^8 \text{ m/s}}{1.0 \text{ m}} = \boxed{3.0 \times 10^8 \text{ Hz}}$

17. **Strategy** Use Eq. (11-6).

    Solution Find the frequency with which the buoy bobs up and down.
    $f = \dfrac{v}{\lambda} = \dfrac{2.5 \text{ m/s}}{7.5 \text{ m}} = \boxed{0.33 \text{ Hz}}$

21. **Strategy** The wave on the string is of the form $y(x, t) = A \sin(\omega t - kx)$. Use the equation, the given information, Eq. (11-7), and the relationship between period and angular frequency to find the amplitude, wavelength, period, and wave speed.

    Solution

    (a) $A = \boxed{4.0 \text{ mm}}$

    (b) $\lambda = \dfrac{2\pi}{k} = \dfrac{2\pi}{6.0 \text{ m}^{-1}} = \boxed{1.0 \text{ m}}$

    (c) $T = \dfrac{2\pi}{\omega} = \dfrac{2\pi}{6.0 \times 10^2 \text{ s}^{-1}} = \boxed{0.010 \text{ s}}$

    (d) $v = \dfrac{\omega}{k} = \dfrac{6.0 \times 10^2 \text{ s}^{-1}}{6.0 \text{ m}^{-1}} = \boxed{100 \text{ m/s}}$

    (e) Since the signs of ωt and kx are opposite, the wave travels $\boxed{\text{in the } +x\text{-direction (to the right)}}$.

23. **Strategy** The equation for a transverse sinusoidal wave moving in the negative x-direction can be written in the form $y(x, t) = A\sin(kx + \omega t)$. Use Eq. (11-7) to find the angular frequency and the wavenumber.

    Solution $A = 0.120$ m, $\lambda = 0.300$ m, $v = 6.40$ m/s, and $y(x, t) = A\sin(\omega t + kx)$.

    $$\omega = \frac{2\pi v}{\lambda} = \frac{2\pi(6.40 \text{ m/s})}{0.300 \text{ m}} = 134 \text{ s}^{-1} \text{ and } k = \frac{2\pi}{\lambda} = \frac{2\pi}{0.300 \text{ m}} = 20.9 \text{ m}^{-1}.$$

    Thus, the equation is $\boxed{y(x, t) = (0.120 \text{ m})\sin[(134 \text{ s}^{-1})t + (20.9 \text{ m}^{-1})x]}$.

25. **Strategy** The maximum y-value is the amplitude. Find the horizontal distance for which the wave repeats to find the wavelength. The wave speed is the distance the wave travels divided by the time interval it took to travel that distance. Use Eq. (11-6) to find the frequency of the wave. The period is the reciprocal of the frequency.

    Solution

    (a) $y_{max} = 2.6$ cm, so $A = \boxed{2.6 \text{ cm}}$.

    (b) $\lambda = \Delta x = 16 \text{ m} - 2 \text{ m} = \boxed{14 \text{ m}}$

    (c) $v = \dfrac{\Delta x}{\Delta t} = \dfrac{7.5 \text{ m} - 5.5 \text{ m}}{0.10 \text{ s}} = \boxed{20 \text{ m/s}}$

    (d) $f = \dfrac{v}{\lambda} = \dfrac{20 \text{ m/s}}{14 \text{ m}} = \boxed{1.4 \text{ Hz}}$

    (e) $T = \dfrac{1}{f} = \dfrac{14 \text{ m}}{20 \text{ m/s}} = \boxed{0.70 \text{ s}}$

29. (a) **Strategy** Substitute $t = 0$, 0.96 s, and 1.92 s into $y(x, t) = (0.80 \text{ mm})\sin[(\pi/5.0 \text{ cm}^{-1})x - (\pi/6.0 \text{ s}^{-1})t]$ and graph the resulting equations.

    Solution The three equations are:
    $y(x, 0) = (0.80 \text{ mm})\sin[(\pi/5.0 \text{ cm}^{-1})x]$, $y(x, 0.96 \text{ s}) = (0.80 \text{ mm})\sin[(\pi/5.0 \text{ cm}^{-1})x - 0.50]$, and

    $y(x, 1.92 \text{ s}) = (0.80 \text{ mm})\sin[(\pi/5.0 \text{ cm}^{-1})x - 1.0]$.
    Find the wavelength.
    $$\lambda = \frac{2\pi}{k} = \frac{2\pi}{\pi/(5.0 \text{ cm})} = 10 \text{ cm}$$

    The amplitude of the wave is $A = 0.80$ mm. The first graph (solid) begins at the origin. The second graph (dashed) is shifted to the right by $(5.0 \text{ cm} \times 0.50)/\pi = 0.80$ cm. The third graph (dotted) is shifted to the right twice as far as the second graph, or 1.6 cm.

    The graphs are shown:

(b) Strategy Substitute $t = 0$, 0.96 s, and 1.96 s into $y(x, t) = (0.50 \text{ mm})\sin[(\pi/5.0 \text{ cm}^{-1})x + (\pi/6.0 \text{ s}^{-1})t]$ and graph the resulting equations.

Solution The three equations are:

$y(x, 0) = (0.50 \text{ mm})\sin[(\pi/5.0 \text{ cm}^{-1})x]$, $y(x, 0.96 \text{ s}) = (0.50 \text{ mm})\sin[(\pi/5.0 \text{ cm}^{-1})x + 0.50]$, and

$y(x, 1.92 \text{ s}) = (0.50 \text{ mm})\sin[(\pi/5.0 \text{ cm}^{-1})x + 1.0]$.

Find the wavelength.

$$\lambda = \frac{2\pi}{k} = \frac{2\pi}{\pi/(5.0 \text{ cm})} = 10 \text{ cm}$$

The amplitude of the wave is $A = 0.50$ mm. The first graph (solid) begins at the origin. The second graph (dashed) is shifted to the left by $(5.0 \text{ cm} \times 0.50)/\pi = 0.80$ cm. The third graph (dotted) is shifted to the left twice as far as the second graph, or 1.6 cm. The graphs are shown:

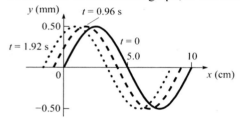

(c) Strategy Refer to the results of parts (a) and (b).

Solution The graphs obtained in part (a) move to the right as time progresses, so $y(x, t) = (0.80 \text{ mm})\sin(kx - \omega t)$ represents a wave traveling in the $+x$-direction. The graphs obtained in part (b) move to the left as time progresses, so $y(x, t) = (0.50 \text{ mm})\sin(kx + \omega t)$ represents a wave traveling in the $-x$-direction.

31. **Strategy** Compute the positions of the peaks of each pulse for the given times. Then use the principle of superposition to graph the shape of the cord for each time.

Solution

t (s)	Short Pulse Position	Tall Pulse Position
0.15	$10 \text{ cm} + (40 \text{ cm/s})(0.15 \text{ s}) = 16 \text{ cm}$	$30 \text{ cm} - (40 \text{ cm/s})(0.15 \text{ s}) = 24 \text{ cm}$
0.25	$10 \text{ cm} + (40 \text{ cm/s})(0.25 \text{ s}) = 20 \text{ cm}$	$30 \text{ cm} - (40 \text{ cm/s})(0.25 \text{ s}) = 20 \text{ cm}$
0.30	$10 \text{ cm} + (40 \text{ cm/s})(0.30 \text{ s}) = 22 \text{ cm}$	$30 \text{ cm} - (40 \text{ cm/s})(0.30 \text{ s}) = 18 \text{ cm}$

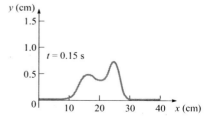

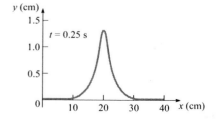

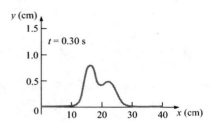

33. Strategy Sketch the sine waves. Use the principle of superposition to find the amplitudes. Let $y_1 = A\sin(\omega t + kx)$

and $y_2 = A\sin(\omega t + kx - \phi)$ and use the trigonometric identity $\sin\alpha + \sin\beta = 2\sin\left(\dfrac{\alpha+\beta}{2}\right)\cos\left(\dfrac{\alpha-\beta}{2}\right)$.

Solution

(a) y (cm)

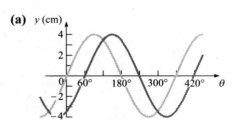

Use the principle of superposition.

$$y = y_1 + y_2 = A\sin(\omega t + kx) + A\sin(\omega t + kx - \phi) = 2A\sin\left(\omega t + kx - \frac{\phi}{2}\right)\cos\frac{\phi}{2} = A'\sin\left(\omega t + kx - \frac{\phi}{2}\right)$$

where $A' = 2A\cos\dfrac{\phi}{2} = 2(4.0\text{ cm})\cos\dfrac{60.0°}{2} = \boxed{6.9\text{ cm}}$.

(b) y (cm)

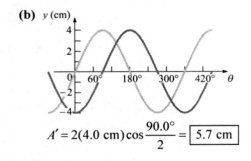

$$A' = 2(4.0\text{ cm})\cos\frac{90.0°}{2} = \boxed{5.7\text{ cm}}$$

35. Strategy Use the principle of superposition and the trigonometric identity

$\sin\alpha + \sin\beta = 2\sin\left(\dfrac{\alpha+\beta}{2}\right)\cos\left(\dfrac{\alpha-\beta}{2}\right)$.

Solution Find the traveling sine wave.

$$y = y_1 + y_2 = A\sin(\omega t + kx) + A\sin(\omega t + kx - \phi) = 2A\sin\left(\omega t + kx - \frac{\phi}{2}\right)\cos\frac{\phi}{2} = A'\sin\left(\omega t + kx - \frac{\phi}{2}\right)$$

where $A' = 2A\cos\dfrac{\phi}{2} = 6.69$ cm. Find ϕ.

$$\cos\frac{\phi}{2} = \frac{A'}{2A},\text{ so }\phi = 2\cos^{-1}\frac{A'}{2A} = 2\cos^{-1}\frac{6.69\text{ cm}}{2(5.00\text{ cm})} = \boxed{96.0°}.$$

37. **Strategy** Refer to the figure. Use $\Delta x = v_x \Delta t$ and the principle of superposition.

Solution The pulse moves $1.80\text{ m} - 1.50\text{ m} = 0.30\text{ m}$ in 0.20 s. So, the speed of the wave is $v = \dfrac{0.30\text{ m}}{0.20\text{ s}} = 1.5$ m/s. When the pulse reaches the right endpoint, it is reflected and inverted. When exactly half of the pulse has been reflected and inverted, the superposition of the incident and reflected waves results in the cancellation of the waves $(y_1 + y_2 = 0)$. Thus, the string looks flat at $t = \dfrac{x}{v} = \dfrac{4.0\text{ m} - 1.5\text{ m}}{1.5\text{ m/s}} = \boxed{1.7\text{ s}}$.

39. **Strategy** The waves are coherent. Use the principle of superposition.

Solution

(a) The resulting wave will have its largest amplitude if the waves interfere constructively. The phase difference is $\boxed{0°}$, and the amplitude is $A_1 + A_2 = 5.0\text{ cm} + 3.0\text{ cm} = \boxed{8.0\text{ cm}}$.

(b) The resulting wave will have its smallest amplitude if the waves interfere destructively. The phase difference is $\boxed{180°}$, and the amplitude is $|A_1 - A_2| = |5.0\text{ cm} - 3.0\text{ cm}| = \boxed{2.0\text{ cm}}$.

(c) $8.0\text{ cm} : 2.0\text{ cm} = \boxed{4:1}$

41. **Strategy** Intensity is proportional to the amplitude squared. For constructive interference, the amplitude of the superposition is the sum of the original amplitudes.

Solution Find A_1/A_2.
$$\frac{A_1}{A_2} = \sqrt{\frac{I_1}{I_2}} = \sqrt{\frac{25}{15}} = \sqrt{\frac{5.0}{3.0}}$$
Find the amplitude of the superposition.
$$A = A_1 + A_2 = A_2\sqrt{\frac{5.0}{3.0}} + A_2 = A_2\left(1 + \sqrt{\frac{5.0}{3.0}}\right)$$
Find the intensity of the superposition.
$$\sqrt{\frac{I}{I_2}} = \frac{A}{A_2} = 1 + \sqrt{\frac{5.0}{3.0}}, \text{ so } I = \left(1 + \sqrt{\frac{5.0}{3.0}}\right)^2 I_2 = \left(1 + \sqrt{\frac{5.0}{3.0}}\right)^2 (15\text{ mW/m}^2) = \boxed{79\text{ mW/m}^2}.$$

45. **Strategy** Use Eqs. (11-2) and (11-13).

Solution
$$f_1 = \frac{v}{2L} \text{ and } v = \sqrt{\frac{TL}{m}}.\ f_1 = \frac{1}{2L}\sqrt{\frac{T_1 L}{m}} = \sqrt{\frac{T_1}{4Lm}}, \text{ so } \left(\frac{f_2}{f_1}\right)^2 = \frac{T_2}{T_1}.$$
Calculate the percentage reduction in the tension.
$$\frac{T_1 - T_2}{T_1} \times 100\% = \frac{T_1 - \left(\frac{f_2}{f_1}\right)^2 T_1}{T_1} \times 100\% = \left[1 - \left(\frac{f_2}{f_1}\right)^2\right] \times 100\% = \left[1 - \left(\frac{1 - 0.040}{1}\right)^2\right] \times 100\% = \boxed{7.8\%}$$

49. **Strategy** The frequencies are given by $f_n = nv/(2L)$. The speed of the transverse waves is related to the tension by $v = \sqrt{T/\mu}$.

Solution

(a) Find the frequency of the fundamental oscillation.

$$f_1 = \frac{v}{2L} = \frac{1}{2L}\sqrt{\frac{T}{\mu}} = \frac{1}{2(1.5\text{ m})}\sqrt{\frac{12\text{ N}}{1.2\times 10^{-3}\text{ kg/m}}} = \boxed{33\text{ Hz}}$$

(b) Find the tension.

$$f_3 = \frac{3v}{2L} = \frac{3}{2L}\sqrt{\frac{T}{\mu}}, \text{ so } T = \frac{4\mu L^2 f_3^2}{9} = \frac{4(1.2\times 10^{-3}\text{ kg/m})(1.5\text{ m})^2(0.50\times 10^3\text{ Hz})^2}{9} = \boxed{300\text{ N}}.$$

53. **(a) Strategy and Solution** All frequencies higher than the fundamental are integral multiples of the fundamental. Since there are no other frequencies between the two given, the fundamental is the difference between those two. Thus, the fundamental frequency is $1040\text{ Hz} - 780\text{ Hz} = \boxed{260\text{ Hz}}$.

(b) **Strategy** Use Eqs. (11-2) and (11-13).

Solution Find the total mass of the string.

$$f_1 = \frac{v}{2L} = \frac{1}{2L}\sqrt{\frac{FL}{m}} = \sqrt{\frac{F}{4mL}}, \text{ so } m = \frac{F}{4f_1^2 L} = \frac{1200\text{ N}}{4(260\text{ Hz})^2(1.6\text{ m})} = \boxed{2.8\text{ g}}.$$

57. **Strategy** The wave speed for the 1.0-Hz waves is twice that for the 2.0-Hz waves, so it takes the 1.0-Hz waves 120 s to reach you. (120 s + 120 s = 240 s is the time it takes the 2.0-Hz waves to reach you; twice as long.)

Solution Compute the distance to the boat.
$$\Delta x = v\Delta t = (1.56\text{ m/s})(120\text{ s}) = \boxed{190\text{ m}}$$

61. **Strategy** Speed is inversely proportional to the time of travel. Form a proportion and use $\Delta x = v\Delta t$.

Solution Relate the speeds to the times of travel.
$$\frac{v_P}{v_S} = \frac{10.0\text{ km/s}}{8.0\text{ km/s}} = \frac{5.00}{4.0} = \frac{t_S}{t_P}$$
Find the time for the S wave to travel from the source to the detector.
$$\Delta t = t_S - t_P = t_S - t_S\left(\frac{4.0}{5.00}\right) = t_S(1 - 0.80), \text{ so } t_S = \frac{2.0\text{ s}}{0.20} = 10\text{ s}.$$
Calculate the distance between the source and the detector.
$$d = v_S t_S = (8.0\text{ km/s})(10\text{ s}) = \boxed{80\text{ km}}$$

65. **Strategy** Even though the wire is cut in two, the linear mass density does not change (half the length and half the mass). According to Eq. (11-4), the speed of waves on a wire is directly proportional to the square root of the tension. According to Eq. (11-13), the frequency of the waves on a wire is directly proportional to the speed of the waves. Therefore, the frequency is directly proportional to the square root of the tension in a wire. Use Newton's second law.

Solution For the single wire:
$$\Sigma F_y = T_1 - mg = 0, \text{ so } T_1 = mg.$$
For the two wires:
$$\Sigma F_y = 2T_2 - mg = 0, \text{ so } T_2 = mg/2.$$
Therefore, $T_1 = 2T_2$. Form a proportion.

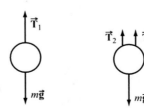

$$\frac{f_2}{f_1} = \sqrt{\frac{T_2}{T_1}} = \sqrt{\frac{T_2}{2T_2}} = \frac{1}{\sqrt{2}}, \text{ so } f_2 = \frac{f_1}{\sqrt{2}}.$$

Thus, the new fundamental frequency of each wire is $660 \text{ Hz}/\sqrt{2} = \boxed{470 \text{ Hz}}$.

69. Strategy Use dimensional analysis.

Solution γ has units $N/m = kg/s^2$. ρ has units kg/m^3. λ has units m. $\gamma/(\lambda \cdot \rho)$ has units

$\dfrac{kg/s^2}{m \cdot kg/m^3} = \dfrac{m^2}{s^2}$. $[\gamma/(\lambda \cdot \rho)]^{1/2}$ has units m/s. So, $\boxed{v \propto \sqrt{\dfrac{\gamma}{\lambda \rho}}}$. Since v depends upon λ, surface waves are

$\boxed{\text{dispersive}}$.

73. Strategy Destructive interference occurs when the path length difference of the two sound waves is an odd multiple of half of the wavelength.

Solution The wavelength is $\lambda = v/f = (340 \text{ m/s})/(680 \text{ Hz}) = 0.50 \text{ m}$. The largest possible path length difference is equal to the distance between the speakers, 1.5 m. $\lambda/2 = 0.25 \text{ m}$, so the path length differences that cause destructive interference are 0.25 m, 0.75 m, and 1.25 m. Let the speakers lie along the x-axis at $x = \pm 0.75 \text{ m}$. Then the path length difference is zero along the y-axis and 1.5 m along the x-axis. As the listener walks along the circle of radius 1 m, the path length difference varies from 0 to 1.5 m. The path length difference equals 0.25 m, 0.75 m, and 1.25 m once for each quadrant of the circle (three occurrences of destructive interference). There are four quadrants, so the listener observes destructive interference at $\boxed{12}$ points along the circle.

77. Strategy Use the principle of superposition.

Solution $\Delta x = 1.80 \text{ m} - 1.50 \text{ m} = 0.30 \text{ m}$ in $\Delta t = 0.20 \text{ s}$, so $v = 0.30 \text{ m}/(0.20 \text{ s}) = 1.5 \text{ m/s}$.
Find the position of the peak at $t = 1.6 \text{ s}$.
$x_{peak} = x_i + vt = 1.5 \text{ m} + (1.5 \text{ m/s})(1.6 \text{ s}) = 3.9 \text{ m}$

The peak of the pulse is nearly to the end of the string. The reflected pulse is below the string, so most of the height of the original pulse is cancelled.

Chapter 12

SOUND

Conceptual Questions

1. The wavelength of the standing waves inside a bassoon is determined by the length of its air chamber. With a fixed fundamental wavelength, the frequency of these waves depends only upon the speed of sound in air, which itself depends significantly upon the air temperature. As a result of thermal contraction and expansion, changes in air temperature also affect the frequency of waves generated on a cello string—albeit much less significantly than for the bassoon.

5. For high-frequency sounds the wavelength is relatively small. The phase difference of the sounds arriving at each ear is then very sensitive to small variations in head position, wind speed, and several other factors. This makes the phase difference method unreliable for high-frequency sounds.

9. The maximum loudness of a stereo is proportional to the maximum intensity level that it can produce. Doubling the power of the stereo's amplifier also doubles its intensity, but it increases the intensity level by only 3 dB.

13. Doubling the pressure amplitude of a sound wave doubles the displacement amplitude, quadruples the intensity, and increases the intensity level by 6 dB.

17. Although the fundamental frequency of the highest note on a piano is approximately 4 kHz, the harmonics of that note occur at higher frequencies. An instrument's timbre is determined by the presence or absence of these harmonics. High quality audio equipment must therefore be able to reproduce frequencies up to the limit of the human audible range in order to faithfully reproduce the sound of an instrument.

Problems

1. **Strategy** Use Eqs. (11-6) and (12-3).

   Solution Find the wavelength of the ultrasonic waves.
   $$\lambda = \frac{v}{f} \text{ and } v = v_0\sqrt{\frac{T}{T_0}}, \text{ so } \lambda = \frac{v}{f} = \frac{v_0}{f}\sqrt{\frac{T}{T_0}} = \frac{331 \text{ m/s}}{1.0\times10^5 \text{ Hz}}\sqrt{\frac{273.15 \text{ K}+15 \text{ K}}{273.15}} = \boxed{3.4 \text{ mm}}.$$

5. **Strategy** Use $\Delta x = v_x\Delta t$ and the speeds of light and sound. For $T = 20.0°C$, $v = 343$ m/s.

   Solution Verify the rule of thumb.
   $$\Delta t_{light} = \frac{\Delta x}{c} = \frac{1.6\times10^3 \text{ m}}{3\times10^8 \text{ m/s}} = 5 \text{ μs is negligible; } \Delta t_{sound} = \frac{\Delta x}{v} = \frac{1.6\times10^3 \text{ m}}{343 \text{ m/s}} = \boxed{4.7 \text{ s}}.$$
   4.7 s = 5 s to one significant figure. The rule of thumb is approximately correct.

7. **Strategy** Use Eq. (12-1).

   Solution Find the speed of sound in mercury.
   $$v = \sqrt{\frac{B}{\rho}} = \sqrt{\frac{2.8\times10^{10} \text{ Pa}}{1.36\times10^4 \text{ kg/m}^3}} = \boxed{1.4 \text{ km/s}}$$

9. **Strategy** Replace each quantity with its SI units and simplify. In (a), use Eq. (12-1). In (b), analyze each combination of ρ and B.

    Solution

    (a) Show that Eq. (12-1) gives the speed of sound in m/s.

    $$v = \sqrt{\frac{B}{\rho}}, \text{ so } \sqrt{\frac{N/m^2}{kg/m^3}} = \sqrt{\frac{(kg \cdot m/s^2)/m^2}{kg/m^3}} = \sqrt{\frac{1/(m \cdot s^2)}{1/m^3}} = \sqrt{\frac{m^2}{s^2}} = m/s.$$

    (b) Show that no other combination of B and ρ of than $\sqrt{B/\rho}$ can give dimensions of speed.

    $$\frac{\rho}{B} \text{ has units } \frac{kg}{m^3} \cdot \frac{m^2}{N} = \frac{kg}{m \cdot kg \cdot m/s^2} = \frac{s^2}{m^2}; \quad \rho B \text{ has units } \frac{kg}{m^3} \cdot \frac{N}{m^2} = \frac{kg^2 \cdot m/s^2}{m^5} = \frac{kg^2}{m^4 \cdot s^2}; \text{ and}$$

    $$\frac{1}{\rho B} \text{ has units } \frac{m^4 \cdot s^2}{kg^2}.$$

    No power of the above three combinations (other than $-1/2$, which gives $\sqrt{B/\rho}$) will give the dimensions of speed; therefore, Eq. (12-1) must be correct except for the possibility of a dimensionless constant.

13. **Strategy** Solve for the intensity in Eq. (12-8). The sound is incoherent, so add the three intensities; then solve for the combined intensity level of the three machines.

    Solution Solve for the intensity.

    $$\beta = (10 \text{ dB}) \log_{10} \frac{I}{I_0}, \text{ so } 10^{\frac{\beta}{10 \text{ dB}}} = \frac{I}{I_0} \text{ and } I = I_0 10^{\frac{\beta}{10 \text{ dB}}}.$$

    Find the combined intensity level.

    $$I_{\text{total}} = I_1 + I_2 + I_3 = I_0 10^{\frac{\beta_1}{10 \text{ dB}}} + I_0 10^{\frac{\beta_2}{10 \text{ dB}}} + I_0 10^{\frac{\beta_3}{10 \text{ dB}}}, \text{ so } \frac{I_{\text{total}}}{I_0} = 10^{\frac{\beta_1}{10 \text{ dB}}} + 10^{\frac{\beta_2}{10 \text{ dB}}} + 10^{\frac{\beta_3}{10 \text{ dB}}}. \text{ Thus,}$$

    $$\beta_{\text{total}} = (10 \text{ dB}) \log_{10} \frac{I_{\text{total}}}{I_0} = (10 \text{ dB}) \log_{10} \left(10^{\frac{\beta_1}{10 \text{ dB}}} + 10^{\frac{\beta_2}{10 \text{ dB}}} + 10^{\frac{\beta_3}{10 \text{ dB}}} \right) = (10 \text{ dB}) \log_{10} \left(10^{\frac{85 \text{ dB}}{10 \text{ dB}}} + 10^{\frac{90 \text{ dB}}{10 \text{ dB}}} + 10^{\frac{93 \text{ dB}}{10 \text{ dB}}} \right)$$

    $$= \boxed{95 \text{ dB}}.$$

    Since 95 dB is comparable to 93 dB, the intensity level of all three machines running is $\boxed{\text{not much different than with only one machine running}}$.

17. **Strategy** Use Eq. (12-8) to find an expression for the intensity. Then use each relationship given for the intensities to obtain the relationships for the intensity levels.

Solution Solve for the intensity.

$$\beta = (10 \text{ dB}) \log \frac{I}{I_0}$$

$$10^{\frac{\beta}{10 \text{ dB}}} = \frac{I}{I_0}$$

$$I = I_0 10^{\frac{\beta}{10 \text{ dB}}}$$

(a) Show that if $I_2 = 10.0 I_1$, $\beta_2 = \beta_1 + 10.0$ dB.

$$I_2 = 10.0 I_1$$

$$I_0 10^{\frac{\beta_2}{10 \text{ dB}}} = 10.0 I_0 10^{\frac{\beta_1}{10 \text{ dB}}}$$

$$10^{\frac{\beta_2}{10 \text{ dB}}} = (10.0) 10^{\frac{\beta_1}{10 \text{ dB}}}$$

$$\log 10^{\frac{\beta_2}{10 \text{ dB}}} = \log \left[(10.0) 10^{\frac{\beta_1}{10 \text{ dB}}} \right] = \log 10.0 + \log 10^{\frac{\beta_1}{10 \text{ dB}}}$$

$$\frac{\beta_2}{10 \text{ dB}} = 1.00 + \frac{\beta_1}{10 \text{ dB}}$$

$$\beta_2 = \beta_1 + 10.0 \text{ dB}$$

(b) Show that if $I_2 = 2.0 I_1$, $\beta_2 = \beta_1 + 3.0$ dB.

$$I_2 = 2.0 I_1$$

$$I_0 10^{\frac{\beta_2}{10 \text{ dB}}} = 2.0 I_0 10^{\frac{\beta_1}{10 \text{ dB}}}$$

$$10^{\frac{\beta_2}{10 \text{ dB}}} = (2.0) 10^{\frac{\beta_1}{10 \text{ dB}}}$$

$$\log 10^{\frac{\beta_2}{10 \text{ dB}}} = \log \left[(2.0) 10^{\frac{\beta_1}{10 \text{ dB}}} \right] = \log 2.0 + \log 10^{\frac{\beta_1}{10 \text{ dB}}}$$

$$\frac{\beta_2}{10 \text{ dB}} = 0.30 + \frac{\beta_1}{10 \text{ dB}}$$

$$\beta_2 = \beta_1 + 3.0 \text{ dB}$$

21. (a) **Strategy** $f_n = nv/(2L)$ for a pipe open at both ends and $v = 343$ m/s for $T = 20.0°C$.

Solution Find the length of the organ pipe.

$$f_1 = \frac{v}{2L}, \text{ so } L = \frac{v}{2f_1} = \frac{343 \text{ m/s}}{2(261.5 \text{ Hz})} = \boxed{65.6 \text{ cm}}$$

(b) **Strategy** The frequency of the organ pipe is proportional to the speed of the waves and the speed is proportional to the square root of temperature, so $f \propto v \propto \sqrt{T}$.

Solution Find the fundamental frequency after the temperature drop.

$$f_{0.0°} = f_{20°} \sqrt{\frac{T_{0.0°}}{T_{20°}}} = (261.5 \text{ Hz}) \sqrt{\frac{273.15 \text{ K} + 0.0 \text{ K}}{273.15 \text{ K} + 20.0 \text{ K}}} = \boxed{252.4 \text{ Hz}}$$

23. **Strategy** $f_n = nv/(2L)$ for a pipe open at both ends.

    Solution Find the length of the organ pipe.

    $$f_1 = \frac{v}{2L}, \text{ so } L = \frac{v}{2f_1} = \frac{331 \text{ m/s}}{2(382 \text{ Hz})} = \boxed{43.3 \text{ cm}}.$$

25. **Strategy** $f_n = nv/(2L)$ for a pipe open at both ends and $v = v_0\sqrt{T/T_0}$.

    Solution Find an expression for the temperature in terms of n.

    $$f_n = \frac{nv}{2L} = \frac{nv_0}{2L}\sqrt{\frac{T}{T_0}}, \text{ so } T = T_0\left(\frac{2Lf_n}{nv_0}\right)^2 = \frac{273.15 \text{ K}}{n^2}\left[\frac{2(2.0 \text{ m})(702 \text{ Hz})}{331 \text{ m/s}}\right]^2 = \frac{19,658 \text{ K}}{n^2}.$$

    The assumed temperature range is 293 K (20°C) to 308 K (35°C). We need to find n such that T falls within this range. By trial and error, n is found to be 8. So, $T = 19,658 \text{ K}/8^2 = 307 \text{ K} = \boxed{34°C}$.

29. **(a) Strategy** The rod is analogous to a pipe open at both ends.

    Solution There is a displacement node (pressure antinode) at the center of the rod and displacement antinodes (pressure nodes) at the ends.

    (b) Strategy For a pipe open at both ends, $\lambda_1 = 2L$. Use Eq. (11-6).

    Solution Calculate the speed of sound in aluminum.
    $$v = \lambda_1 f_1 = 2Lf_1 = 2(1.0 \text{ m})(2.55\times10^3 \text{ Hz}) = \boxed{5100 \text{ m/s}}.$$

    (c) Strategy The frequency of the sound wave in the rod is the same as the frequency of the sound wave in the air. Use Eq. (11-6).

    Solution Find the wavelength of the sound wave in the air.
    $$\lambda = \frac{v}{f} = \frac{334 \text{ m/s}}{2.55\times10^3 \text{ Hz}} = \boxed{13.1 \text{ cm}}$$

    (d) Strategy and Solution The longitudinal motion of the rod is symmetrical about a central axis, so the ends move in opposite directions and, thus, they are out of phase.

33. **(a) Strategy** Use Eqs. (11-2) and (11-13).

    Solution Solve for the tension in the string.
    $$v = \sqrt{FL/m} \text{ and } f_1 = v/(2L), \text{ so } F = 4mLf_1^2 = 4(0.300\times10^{-3} \text{ kg})(0.655 \text{ m})(330.0 \text{ Hz})^2 = \boxed{85.6 \text{ N}}.$$

    (b) Strategy and Solution The waves travel on the string with a speed of
    $$v = 2Lf_1 = 2(0.655 \text{ m})(330.0 \text{ Hz}) = \boxed{432 \text{ m/s}}.$$

    (c) Strategy and Solution Since the other musician is lowering the frequency of the whistle, the frequency being played is 330.0 Hz + 5 Hz = $\boxed{335 \text{ Hz}}$ when beats are first heard.

    (d) Strategy Use Eq. (12-10b) for a pipe closed at one end.

    Solution Find the length of the slide whistle.
    $$f = \frac{v}{4L}, \text{ so } L = \frac{v}{4f} = \frac{343 \text{ m/s}}{4(335 \text{ Hz})} = \boxed{0.256 \text{ m}}.$$

37. **Strategy** Since the observer is moving and the source is stationary, use Eq. (12-13).

    Solution As Mandy walks toward one siren (1), $v_o < 0$. As she recedes from the other siren (2), $v_o > 0$. Find the beat frequency heard by Mandy.

    $$f_1 - f_2 = \left(1 + \frac{|v_o|}{v}\right)f_s - \left(1 - \frac{|v_o|}{v}\right)f_s = \frac{2|v_o|f_s}{v} = \frac{2(1.56 \text{ m/s})(698 \text{ Hz})}{343 \text{ m/s}} = \boxed{6.35 \text{ Hz}}$$

39. **Strategy** Since the observer is moving and the source is stationary, use Eq. (12-13).

    Solution Compute the frequencies of the sound observed by the moving observer.

    (a) The observer is moving toward a stationary source $(v_o < 0)$.

    $$f_o = \left(1 - \frac{v_o}{v}\right)f_s = [1 - (-0.50)](1.0 \text{ kHz}) = \boxed{1.5 \text{ kHz}}$$

    (b) The observer is now moving away from the source $(v_o > 0)$.

    $$f_o = (1 - 0.50)(1.0 \text{ kHz}) = \boxed{500 \text{ Hz}}$$

41. **Strategy** Since both source and observer are moving, use Eq. (12-14).

    Solution Compute the frequencies of the sound observed by the moving observer.

    (a) A source and an observer are traveling toward each other $(v_s > 0, v_o < 0)$.

    $$f_o = \frac{1 - \frac{v_o}{v}}{1 - \frac{v_s}{v}}f_s = \frac{1 - (-0.50)}{1 - 0.50}(1.0 \text{ kHz}) = \boxed{3.0 \text{ kHz}}$$

    (b) A source and an observer are traveling away from each other $(v_s < 0, v_o > 0)$.

    $$f_o = \frac{1 - 0.50}{1 + 0.50}(1.0 \text{ kHz}) = \boxed{330 \text{ Hz}}$$

    (c) A source and an observer are traveling in the same direction $(v_s < 0, v_o < 0)$.

    $$f_o = \frac{1 + 0.50}{1 + 0.50}(1.0 \text{ kHz}) = \boxed{1.0 \text{ kHz}}$$

45. (a) **Strategy** The distance traveled (round trip) by the sound of the firing pistol in time Δt is $v\Delta t$. The distance between the ship and one side of the fjord is half this distance. Use Eq. (12-3).

    Solution Find the distance between the ship and one side of the fjord.

    $$d = \frac{1}{2}v\Delta t = \frac{1}{2}\Delta t v_0 \sqrt{\frac{T}{T_0}} = \frac{1}{2}(4.0 \text{ s})(331 \text{ m/s})\sqrt{\frac{273.15 \text{ K} + 5.0 \text{ K}}{273.15 \text{ K}}} = \boxed{670 \text{ m}}$$

    (b) **Strategy** Let the distance to the closer side of the fjord be $d_1 = \frac{1}{2}vt_1$; then the distance to the other side of the fjord is $d_2 = \frac{1}{2}vt_2$.

    Solution Find the time interval between the two echoes.

    $$\Delta t = t_2 - t_1 = \frac{2d_2}{v} - t_1 = \frac{2d_2}{2d_1/t_1} - t_1 = \frac{d_2}{d_1}t_1 - t_1 = \left(\frac{1.80 \text{ km} - 0.668 \text{ km}}{0.668 \text{ km}} - 1\right)(4.0 \text{ s}) = \boxed{2.8 \text{ s}}$$

47. Strategy The distance traveled (round trip) by the sound wave in time Δt is $v\Delta t$. The depth d of the lake is half this distance.

Solution Find the depth of the lake.

$$d = \frac{1}{2}v\Delta t = \frac{1}{2}(0.540 \text{ s})(1493 \text{ m/s}) = \boxed{403 \text{ m}}$$

49. Strategy First treat the moth as a receiver moving away from the source (the bat); then treat the moth as the source moving away from an observer (the bat). Use Eqs. (12-3) and (12-14).

Solution Moth as receiver (v_m and $v_b > 0$; in the direction of propagation):

$$f_1 = \frac{1 - \frac{v_m}{v}}{1 - \frac{v_b}{v}} f_s$$

Moth as source (v_m and $v_b < 0$; opposite the direction of propagation):

$$f_2 = \frac{1 + \frac{v_b}{v}}{1 + \frac{v_m}{v}} f_1 = \frac{\left(1 + \frac{v_b}{v}\right)\left(1 - \frac{v_m}{v}\right)}{\left(1 + \frac{v_m}{v}\right)\left(1 - \frac{v_b}{v}\right)} f_s = \frac{(v + v_b)(v - v_m)}{(v + v_m)(v - v_b)} f_s$$

Find v at $10.0°$C.

$$v = v_0\sqrt{\frac{T}{T_0}} = (331 \text{ m/s})\sqrt{\frac{273.15 \text{ K} + 10.0 \text{ K}}{273.15 \text{ K}}} = 337 \text{ m/s}$$

Calculate the frequency, f_2.

$$f_2 = \frac{(337 \text{ m/s} + 4.40 \text{ m/s})(337 \text{ m/s} - 1.20 \text{ m/s})}{(337 \text{ m/s} + 1.20 \text{ m/s})(337 \text{ m/s} - 4.40 \text{ m/s})}(82.0 \text{ kHz}) = \boxed{83.6 \text{ kHz}}$$

53. (a) Strategy Use Eqs. (11-2) and (11-13).

Solution Find f_2.

$$v = \sqrt{\frac{FL}{m}} \text{ and } f_2 = \frac{v}{L}, \text{ so } f_2 = \sqrt{\frac{F}{mL}} = \sqrt{\frac{7.00 \text{ N}}{(0.230 \times 10^{-3} \text{ kg})(0.300 \text{ m})}} = \boxed{319 \text{ Hz}}.$$

(b) Strategy The frequency in air is the same as the frequency for the string. Use Eq. (11-6).

Solution The frequency of the sound in the surrounding air is $\boxed{319 \text{ Hz}}$. The wavelength of the sound in the surrounding air is $\lambda = \frac{v}{f_2} = \frac{350 \text{ m/s}}{319 \text{ Hz}} = \boxed{1.1 \text{ m}}$.

57. (a) Strategy and Solution The intensity is inversely proportional to distance, so $I_1 \propto I_0/d^2$, where I_1 and I_0 are the intensity at the object causing the reflection and the original intensity, respectively, and d is the distance between the bat and the object. Similarly, for the intensity of the echo, $I_2 \propto I_1/d^2$ where I_2 is the intensity of the echo at the bat. Therefore, $I_2 \propto \dfrac{I_1}{d^2} \propto \dfrac{I_0/d^2}{d^2} \propto \dfrac{1}{d^4}$.

(b) Strategy Use the result of part (a).

Solution Find the percent increase in the intensity of the echo in terms of the distances.

$$\frac{\Delta I}{I_0} \times 100\% = \frac{I - I_0}{I_0} \times 100\% = \frac{\frac{1}{d^4} - \frac{1}{d_0^4}}{\frac{1}{d_0^4}} \times 100\% = \left[\left(\frac{d_0}{d}\right)^4 - 1\right] \times 100\%$$

Compute the percent increase for each object.

First object: $\left[\left(\dfrac{0.60}{0.50}\right)^4 - 1\right] \times 100\% = \boxed{110\%}$; second object: $\left[\left(\dfrac{1.10}{1.00}\right)^4 - 1\right] \times 100\% = \boxed{46\%}$

61. Strategy The distance traveled (round trip) by the sound pulse in time Δt is $v\Delta t$. The distance to the ocean floor is half this distance.

Solution Find the elapsed time between an emitted pulse and the return of its echo at the correct depth d.

$$2d = v\Delta t, \text{ so } \Delta t = \frac{2d}{v} = \frac{2(40.0 \text{ fathoms})\left(1.83 \; \frac{m}{\text{fathom}}\right)}{1533 \text{ m/s}} = \boxed{0.0955 \text{ s}}.$$

65. Strategy The greatest common factor of the frequencies is the fundamental. Compute ratios of the frequencies and use these ratios to determine the greatest common factor of the frequencies.

Solution Compute the ratios.
$$\frac{588}{392} = \frac{3}{2} \text{ and } \frac{980}{392} = \frac{5}{2}, \text{ so } \frac{392}{2} = 196 \times 1, \ 196 \times 3 = 588, \text{ and } 196 \times 5 = 980.$$
The fundamental frequency is $\boxed{196 \text{ Hz}}$.

69. Strategy Since the combined sound intensity is incoherent, the sound intensity of one of the violins is 1/8 of the combined total $(I_2 = I_1/8)$. Use Eq. (12-9).

Solution Find the sound intensity level for one violin.
$$\beta_2 - \beta_1 = (10 \text{ dB})\log\frac{I_2}{I_1} = (10 \text{ dB})\log\frac{1}{8} = -(10 \text{ dB})\log 8, \text{ so}$$
$$\beta_2 = \beta_1 - (10 \text{ dB})\log 8 = 38.0 \text{ dB} - (10 \text{ dB})\log 8 = \boxed{29.0 \text{ dB}}.$$

REVIEW AND SYNTHESIS: CHAPTERS 9–12

Review Exercises

1. **Strategy** The magnitude of the buoyant force on an object in water is equal to the weight of the water displaced by the object.

   Solution

   (a) Lead is much denser than aluminum, so for the same mass, its volume is much less. Therefore, aluminum has the larger buoyant force acting on it; since it is less dense it occupies more volume.

   (b) Steel is denser than wood. Even though the wood is floating, it displaces more water than does the steel. Therefore, wood has the larger buoyant force acting on it; since it displaces more water than the steel.

   (c) Lead: $\rho_w g V_{Pb} = \rho_w g \dfrac{m_{Pb}}{\rho_{Pb}} = (1.00 \times 10^3 \text{ kg/m}^3)(9.80 \text{ m/s}^2) \dfrac{1.0 \text{ kg}}{11{,}300 \text{ kg/m}^3} = \boxed{0.87 \text{ N}}$

   Aluminum: $\rho_w g V_{Al} = \rho_w g \dfrac{m_{Al}}{\rho_{Al}} = (1.00 \times 10^3 \text{ kg/m}^3)(9.80 \text{ m/s}^2) \dfrac{1.0 \text{ kg}}{2702 \text{ kg/m}^3} = \boxed{3.6 \text{ N}}$

   Steel: $\rho_w g V_{Steel} = \rho_w g \dfrac{m_{Steel}}{\rho_{Steel}} = (1.00 \times 10^3 \text{ kg/m}^3)(9.80 \text{ m/s}^2) \dfrac{1.0 \text{ kg}}{7860 \text{ kg/m}^3} = \boxed{1.2 \text{ N}}$

   Wood: $mg = (1.0 \text{ kg})(9.80 \text{ m/s}^2) = \boxed{9.8 \text{ N}}$ (Since the wood is floating, the buoyant force is equal to its weight.)

5. **Strategy** Use Eqs. (10-20a), (10-21), and (10-22), and Newton's second law.

   Solution The normal force on the 1.0-kg block m_1 is $N = m_1 g$. So, the force of friction on m_1 is $f = \mu N = \mu m_1 g$.

   Find the maximum acceleration that the top block can experience before it starts to slip.

   $\Sigma F = f = m_1 a$, so $a = \dfrac{f}{m_1} = \dfrac{\mu m_1 g}{m_1} = \mu g$.

   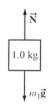

   For SHM, the maximum acceleration is $a_m = \omega^2 A = \dfrac{k}{m} A$, which in this case is equal to

   $\dfrac{kA}{m_1 + m_2}$. Equate the accelerations and solve for A.

   $\dfrac{kA}{m_1 + m_2} = \mu g$, so $A = \dfrac{\mu(m_1 + m_2)g}{k}$.

   The maximum speed is $v_m = \omega A$. Compute the maximum speed that this set of blocks can have without the top block slipping.

   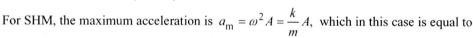

   $v_m = \omega A = \sqrt{\dfrac{k}{m_1 + m_2}} \left[\dfrac{\mu(m_1 + m_2)g}{k} \right] = \mu g \sqrt{\dfrac{m_1 + m_2}{k}} = 0.45(9.80 \text{ m/s}^2)\sqrt{\dfrac{1.0 \text{ kg} + 5.0 \text{ kg}}{150 \text{ N/m}}} = \boxed{0.88 \text{ m/s}}$

9. **(a) Strategy** Use Eqs. (11-2) and (11-13).

   Solution Find the tension in the guitar string.

   $$f_1 = \frac{v}{2L} = \frac{1}{2L}\sqrt{\frac{FL}{m}} = \sqrt{\frac{F}{4Lm}}, \text{ so } F = 4Lmf_1^2 = 4(0.655 \text{ m})(0.00331 \text{ kg})(82 \text{ Hz})^2 = \boxed{58 \text{ N}}.$$

   (b) Strategy Use Eqs. (11-4) and (11-13).

   Solution Find the length of the lowest frequency string when it is fingered at the fifth fret.

   $$f_1 = \frac{v}{2L} = \frac{1}{2L}\sqrt{\frac{F}{\mu}}, \text{ so } L = \frac{1}{2f_1}\sqrt{\frac{F}{\mu}} = \frac{1}{2(110 \text{ Hz})}\sqrt{\frac{58 \text{ N}}{0.00331 \text{ kg}/(0.655 \text{ m})}} = \boxed{49 \text{ cm}}.$$

13. **Strategy** The frequency of the sound is increased by a factor equal to the number of holes in the disk. Use Eq. (11-6).

   Solution The frequency of the sound is

   $$f = 25(60.0 \text{ Hz}) = \boxed{1500 \text{ Hz}}.$$

   Compute the wavelength that corresponds to this frequency.

   $$\lambda = \frac{v}{f} = \frac{343 \text{ m/s}}{1.50 \times 10^3 \text{ Hz}} = \boxed{22.9 \text{ cm}}$$

17. **Strategy** The source is moving in the direction of propagation of the sound, so $v_s > 0$. The observer is moving in the direction opposite the propagation of the sound, so $v_o < 0$. Use Eq. (12-14).

   Solution Find the frequency heard by the passenger in the oncoming boat.

   $$f_o = \frac{v - v_o}{v - v_s} f_s = \frac{343 \text{ m/s} - (-15.6 \text{ m/s})}{343 \text{ m/s} - 20.1 \text{ m/s}}(312 \text{ Hz}) = \boxed{346 \text{ Hz}}$$

21. **Strategy** Use the equations describing waves on a string fixed at both ends.

   Solution

   (a) The wavelength of the fundamental mode is $\lambda = 2L = 2(0.640 \text{ m}) = \boxed{1.28 \text{ m}}$.

   (b) The wave speed on the string is $v = \lambda f = (1.28 \text{ m})(110.0 \text{ Hz}) = \boxed{141 \text{ m/s}}$.

   (c) Find the linear mass density of the string.

   $$v = \sqrt{\frac{F}{\mu}}, \text{ so } \mu = \frac{F}{v^2} = \frac{133 \text{ N}}{(140.8 \text{ m/s})^2} = \boxed{6.71 \text{ g/m}}.$$

   (d) Find the maximum speed of a point on the string.

   $$v = \omega A = 2\pi f A = 2\pi(110.0 \text{ Hz})(0.00230 \text{ m}) = \boxed{0.253 \text{ m/s}}$$

   (e) The frequency in air is that same as that in the bridge and body of the guitar, which is the same as the frequency of the string: $\boxed{110.0 \text{ Hz}}$.

   (f) The speed of sound in air is $v = 331 \text{ m/s} + (0.60 \text{ m/s})(20) = 343 \text{ m/s}$.

   The wavelength of the sound wave in air is $\lambda = \frac{v}{f} = \frac{343 \text{ m/s}}{110.0 \text{ Hz}} = \boxed{3.12 \text{ m}}$.

25. Strategy Use Newton's second law and Hooke's law.

Solution

(a) Find the tension in the string when hanging straight down.

$\Sigma F = T_1 - mg = 0$, so $T_1 = mg$.

Find the tension in the string while it is swinging.

$\Sigma F_y = T_2 \cos\theta - mg = 0$, so $T_2 = \dfrac{mg}{\cos\theta}$.

Find the stretch of the string.

$Y = \dfrac{\Delta T/A}{\Delta L/L}$, so

$$\Delta L = \frac{L\Delta T}{AY} = \frac{Lmg}{AY}\left(\frac{1}{\cos\theta} - 1\right) = \frac{(2.200\text{ m})(0.411\text{ kg})(9.8\text{ m/s}^2)}{\pi(0.00125\text{ m})^2(4.00\times10^9\text{ Pa})}\left(\frac{1}{\cos 65.0^\circ} - 1\right) = \boxed{6.17\times10^{-4}\text{ m}}.$$

(b) Find the kinetic energy.

$$\Sigma F_x = T_2 \sin\theta = ma_x = \frac{mv^2}{r} = \frac{2K}{r}, \text{ so}$$

$$K = \frac{rT_2\sin\theta}{2} = \frac{L\sin\theta\frac{mg}{\cos\theta}\sin\theta}{2} = \frac{mgL\tan\theta\sin\theta}{2} = \frac{(0.411\text{ kg})(9.80\text{ m/s}^2)(2.200\text{ m})\tan 65.0^\circ\sin 65.0^\circ}{2}$$

$$= \boxed{8.61\text{ J}}$$

(c) Find the time it takes a transverse wave pulse to travel the length of the string. Neglect the small change due to the stretch when finding the linear mass density.

$$v = \sqrt{\frac{T_2}{\mu}} = \sqrt{\frac{T_2}{\rho A}} \text{ and } L = vt, \text{ so}$$

$$t = \frac{L+\Delta L}{v} = (L+\Delta L)\sqrt{\frac{\rho A}{T_2}} = (L+\Delta L)\sqrt{\frac{\rho A\cos\theta}{mg}} = (2.2006\text{ m})\sqrt{\frac{(1150\text{ kg/m}^3)\pi(0.00125\text{ m})^2\cos 65.0^\circ}{(0.411\text{ kg})(9.80\text{ m/s}^2)}}$$

$$= \boxed{0.0536\text{ s}}$$

MCAT Review

1. **Strategy** The buoyant force on the brick is equal in magnitude to the weight of the volume of water it displaces.

   Solution The brick is completely submerged, so its volume is equal to that of the displaced water. The weight of the displaced water is $30\text{ N} - 20\text{ N} = 10\text{ N}$. Find the volume of the brick.

   $$\rho_w g V_w = \rho_w g V_{brick} = 10\text{ N, so } V_{brick} = \frac{10\text{ N}}{\rho_w g} = \frac{10\text{ N}}{(1000\text{ kg/m}^3)(10\text{ m/s}^2)} = 1 \times 10^{-3}\text{ m}^3$$

   The correct answer is $\boxed{A}$.

2. **Strategy** The expansion of the cable obeys $F = k\Delta L$.

   Solution Compute the expansion of the cable.
   $$\Delta L = \frac{F}{k} = \frac{5000\text{ N}}{5.0 \times 10^6\text{ N/m}} = 10^{-3}\text{ m}$$
   The correct answer is $\boxed{A}$.

3. **Strategy** Solve for the intensity in the definition of sound level.

   Solution Find the intensity of the fire siren.
   $$\text{SL} = 10\log_{10}\frac{I}{I_0}, \text{ so } I = I_0 10^{\text{SL}/10} = (1.0 \times 10^{-12}\text{ W/m}^2)10^{100/10} = 1.0 \times 10^{-2}\text{ W/m}^2.$$

   The correct answer is $\boxed{D}$.

4. **Strategy and Solution** The two centimeters of liquid with a specific gravity of 0.5 are equivalent to one centimeter of water; that is, the 6-cm column of liquid is equivalent to a 5-cm column of water. Therefore, the new gauge pressure at the base of the column is five-fourths the original. The correct answer is $\boxed{C}$.

5. **Strategy** The wavelength is inversely proportional to the index n. Let $\lambda_n = 8$ m and $\lambda_{n+2} = 4.8$ m.

   Solution Find n.
   $$L = \frac{n\lambda_n}{4} = \frac{(n+2)\lambda_{n+2}}{4}, \text{ so } \frac{\lambda_n}{\lambda_{n+2}} = \frac{n+2}{n} = 1 + \frac{2}{n}, \text{ or } n = 2\left(\frac{\lambda_n}{\lambda_{n+2}} - 1\right)^{-1}.$$
   Compute L.
   $$L = \frac{n\lambda_n}{4} = 2\left(\frac{\lambda_n}{\lambda_{n+2}} - 1\right)^{-1}\frac{\lambda_n}{4} = \frac{1}{2}\left(\frac{8\text{ m}}{4.8\text{ m}} - 1\right)^{-1}(8\text{ m}) = 6\text{ m}$$
   The correct answer is $\boxed{C}$.

6. **Strategy** The wave may interfere within the range of possibility of totally constructive or totally destructive interference.

   Solution For totally constructive interference, the amplitude of the combined waves is $5 + 3 = 8$ units. For totally destructive interference, the amplitude of the combined waves is $5 - 3 = 2$ units. The correct answer is $\boxed{B}$.

7. **Strategy and Solution** As the bob repeatedly swings to and fro, it speeds up and slows down, as well as changes direction. Therefore, its linear acceleration must change in both magnitude and direction.

   The correct answer is $\boxed{D}$.

8. **Strategy** Since $K \propto v^2$ and $v \propto r^{-4}$, $K \propto r^{-8}$.

   Solution Compute the ratio of kinetic energies.

   $$\frac{K_2}{K_1} = \left(\frac{2}{1}\right)^{-8} = \frac{1}{256} = \frac{1}{4^4} \text{ or } K_2 : K_1 = 1 : 4^4.$$

   The correct answer is $\boxed{B}$.

9. **Strategy** The buoyant force on a ball is equal to the weight of the volume of water displaced by that ball.

   Solution Since B_1 is not fully submerged and B_2 and B_3 are, the buoyant force on B_1 is less than the buoyant forces on the other two. Since B_2 and B_3 are fully submerged, the buoyant forces on each are the same.

   The correct answer is $\boxed{B}$.

10. **Strategy and Solution** Ball 1 is floating, so its density is less than that of the water. Since Ball 2 is submerged within the water, its density is the same as that of the water. Ball 3 is sitting on the bottom of the tank, so its density is greater than that of the water. The correct answer is $\boxed{A}$.

11. **Strategy** The supporting force of the bottom of the tank is equal to the weight of Ball 3 less the buoyant force of the water.

    Solution Compute the supporting force on Ball 3.

    $$\rho_{\text{Ball 3}} g V_{\text{Ball 3}} - \rho_{\text{water}} g V_{\text{Ball 3}} = g V_{\text{Ball 3}} (\rho_{\text{Ball 3}} - \rho_{\text{water}})$$
    $$= (9.80 \text{ m/s}^2)(1.0 \times 10^{-6} \text{ m}^3)(7.8 \times 10^3 \text{ kg/m}^3 - 1.0 \times 10^3 \text{ kg/m}^3) = 6.7 \times 10^{-2} \text{ N}$$

    The correct answer is $\boxed{B}$.

12. **Strategy** The relationship between the fraction of a floating object's volume that is submerged to the ratio of the object's density to the fluid in which it floats is $V_f / V_o = \rho_o / \rho_f$. So, the fraction of the object that is not submerged is $V_{ns} = 1 - V_f / V_o = 1 - \rho_o / \rho_f$.

    Solution Find the fraction of the volume of Ball 1 that is above the surface of the water.

    $$1 - \frac{\rho_{\text{Ball 1}}}{\rho_{\text{water}}} = 1 - \frac{8.0 \times 10^2 \text{ kg/m}^3}{1.0 \times 10^3 \text{ kg/m}^3} = \frac{1}{5}$$

    The correct answer is $\boxed{D}$.

13. **Strategy** The pressure difference at a depth d in water is given by $\rho_{water}gd$.

   Solution Compute the approximate difference in pressure between the two balls.
   $$\rho_{water}gd = (1.0 \times 10^3 \text{ kg/m}^3)(9.80 \text{ m/s}^2)(0.20 \text{ m}) = 2.0 \times 10^3 \text{ N/m}^2$$
   The correct answer is $\boxed{C}$.

14. **Strategy** For the force exerted on Ball 3 by the bottom of the tank to be zero, Ball 3 must have the same density as the water.

   Solution Find the volume of the hollow portion of Ball 3, V_H.

   $$\rho_{Ball\ 3} = \frac{m_{Ball\ 3}}{V_{Ball\ 3}} = \frac{\rho_{Fe}V_{Fe}}{V_{Ball\ 3}} = \rho_{water} \text{ and } V_H = V_{Ball\ 3} - V_{Fe}, \text{ so}$$

   $$V_H = V_{Ball\ 3} - \frac{\rho_{water}}{\rho_{Fe}}V_{Ball\ 3} = V_{Ball\ 3}\left(1 - \frac{\rho_{water}}{\rho_{Fe}}\right) = (1.0 \times 10^{-6} \text{ m}^3)\left(1 - \frac{1.0 \times 10^3 \text{ kg/m}^3}{7.8 \times 10^3 \text{ kg/m}^3}\right) = 0.87 \times 10^{-6} \text{ m}^3.$$

   The correct answer is $\boxed{C}$.

Chapter 13

TEMPERATURE AND THE IDEAL GAS

Conceptual Questions

1. The development of standard temperature scales requires use of the zeroth law of thermodynamics. This law tells us that if a thermometer is in thermodynamic equilibrium with both a test object and an object used to define a standard, then the test and standard objects must be in thermodynamic equilibrium with each other. If this law did not hold, the temperature scale on a thermometer would have no relation to the actual temperature of the test object, and therefore, no standard could be defined.

5. The thermal expansion coefficients of silver and brass only differ by about 5%. Thus, a bimetallic strip made from these materials would bend only slightly during expansion, contrary to its intended purpose.

9. The SI units of mass density and number density are kg/m^3 and m^{-3}, respectively. An equal number density does not imply an equal mass density because the mass of an individual atom may be different in each gas.

13. The pressure of the air outside the balloon decreases as its distance above the Earth increases. The balloon expands until the pressures inside and outside are equal.

17. We could use any two of the quantities to decide whether the gas is dilute. The gas is dilute if the number density is low enough, or equivalently if the average intermolecular distance is large enough compared to the molecular diameter. For example, if the mean free path is much larger than the diameter, then we know that the intermolecular distance is large enough and the number density small enough so that the gas is dilute.

Problems

1. **Strategy** Use Eqs. (13-2b) and (13-3).

   Solution Convert the temperature.

   (a) $T_C = \dfrac{T_F - 32°F}{1.8°F/°C} = \dfrac{84°F - 32°F}{1.8°F/°C} = \boxed{29°C}$

   (b) $T = 29\ K + 273.15\ K = \boxed{302\ K}$

5. **Strategy and Solution** There are $78 + 114 = 192$ degrees C and 144 degrees J between the freezing and boiling points of ethyl alcohol. Thus, the conversion factor is $144/192 = 0.750$. So, $T_J = (0.750°J/°C)T_C + A$, where A is the offset to be determined. Find A by setting both temperatures equal to their respective boiling temperatures. $144°J = (0.750°J/°C)(78°C) + A$, so $A = 144°J - (0.750°J/°C)(78°C) = 85.5°J$.

   Thus, the conversion from °J to °C is given by $\boxed{T_J = (0.750°J/°C)T_C + 85.5°J}$.

9. **Strategy** Since each concrete slab expands along its entire length, only half of the expansion is considered for a particular gap. Two sections meet at a gap, so the gap should be as wide as the expansion of one concrete slab. Use Eq. (13-4).

   Solution Find the sizes of the expansion gaps.

   (a) $\Delta L = L_0 \alpha \Delta T = (15\ m)(12 \times 10^{-6}\ K^{-1})(40.0°C - 20.0°C) = \boxed{3.6\ mm}$

(b) $\Delta L = L_0 \alpha \Delta T = (15 \text{ m})(12 \times 10^{-6} \text{ K}^{-1})(-20.0°\text{C} - 20.0°\text{C}) = -7.2 \text{ mm}$

gap width $= 7.2 \text{ mm} + 3.6 \text{ mm} = \boxed{10.8 \text{ mm}}$

11. **Strategy** The hole expands just as if it were a solid brass disk. Use Eq. (13-6).

    Solution Find the increase in area of the hole.
    $$\Delta A = 2\alpha A_0 \Delta T = 2(1.9 \times 10^{-5} \text{ °C}^{-1})(1.00 \text{ mm}^2)(30.0°\text{C} - 20.0°\text{C}) = \boxed{3.8 \times 10^{-4} \text{ mm}^2}$$

13. **Strategy** A decrease in volume for a fixed mass increases the density, and vice versa. Use Eq. (13-7).

    Solution

    (a) $\dfrac{\Delta V}{V_0} = \beta \Delta T$ and $\dfrac{\Delta \rho}{\rho} = -\dfrac{\Delta V}{V_0}$. Thus, $\dfrac{\Delta \rho}{\rho} = -\beta \Delta T$, so $\Delta \rho = -\beta \rho \Delta T$.

    (b) Compute the fractional change in density.
    $$\frac{\Delta \rho}{\rho} = -\beta \Delta T = -(57 \times 10^{-6} \text{ K}^{-1})(-10.0°\text{C} - 32°\text{C}) = \boxed{2.4 \times 10^{-3}}$$

17. **Strategy** Use Eq. (13-4). The internal radius of the ring expands as if it were a solid piece of brass.

    Solution Find the temperature at which the internal radius of the ring is 1.0010 cm.
    $\dfrac{\Delta L}{L_0} = \alpha \Delta T$, so $\dfrac{\Delta L}{\alpha L_0} = T - T_0$. Compute the temperature.
    $$T = \frac{\Delta L}{\alpha L_0} + T_0 = \frac{1.0010 \text{ cm} - 1.0000 \text{ cm}}{(19 \times 10^{-6} \text{ K}^{-1})(1.0000 \text{ cm})} + 22.0°\text{C} = \boxed{75°\text{C}}$$

21. **Strategy** The diameter of the hole expands as if it were a solid piece of copper. Use Eq. (13-4).

    Solution Find the temperature at which the diameter of the hole is 1.0000 cm.
    $$\frac{\Delta L}{L_0} = \alpha \Delta T = \alpha(T - T_0), \text{ so } T = \frac{\Delta L}{\alpha L_0} + T_0 = \frac{1.0000 \text{ cm} - 0.9980 \text{ cm}}{(16 \times 10^{-6} \text{ K}^{-1})(0.9980 \text{ cm})} + 20.0°\text{C} = \boxed{150°\text{C}}.$$

25. **Strategy** Use Eqs. (13-4) and (13-6) and the given initial and final areas.

    Solution Find the fractional change.
    $$\frac{\Delta A}{A_0} = \frac{A - A_0}{A_0} = \frac{(s_0 + \Delta s)^2 - s_0^2}{s_0^2} = \frac{s_0^2 + 2s_0 \Delta s + (\Delta s)^2 - s_0^2}{s_0^2} = \frac{2s_0 \Delta s + (\Delta s)^2}{s_0^2} = \frac{\Delta s(2s_0 + \Delta s)}{s_0^2}$$
    Now, since $s_0 \gg \Delta s$, we have
    $$\frac{\Delta A}{A_0} \approx \frac{\Delta s(2s_0 + 0)}{s_0^2} = \frac{2s_0 \Delta s}{s_0^2} = \frac{2\Delta s}{s_0} = 2\alpha \Delta T \text{ since } \frac{\Delta s}{s_0} = \alpha \Delta T.$$

29. **Strategy** Add the molecular masses of each element in carbon dioxide.

    Solution Find the mass of carbon dioxide in kg.
    $$\text{mass of CO}_2 \text{ in kg} = m_C + 2m_O = [12.011 \text{ u} + 2(15.9994 \text{ u})](1.6605 \times 10^{-27} \text{ kg/u}) = \boxed{7.31 \times 10^{-26} \text{ kg}}$$

33. **Strategy** Divide the total mass by the molar mass of sucrose to find the number of moles. Then use Eq. (13-11) to find the number of hydrogen atoms.

Solution $m_{C_{12}H_{22}O_{11}} = 12(12.011 \text{ g/mol}) + 22(1.00794 \text{ g/mol}) + 11(15.9994 \text{ g/mol}) = 342.30 \text{ g/mol}$

There are 342.30 grams of sucrose per mole, so there are $\dfrac{684.6 \text{ g}}{342.30 \text{ g/mol}} = 2.000$ mol of sucrose.

There are $2.000(22) = 44.00$ moles of hydrogen. Find the number of hydrogen atoms.

$N = nN_A = (44.00 \text{ mol})(6.022 \times 10^{23} \text{ mol}^{-1}) = \boxed{2.650 \times 10^{25} \text{ atoms}}$

35. **Strategy** Divide the total mass of methane by its molar mass.

Solution Find the number of moles.

$n_{CH_4} = \dfrac{\text{mass of } CH_4}{\text{molar mass of } CH_4} = \dfrac{144.36 \text{ g}}{12.011 \text{ g/mol} + 4(1.00794 \text{ g/mol})} = \boxed{8.9985 \text{ mol}}$

37. **Strategy** Use Eq. (13-10).

Solution Find the number of air molecules.

$N = \dfrac{\rho V}{m} = (1.2 \text{ kg/m}^3)(1.0 \text{ cm}^3)\left(\dfrac{1 \text{ m}^3}{10^6 \text{ cm}^3}\right)\left(\dfrac{1}{29.0 \text{ u}}\right)\left(\dfrac{1 \text{ u}}{1.66 \times 10^{-27} \text{ kg}}\right) = \boxed{2.5 \times 10^{19} \text{ molecules}}$

41. **Strategy** Use the macroscopic form of the ideal gas law, Eq. (13-16).

Solution Find the new temperature of the air.

$PV = nRT$, so $\dfrac{T_f}{T_i} = \dfrac{\frac{P_f V_f}{nR}}{\frac{P_i V_i}{nR}} = \dfrac{P_f V_f}{P_i V_i}$, or

$T_f = \dfrac{P_f V_f T_i}{P_i V_i} = \dfrac{20.0 P_i (0.111 V_i) T_i}{P_i V_i} = 20.0(0.111)(30 \text{ K} + 273.15 \text{ K}) = 673 \text{ K} = \boxed{400°C}$.

45. **Strategy** The volume and moles of the gas are constant. Use Gay-Lussac's law.

Solution Find the pressure at the higher temperature.

$P \propto T$, so $P_f = \dfrac{T_f}{T_i} P_i = \dfrac{70.0 \text{ K} + 273.15 \text{ K}}{20.0 \text{ K} + 273.15 \text{ K}}(115 \text{ kPa}) = \boxed{135 \text{ kPa}}$.

49. **Strategy** The number of moles of the gas is constant. Use the ideal gas law.

Solution Find the volume of the hydrogen.

$\dfrac{P_f V_f}{T_f} = \dfrac{P_i V_i}{T_i}$, so $V_f = \dfrac{P_i V_i T_f}{P_f T_i} = \dfrac{(1.00 \times 10^5 \text{ N/m}^2)(5.0 \text{ m}^3)(273.15 \text{ K} - 13 \text{ K})}{(0.33 \times 10^3 \text{ N/m}^2)(273.15 \text{ K} + 27 \text{ K})} = \boxed{1.3 \times 10^3 \text{ m}^3}$.

53. **Strategy** Use the microscopic form of the ideal gas law, Eq. (13-13).

Solution Find the number of air molecules released.

$N = \dfrac{PV}{kT}$, and V, k, and T are constant, so

$$\Delta N = \frac{V\Delta P}{kT} = \frac{(1.0 \text{ m}^3)(15.0 \text{ atm} - 20.0 \text{ atm})(1.013\times10^5 \text{ Pa/atm})}{(1.38\times10^{-23} \text{ J/K})(273 \text{ K})} = -1.3\times10^{26}.$$

$\boxed{1.3\times10^{26}}$ air molecules were released.

55. Strategy The number of moles of the gas is constant and $V = \frac{1}{6}\pi d^3$. Use the ideal gas law and Eq. (9-3).

Solution Find the diameter of the bubble when it reaches the surface.

$$\frac{P_f V_f}{T_f} = \frac{P_i V_i}{T_i}, \text{ so } V_f = \frac{1}{6}\pi d_f^3 = \frac{P_i T_f}{P_f T_i}V_i = \frac{(P_f + \rho gh)T_f}{P_f T_i}\left(\frac{1}{6}\pi d_i^3\right). \text{ Solve for } d_f.$$

$$d_f = d_i \sqrt[3]{\frac{(P_f + \rho gh)T_f}{P_f T_i}}$$

$$= (1.00 \text{ mm})\sqrt[3]{\frac{[(1.0 \text{ atm})(1.013\times10^5 \text{ Pa/atm}) + (1.0\times10^3 \text{ kg/m}^3)(9.80 \text{ m/s}^2)(80.0 \text{ m})](273.15 \text{ K} + 18 \text{ K})}{(1.0 \text{ atm})(1.013\times10^5 \text{ Pa/atm})(273.15 \text{ K} + 4 \text{ K})}}$$

$$= \boxed{2.1 \text{ mm}}$$

57. Strategy The number of moles of air is constant. Assume that the temperature is constant.

$$V_f = \frac{\Delta V}{\Delta t}\Delta t \text{ where } \frac{\Delta V}{\Delta t} = 0.500 \text{ L/s} = 5.00\times10^2 \text{ cm}^3/\text{s} \text{ and } P_f = P_i + \rho gd. \text{ Use Boyle's law.}$$

Solution Find how long the tank of air will last for each depth.

(a) $$\frac{V_f}{V_i} = \frac{\frac{\Delta V}{\Delta t}\Delta t}{V_i} = \frac{P_{tank}}{P_f} = \frac{P_{tank}}{P_{atm} + \rho gd}, \text{ so}$$

$$\Delta t = \frac{P_{tank}V_i}{\frac{\Delta V}{\Delta t}(P_{atm} + \rho gd)}$$

$$= \frac{(1.0\times10^7 \text{ Pa})(0.010 \text{ m}^3)}{(5.00\times10^2 \text{ cm}^3/\text{s})(10^{-6} \text{ m}^3/\text{cm}^3)(60 \text{ s/min})\left[1.013\times10^5 \text{ Pa} + (1.0\times10^3 \text{ kg/m}^3)(9.80 \text{ m/s}^2)(2.0 \text{ m})\right]}$$

$$= \boxed{28 \text{ min}}$$

(b) $$\Delta t = \frac{(1.0\times10^7 \text{ Pa})(0.010 \text{ m}^3)}{(5.00\times10^2 \text{ cm}^3/\text{s})(10^{-6} \text{ m}^3/\text{cm}^3)(60 \text{ s/min})\left[1.013\times10^5 \text{ Pa} + (1.0\times10^3 \text{ kg/m}^3)(9.80 \text{ m/s}^2)(20.0 \text{ m})\right]}$$

$$= \boxed{11 \text{ min}}$$

61. Strategy The total translational kinetic energy of the gas molecules is equal to the number of molecules times the average translational kinetic energy per molecule. Use Eqs. (13-13) and (13-20).

Solution Find the total translational kinetic energy of the gas molecules.

$$K_{total} = N\langle K_{tr}\rangle = N\left(\frac{3}{2}kT\right) = \frac{3}{2}PV = \frac{3}{2}(1.013\times10^5 \text{ Pa})(0.00100 \text{ m}^3) = \boxed{152 \text{ J}}$$

65. **Strategy** The total internal kinetic energy of the ideal gas is equal to the number of molecules times the average kinetic energy per molecule. Use Eq. (13-20).

   Solution Find the total internal kinetic energy of the ideal gas.

$$K_{\text{total}} = N\langle K_{\text{tr}}\rangle = N\left(\frac{3}{2}kT\right) = \frac{3}{2}nRT = \frac{3}{2}(1.0 \text{ mol})[8.314 \text{ J/(mol·K)}](273.15 \text{ K} + 0.0 \text{ K}) = \boxed{3.4 \text{ kJ}}$$

69. **Strategy** Use Eq. (13-22).

   Solution Find the rms speeds of the molecules.

   (a) $v_{\text{rms}} = \sqrt{\dfrac{3kT}{m}} = \sqrt{\dfrac{3(1.38\times10^{-23} \text{ J/K})(273.15 \text{ K} + 0.0 \text{ K})}{2(14.00674 \text{ u})(1.66\times10^{-27} \text{ kg/u})}} = \boxed{493 \text{ m/s}}$

   (b) $v_{\text{rms}} = \sqrt{\dfrac{3(1.38\times10^{-23} \text{ J/K})(273.15 \text{ K} + 0.0 \text{ K})}{2(15.9994 \text{ u})(1.66\times10^{-27} \text{ kg/u})}} = \boxed{461 \text{ m/s}}$

   (c) $v_{\text{rms}} = \sqrt{\dfrac{3(1.38\times10^{-23} \text{ J/K})(273.15 \text{ K} + 0.0 \text{ K})}{[12.011 \text{ u} + 2(15.9994 \text{ u})](1.66\times10^{-27} \text{ kg/u})}} = \boxed{393 \text{ m/s}}$

73. **Strategy** Use Eq. (13-20).

   Solution Find the temperature of the ideal gas.

$$\langle K_{\text{tr}}\rangle = \frac{3}{2}kT, \text{ so } T = \frac{2\langle K_{\text{tr}}\rangle}{3k} = \frac{2(4.60\times10^{-20} \text{ J})}{3(1.38\times10^{-23} \text{ J/K})} = \boxed{2220 \text{ K}}.$$

75. **Strategy and Solution** The average translational energy of a molecule in an ideal gas is $\langle K_{\text{tr}}\rangle = \dfrac{1}{2}m\langle v^2\rangle$. The rms speed is $v_{\text{rms}} = \sqrt{\langle v^2\rangle}$, so $\dfrac{1}{2}mv_{\text{rms}}^2 = \langle K_{\text{tr}}\rangle$. From Eq. (13-19), we know that $\langle K_{\text{tr}}\rangle = \dfrac{3PV}{2N}$, and using the ideal gas law, $PV = nRT$, we have $\dfrac{1}{2}mv_{\text{rms}}^2 = \dfrac{3}{2}\left(\dfrac{PV}{N}\right) = \dfrac{3}{2}\left(\dfrac{nRT}{N}\right)$, so $v_{\text{rms}} = \sqrt{3\left(\dfrac{n}{mN}\right)RT} = \sqrt{\dfrac{3RT}{M}}$.

77. **Strategy** Form a proportion with the two reaction rates and solve for the temperature increase. Use Eq. (13-24).

Solution Find the temperature increase.

$$\frac{1.035}{1} = \frac{e^{-\frac{E_a}{kT_2}}}{e^{-\frac{E_a}{kT_1}}} = e^{\frac{E_a}{k}\left(\frac{1}{T_1}-\frac{1}{T_2}\right)}$$

$$\ln 1.035 = \frac{E_a}{k}\left(\frac{1}{T_1}-\frac{1}{T_2}\right)$$

$$\frac{k \ln 1.035}{E_a} = \frac{1}{T_1}-\frac{1}{T_2}$$

$$\frac{1}{T_2} = \frac{1}{T_1}-\frac{k \ln 1.035}{E_a}$$

$$T_2 = \left(\frac{1}{T_1}-\frac{k \ln 1.035}{E_a}\right)^{-1}$$

$$\Delta T = \left(\frac{1}{T_1}-\frac{k \ln 1.035}{E_a}\right)^{-1} - T_1$$

$$= \left[\frac{1}{273.15\text{ K}+10.00\text{ K}} - \frac{(1.38\times10^{-23}\text{ J/K})\ln 1.035}{2.81\times10^{-19}\text{ J}}\right]^{-1} - (273.15\text{ K}+10.00\text{ K}) = \boxed{0.14°\text{C}}$$

81. **Strategy** Use Eq. (13-26).

Solution Estimate the time it takes a sucrose molecule to move 5.00 mm in one direction.

$$x_{rms} = \sqrt{2Dt},\text{ so }t = \frac{x_{rms}^2}{2D} = \frac{(5.00\times10^{-3}\text{ m})^2}{2(5.0\times10^{-10}\text{ m}^2/\text{s})} = \boxed{2.5\times10^4\text{ s}}.$$

85. **Strategy** Find the temperature at which the radius of the steel sphere and the internal radius of the brass ring are the same. Use Eq. (13-5).

Solution Set the final lengths of the radii equal and solve for the final temperature.

$$L_{b0} + L_{b0}\alpha_b\Delta T = L_{s0} + L_{s0}\alpha_s\Delta T$$
$$L_{b0}\alpha_b(T_f - T_i) = L_{s0} - L_{b0} + L_{s0}\alpha_s(T_f - T_i)$$
$$T_f(L_{b0}\alpha_b - L_{s0}\alpha_s) = L_{s0} - L_{b0} + T_i(L_{b0}\alpha_b - L_{s0}\alpha_s)$$

$$T_f = \frac{L_{s0} - L_{b0}}{L_{b0}\alpha_b - L_{s0}\alpha_s} + T_i$$

$$T_f = \frac{1.0010\text{ cm}-1.0000\text{ cm}}{(1.0000\text{ cm})(19\times10^{-6}\text{ K}^{-1})-(1.0010\text{ cm})(12\times10^{-6}\text{ K}^{-1})} + 22.0°\text{C} = \boxed{165°\text{C}}$$

89. **(a) Strategy** Use Eq. (13-20).

Solution Compute the average kinetic energy of the air molecules.

$$\langle K_{tr}\rangle = \frac{3}{2}kT = \frac{3}{2}(1.38\times10^{-23}\text{ J/K})\left[\frac{5}{9}(98.6-32.0)\text{ K}+273.15\text{ K}\right] = \boxed{6.42\times10^{-21}\text{ J}}$$

(b) Strategy Since the average kinetic energy is directly proportional to the absolute temperature, find the percent change in the absolute temperature.

Solution Find the percentage by which the kinetic energy of the molecules increased.

$$\frac{\Delta K}{K_0} \times 100\% = \frac{\Delta T}{T_0} \times 100\% = \frac{\frac{5}{9}(100.0 - 98.6)\ \text{K}}{\frac{5}{9}(98.6 - 32.0)\ \text{K} + 273.15\ \text{K}} \times 100\% = \boxed{0.25\%}$$

93. **Strategy and Solution** The average of the test scores is $\dfrac{83 + 62 + 81 + 77 + 68 + 92 + 88 + 83 + 72 + 75}{10} = \boxed{78.1}$.

The rms value is $\sqrt{\dfrac{83^2 + 62^2 + 81^2 + 77^2 + 68^2 + 92^2 + 88^2 + 83^2 + 72^2 + 75^2}{10}} = \boxed{78.6}$. The most probable value

is $\boxed{83}$, since it appears twice as often as any other score.

97. **Strategy** Assume that each air molecule is at the center of a sphere (with volume V/N) of diameter d. Then the average distance between air molecules is approximately d. Use the microscopic form of the ideal gas law, Eq. (13-13).

Solution Estimate the average distance between air molecules.

$$PV = P\left(N\frac{1}{6}\pi d^3\right) = NkT, \text{ so } d = \left(\frac{6kT}{\pi P}\right)^{1/3} = \left[\frac{6(1.38 \times 10^{-23}\ \text{J/K})(273.15\ \text{K} + 0.0\ \text{K})}{\pi(1.00\ \text{atm})(1.013 \times 10^5\ \text{Pa/atm})}\right]^{1/3} \approx \boxed{4\ \text{nm}}.$$

101. **Strategy** Use Eq. (13-4).

Solution Find the approximate temperature for the SR-71 while it is in flight.

$$\frac{\Delta L}{L_0} = \alpha \Delta T, \text{ so } T_f = T_i + \frac{\Delta L}{\alpha L_0} = 20°\text{C} + \frac{0.20\ \text{m}}{(10.1 \times 10^{-6}\ \text{K}^{-1})(32.70\ \text{m})} = \boxed{630°\text{C}}.$$

105. **Strategy** Use Eq. (13-22).

Solution Find the decrease in the rms speed of the air molecules.

$$\Delta v_{\text{rms}} = \sqrt{\frac{3kT_f}{m}} - \sqrt{\frac{3kT_i}{m}}$$

$$= \sqrt{\frac{3(1.38 \times 10^{-23}\ \text{J/K})(10.0\ \text{K} + 273.15\ \text{K})}{[0.750(2)(14.00674\ \text{u}) + 0.250(2)(15.9994\ \text{u})](1.66 \times 10^{-27}\ \text{kg/u})}}$$

$$- \sqrt{\frac{3(1.38 \times 10^{-23}\ \text{J/K})(40.0\ \text{K} + 273.15\ \text{K})}{[0.750(2)(14.00674\ \text{u}) + 0.250(2)(15.9994\ \text{u})](1.66 \times 10^{-27}\ \text{kg/u})}}$$

$$= -25\ \text{m/s}$$

The rms speed will have decreased by $\boxed{25\ \text{m/s}}$.

109. **Strategy** Use ideal gas law and Hooke's law.

Solution Find the pressure of the gas.

$PV = nRT$, so $P_{\text{gas}} = \dfrac{nRT}{V}$. The force with which the piston pushes on the spring is equal to

$F = (P_{\text{gas}} - P_{\text{atm}})A_{\text{piston}}$. Set this equal to $F = k\Delta x$ to find the spring constant.

$k\Delta x = (P_{\text{gas}} - P_{\text{atm}})A_{\text{piston}} = \left(\dfrac{nRT}{V} - P_{\text{atm}}\right)A_{\text{piston}}$, so

$$k = \left[\frac{(6.50 \times 10^{-2}\ \text{mol})[8.314\ \text{J/(mol·K)}](20.0\ \text{K} + 273.15\ \text{K})}{(0.120\ \text{m} + 0.0540\ \text{m})\pi(0.0800/2\ \text{m})^2} - 1.013 \times 10^5\ \text{Pa}\right]\frac{\pi(0.0800/2\ \text{m})^2}{0.0540\ \text{m}} = \boxed{7.4 \times 10^3\ \text{N/m}}.$$

Chapter 14

HEAT

Conceptual Questions

1. Heat flows from the hotter to the colder object.

5. The air between the layers acts as very good thermal insulation.

9. The fins increase the area of the surface responsible for radiating heat from the engine to the air so that the engine cools more efficiently.

13. Water contained in the sauce and cheese of a pizza has a higher specific heat and thermal conductivity than the crust. Thus, for a given time interval, the amount of heat energy transferred to the roof of the mouth by the sauce and cheese is greater than that which is transferred to the hand by the crust.

17. The rate at which heat flows via conduction, convection, and radiation depends upon the temperature difference between the hot and cold bodies. By adding the milk to the coffee immediately, the temperature difference between the coffee and the surrounding air is minimized, so the rate of heat flow is smaller. Additionally, when the color of the coffee is lighter, the radiative heat loss will be further reduced.

21. In winter, the walls of a room are at a lower temperature than the air within the room. In summer, the walls are warmer than the inside air. The amount of heat lost from a person standing within the room via conduction and convection is constant throughout the year since the temperature of the air is held fixed. The temperature difference between the body and the walls is greatest in winter. As a result, more heat is lost via radiative transfer in winter and the room therefore feels cooler.

Problems

1. (a) **Strategy** The gravitational potential energy of the 1.4 kg of water is converted to internal energy in the 6.4-kg system.

   Solution Compute the increase in internal energy.
   $$U = mgh = (1.4 \text{ kg})(9.80 \text{ m/s}^2)(2.5 \text{ m}) = \boxed{34 \text{ J}}$$

   (b) **Strategy and Solution** Yes; the increase in internal energy increases the average kinetic energy of the water molecules, thus the temperature is slightly increased.

3. **Strategy** The amount of internal energy generated is equal to the decrease in kinetic energy of the bullet.

   Solution Compute the amount of internal energy generated.
   $$|\Delta K| = \frac{1}{2}mv_i^2 = \frac{1}{2}(0.0200 \text{ kg})(7.00 \times 10^2 \text{ m/s})^2 = \boxed{4.90 \text{ kJ}}$$

5. **(a) Strategy** The decrease in gravitational potential energy of the child is equal to the amount of internal energy generated.

    Solution Compute the amount of internal energy generated.
    $$U = mgh = (15 \text{ kg})(9.80 \text{ m/s}^2)(1.7 \text{ m}) = \boxed{250 \text{ J}}$$

    (b) Strategy and Solution Friction warms the slide and the child, and the air molecules are deflected by the child's body. The energy goes into $\boxed{\text{all three}}$.

9. **Strategy** The conversion factor is $1 \text{ kW} \cdot \text{h} = 3.600 \text{ MJ}$.

    Solution Convert 1.00 kJ to kilowatt-hours.
    $$(1.00 \times 10^3 \text{ J}) \frac{1 \text{ kW} \cdot \text{h}}{3.600 \times 10^6 \text{ J}} = \boxed{2.78 \times 10^{-4} \text{ kW} \cdot \text{h}}$$

11. **Strategy** The heat capacity of an object is equal to its mass times its specific heat.

    Solution Find the heat capacity of the 5.00-g gold ring.
    $$C = mc = (0.00500 \text{ kg})[0.128 \text{ kJ/(kg} \cdot \text{K)}] = \boxed{6.40 \times 10^{-4} \text{ kJ/K}}$$

13. **Strategy** Use Eq. (14-4).

    Solution Find the amount of heat that must flow into the water.
    $$Q = mc\Delta T = (2.0 \times 10^{-3} \text{ m}^3)(1.0 \times 10^3 \text{ kg/m}^3)[4186 \text{ J/(kg} \cdot \text{K)}](80.0 - 20.0) \text{ K} = \boxed{0.50 \text{ MJ}}$$

15. **Strategy** The 3.3% of the energy from the food is converted to gravitational potential energy of the high jumper.

    Solution Find the height the athlete could jump.
    $$U = mgh, \text{ so } h = \frac{U}{mg} = \frac{(3.00 \times 10^6 \text{ cal})(4.186 \text{ J/cal})(0.033)}{(60.0 \text{ kg})(9.80 \text{ m/s}^2)} = \boxed{700 \text{ m}}.$$

17. **Strategy** The heat capacity of an object is equal to its mass times its specific heat. The mass of an object is equal to its density times its volume.

    Solution Find the heat capacities.

    (a) $C = mc = \rho Vc = (2702 \text{ kg/m}^3)(1.00 \text{ m}^3)[0.900 \text{ kJ/(kg} \cdot \text{K)}] = \boxed{2430 \text{ kJ/K}}$

    (b) $C = mc = \rho Vc = (7860 \text{ kg/m}^3)(1.00 \text{ m}^3)[0.44 \text{ kJ/(kg} \cdot \text{K)}] = \boxed{3500 \text{ kJ/K}}$

19. **Strategy** Use Eq. (14-4) to find the heat required.

    Solution The heat capacity of the system is $C = m_{\text{Al}}c_{\text{Al}} + m_{\text{w}}c_{\text{w}}$.
    $$Q = mc\Delta T = \{(0.400 \text{ kg})[0.900 \text{ kJ/(kg} \cdot \text{K)}] + (2.00 \text{ kg})[4.186 \text{ kJ/(kg} \cdot \text{K)}]\}(100.0°\text{C} - 15.0°\text{C}) = \boxed{742 \text{ kJ}}$$

21. Strategy Use Eq. (14-4).

Solution Find the specific heat of lead.

$$c = \frac{Q}{m\Delta T} = \frac{0.88 \text{ kJ}}{(0.35 \text{ kg})(20.0 \text{ K})} = \boxed{0.13 \text{ kJ/(kg} \cdot \text{K)}}$$

25. Strategy The energy required to increase the internal energy of the gas is equal to $nC_V \Delta T$ where $C_V = 20.4$ J/(mol·K) for H_2. Use the ideal gas law to find the number of moles of H_2.

Solution Find the energy required.

$$\Delta U = nC_V \Delta T = \frac{PV}{RT} C_V \Delta T = \frac{(10.0 \text{ atm})(1.013 \times 10^5 \text{ Pa/atm})(250 \text{ L})(10^{-3} \text{ m}^3/\text{L})[20.4 \text{ J/(mol} \cdot \text{K})](25.0 \text{ K})}{[8.314 \text{ J/(mol} \cdot \text{K})](273.15 \text{ K} + 0.0 \text{ K})}$$

$$= \boxed{57 \text{ kJ}}$$

29. (a) Strategy Phase changes are indicated by the graph where the temperature is constant while heat is added.

Solution Initially, the substance is solid. As the temperature increases, the substance changes from the solid to the liquid phase, then from the liquid phase to the gas phase. There are two phase changes shown by the graph: from $\boxed{\text{B to C, solid to liquid}}$; and from $\boxed{\text{D to E, liquid to gas}}$.

(b) Strategy and Solution The beginning of the phase change from solid to liquid indicates the melting point of the substance. That point is labeled by the letter $\boxed{\text{B}}$.

(c) Strategy and Solution The beginning of the phase change from liquid to gas indicates the boiling point of the substance. That point is labeled by the letter $\boxed{\text{D}}$.

31. Strategy The sum of the heat flows is zero. Use Eqs. (14-4) and (14-9).

Solution Find the heat of fusion of water.

$$0 = Q_{\text{ice}} + Q_{\text{w}} + Q_{\text{c}} = m_{\text{ice}} L_{\text{f}} + m_{\text{ice}} c_{\text{w}} \Delta T_{\text{ice}} + m_{\text{w}} c_{\text{w}} \Delta T_{\text{w}} + m_{\text{c}} c_{\text{c}} \Delta T_{\text{c}}, \text{ so}$$

$$L_{\text{f}} = -\frac{c_{\text{w}}(m_{\text{ice}} \Delta T_{\text{ice}} + m_{\text{w}} \Delta T_{\text{w}}) + m_{\text{c}} c_{\text{c}} \Delta T_{\text{c}}}{m_{\text{ice}}}$$

$$= -\frac{[4.186 \text{ J/(g} \cdot \text{K})][(30.0 \text{ g})(8.5 \text{ K}) + (2.00 \times 10^2 \text{ g})(-11.5 \text{ K})] + (3.00 \times 10^2 \text{ g})[0.380 \text{ J/(g} \cdot \text{K})](-11.5 \text{ K})}{30.0 \text{ g}}$$

$$= \boxed{330 \text{ J/g}}$$

33. Strategy The sum of the heat flows is zero. Find the mass of the water required to melt the ice and leave the temperature of the drink at 5.0°C. Use Eqs. (14-4) and (14-9).

Solution Find the mass of water to be added to the cup.

$$0 = Q_{\text{w}} + Q_{\text{ice}}$$

$$0 = m_{\text{w}} c_{\text{w}} \Delta T_{\text{w}} + m_{\text{ice}} L_{\text{f}} + m_{\text{ice}} c_{\text{ice}} \Delta T_1 + m_{\text{ice}} c_{\text{w}} \Delta T_2$$

$$-m_{\text{w}} c_{\text{w}} \Delta T_{\text{w}} = m_{\text{ice}}(L_{\text{f}} + c_{\text{ice}} \Delta T_1 + c_{\text{w}} \Delta T_2)$$

$$m_{\text{w}} = \frac{m_{\text{ice}}(L_{\text{f}} + c_{\text{ice}} \Delta T_1 + c_{\text{w}} \Delta T_2)}{-c_{\text{w}} \Delta T_{\text{w}}}$$

$$m_{\text{w}} = -\frac{(50.0 \text{ g} + 50.0 \text{ g})\{333.7 \text{ J/g} + [2.1 \text{ J/(g} \cdot \text{K})](15.0 \text{ K}) + [4.186 \text{ J/(g} \cdot \text{K})](5.0 \text{ K})\}}{[4.186 \text{ J/(g} \cdot \text{K})](-20.0 \text{ K})} = \boxed{461 \text{ g}}$$

35. **Strategy** Heat flows from the water to the ice, melting some of it. Find the mass of ice required to lower the temperature of the water to 0.0°C. Use Eqs. (14-4) and (14-9).

    Solution Find the required mass of ice.
    $0 = Q_{ice} + Q_w = m_{ice} L_f + m_w c_w \Delta T_w$, so
    $$m_{ice} = -\frac{m_w c_w \Delta T_w}{L_f} = -\frac{(5.00 \times 10^2 \text{ mL})(1.00 \text{ g/mL})[4.186 \text{ J/(g·K)}](-25.0 \text{ K})}{333.7 \text{ J/g}} = \boxed{157 \text{ g}}.$$

37. **Strategy** The sum of the heat flows is zero. The tea is basically water. The mass of the tea is found by multiplying the density of water by the volume of the tea. Do not neglect the temperature change of the glass. Use Eqs. (14-4) and (14-9).

    Solution Find the mass of the ice required to cool the tea to 10.0°C. Let ΔT be the temperature change of the tea and the glass.
    $$0 = Q_t + Q_{ice} + Q_g$$
    $$0 = \rho_w V_t c_w \Delta T + m_{ice} L_f + m_{ice} c_{ice} \Delta T_1 + m_{ice} c_w \Delta T_2 + m_g c_g \Delta T$$
    $$0 = (\rho_w V_t c_w + m_g c_g) \Delta T + m_{ice} (L_f + c_{ice} \Delta T_1 + c_w \Delta T_2)$$
    $$m_{ice} = \frac{-(\rho_w V_t c_w + m_g c_g) \Delta T}{L_f + c_{ice} \Delta T_1 + c_w \Delta T_2}$$
    $$= \frac{-\{(1.00 \times 10^3 \text{ kg/m}^3)(2.00 \times 10^{-4} \text{ m}^3)[4.186 \text{ kJ/(kg·K)}] + (0.35 \text{ kg})[0.837 \text{ kJ/(kg·K)}]\}(-85.0 \text{ K})}{333.7 \text{ kJ/kg} + [2.1 \text{ kJ/(kg·K)}](10.0 \text{ K}) + [4.186 \text{ kJ/(kg·K)}](10.0 \text{ K})}$$
    $$= \boxed{242 \text{ g}}$$

    The percentage change from the answer for Problem 40 is $\frac{242 \text{ g} - 179 \text{ g}}{179 \text{ g}} \times 100\% = \boxed{35\%}$.

39. **Strategy** Heat flows from the aluminum into the ice. Use Eqs. (14-4) and (14-9).

    Solution Find the mass of aluminum required to melt 10.0 g of ice.
    $Q_{Al} + Q_{ice} = m_{Al} c_{Al} \Delta T_{Al} + m_{ice} L_f = 0$, so
    $$m_{Al} = -\frac{m_{ice} L_f}{c_{Al} \Delta T_{Al}} = -\frac{(10.0 \text{ g})(333.7 \text{ J/g})}{[0.900 \text{ J/(g·K)}](0.0°C - 80.0°C)} = \boxed{46.3 \text{ g}}.$$

41. **Strategy** Use Eq. (14-4) for vaporization. Q is equal to the rate of heat loss per square meter times the area times the time (1 h = 3600 s).

    Solution Find the mass of water lost through transpiration.
    $$m = \frac{Q}{L_v} = \frac{(250 \text{ W/m}^2)(0.005 \text{ m}^2)(3600 \text{ s})}{2256 \text{ J/g}} = \boxed{2 \text{ g}}$$

45. **Strategy** The heat supplied heats the substance to its melting point, melts it, then raises the temperature of the resulting liquid to 327°C. Use Eqs. (14-4) and (14-9).

    Solution Compute the heat of fusion.
    $$Q = m L_f + m c \Delta T, \text{ so } L_f = -c\Delta T + \frac{Q}{m} = -[0.129 \text{ kJ/(kg·K)}](327 - 21) \text{ K} + \frac{31.15 \text{ kJ}}{0.500 \text{ kg}} = \boxed{22.8 \text{ kJ/kg}}.$$

49. Strategy Use Eq. (14-12).

Solution Compute the thermal resistance for each material.

(a) $R = \dfrac{d}{\kappa A} = \dfrac{2.0 \times 10^{-2} \text{ m}}{[0.17 \text{ W/(m} \cdot \text{K)}](1.0 \text{ m}^2)} = \boxed{0.12 \text{ K/W}}$

(b) $R = \dfrac{2.0 \times 10^{-2} \text{ m}}{[80.2 \text{ W/(m} \cdot \text{K)}](1.0 \text{ m}^2)} = \boxed{2.5 \times 10^{-4} \text{ K/W}}$

(c) $R = \dfrac{2.0 \times 10^{-2} \text{ m}}{[401 \text{ W/(m} \cdot \text{K)}](1.0 \text{ m}^2)} = \boxed{5.0 \times 10^{-5} \text{ K/W}}$

51. Strategy From the given information, $\Delta T = \mathcal{P}_1 R_1 = \mathcal{P}_2 R_2 = \mathcal{P}(R_1 + R_2)$.

Solution Find the rate of heat flow per unit area.
$$\frac{R_1}{R_2} = \frac{\mathcal{P}_2}{\mathcal{P}_1} \text{ and } \mathcal{P} = \frac{\mathcal{P}_2 R_2}{R_1 + R_2}. \text{ Thus, } \mathcal{P} = \frac{\mathcal{P}_2}{\frac{R_1}{R_2} + 1} = \frac{\mathcal{P}_2}{\frac{\mathcal{P}_2}{\mathcal{P}_1} + 1} = \frac{20.0 \text{ W/m}^2}{\frac{20.0}{10.0} + 1} = \boxed{6.67 \text{ W/m}^2}.$$

53. Strategy Use Fourier's law of heat conduction, Eq. (14-10).

Solution Find the lowest temperature the dog can withstand without increasing its heat output.
$$\mathcal{P} = \kappa A \frac{\Delta T}{d} = \kappa A \frac{T_i - T_o}{d}, \text{ so } T_o = T_i - \frac{d\mathcal{P}}{\kappa A} = 38°C - \frac{(0.050 \text{ m})(51 \text{ W})}{[0.026 \text{ W/(m} \cdot \text{K)}](1.31 \text{ m}^2)} = \boxed{-37°C}.$$

57. Strategy Use Fourier's law of heat conduction, Eq. (14-10).

Solution

(a) $\mathcal{P} = \kappa A \dfrac{\Delta T}{d} = [401 \text{ W/(m} \cdot \text{K)}](1.0 \times 10^{-6} \text{ m}^2) \dfrac{104°C - 24°C}{0.10 \text{ m}} = \boxed{0.32 \text{ W}}$

(b) $\dfrac{\Delta T}{d} = \dfrac{104°C - 24°C}{0.10 \text{ m}} = \boxed{800 \text{ K/m}}$

(c) The effective length has doubled.
$$\mathcal{P} = [401 \text{ W/(m} \cdot \text{K)}](1.0 \times 10^{-6} \text{ m}^2) \frac{104°C - 24°C}{0.20 \text{ m}} = \boxed{0.16 \text{ W}}$$

(d) The effective area has doubled.
$$\mathcal{P} = [401 \text{ W/(m} \cdot \text{K)}](2.0 \times 10^{-6} \text{ m}^2) \frac{104°C - 24°C}{0.10 \text{ m}} = \boxed{0.64 \text{ W}}$$

(e) Since the bars are identical, the temperature at the junction will be midway between the temperatures of the baths.
$$\frac{104°C + 24°C}{2} = \boxed{64°C}$$

61. **Strategy** Use Stefan's law of radiation, Eq. (14-16).

    Solution Compute the power radiated by the bulb.
    $$\mathcal{P} = e\sigma A T^4 = 0.32[5.670\times10^{-8} \text{ W}/(\text{m}^2\cdot\text{K}^4)](1.00\times10^{-4} \text{ m}^2)(3.00\times10^3 \text{ K})^4 = \boxed{150 \text{ W}}$$

65. **Strategy** Use Stefan's law of radiation, Eq. (14-16). Form a proportion. Assume the cross-sectional area A is constant.

    Solution Find the temperature of the filament.
    $$\mathcal{P} = e\sigma A T^4, \text{ so } \frac{\mathcal{P}_{58}}{\mathcal{P}_{60}} = \frac{T_{58}^4}{T_{60}^4} \text{ and } T_{58} = \left(\frac{\mathcal{P}_{58}}{\mathcal{P}_{60}}\right)^{1/4} T_{60} = \left(\frac{58.0}{60.0}\right)^{1/4}(2820 \text{ K}) = \boxed{2800 \text{ K}}.$$

69. **Strategy** Approximate the pots as cubes of similar volume. Use Eq. (14-18).

    Solution Find the net rate of radiative heat loss from the two pots.
    $$s^3 = V, \text{ so } s = V^{1/3} \text{ and } 6s^2 = A = 6V^{2/3}.$$
    Coffeepot:
    $$\mathcal{P}_{\text{net}} = e\sigma A(T^4 - T_s^4)$$
    $$= 0.12[5.670\times10^{-8} \text{ W}/(\text{m}^2\cdot\text{K}^4)][6(1.00 \text{ L})^{2/3}(10^{-3} \text{ m}^3/\text{L})^{2/3}][(98 \text{ K} + 273.15 \text{ K})^4 - (25 \text{ K} + 273.15 \text{ K})^4]$$
    $$= \boxed{4.5 \text{ W}}$$
    Teapot:
    $$\mathcal{P}_{\text{net}} = 0.65[5.670\times10^{-8} \text{ W}/(\text{m}^2\cdot\text{K}^4)][6(1.00 \text{ L})^{2/3}(10^{-3} \text{ m}^3/\text{L})^{2/3}][(98 \text{ K} + 273.15 \text{ K})^4 - (25 \text{ K} + 273.15 \text{ K})^4]$$
    $$= \boxed{24 \text{ W}}$$

71. **(a) Strategy** The power absorbed by the leaf must equal that radiated away. Power is equal to intensity times area. Use Stefan's law of radiation, Eq. (14-16).

    Solution Absorbed:
    $$I_{\text{top}}A + e\sigma A T_s^4 = 0.700(9.00\times10^2 \text{ W}/\text{m}^2)(5.00\times10^{-3} \text{ m}^2) +$$
    $$(1)[5.670\times10^{-8} \text{ W}/(\text{m}^2\cdot\text{K}^4)](5.00\times10^{-3} \text{ m}^2)(273.15 \text{ K} + 25.0 \text{ K})^4$$
    $$= 5.39 \text{ W}$$
    Find the temperature of the leaf. The area is now $2(5.00\times10^{-3} \text{ m}^2) = 10.0\times10^{-3} \text{ m}^2$ (both sides of the leaf).
    $$\mathcal{P} = e\sigma A T^4, \text{ so}$$
    $$T = \left(\frac{\mathcal{P}}{e\sigma A}\right)^{1/4} = \left[\frac{5.39 \text{ W}}{(1)[5.670\times10^{-8} \text{ W}/(\text{m}^2\cdot\text{K}^4)](10.0\times10^{-3} \text{ m}^2)}\right]^{1/4} = 312 \text{ K} - 273 \text{ K} = \boxed{39^\circ\text{C}}.$$

    (b) Strategy Since the bottom of the leaf absorbs and emits at the same rate, it can be ignored.

    Solution Find the power per unit area that must be lost by other methods.
    $$\frac{\mathcal{P}_{\text{abs,Sun}}}{A} = \frac{\mathcal{P}_{\text{rad}}}{A} + \frac{\mathcal{P}_{\text{other}}}{A} = e\sigma T^4 + \frac{\mathcal{P}_{\text{other}}}{A}, \text{ so}$$
    $$\frac{\mathcal{P}_{\text{other}}}{A} = \frac{\mathcal{P}_{\text{abs,Sun}}}{A} - e\sigma T^4 = 0.700(9.00\times10^2 \text{ W}/\text{m}^2) - (1)[5.670\times10^{-8} \text{ W}/(\text{m}^2\cdot\text{K}^4)](273.15 \text{ K} + 25.0 \text{ K})^4$$
    $$= 182 \text{ W}/\text{m}^2$$

    Thus, the power per unit area that must be lost by other methods is $\boxed{182 \text{ W}/\text{m}^2}$.

73. Strategy Use Eq. (14-4) and the relationship between power and intensity.

Solution The energy provided by the sunlight is converted to heat in the water. The energy provided is $\mathscr{P}\Delta t = IA\Delta t$. Compute the time to heat the water.

$$Q = mc\Delta T = IA\Delta t, \text{ so } \Delta t = \frac{mc\Delta T}{IA} = \frac{(1.0 \text{ L})(1000 \text{ g/L})[4.186 \text{ J}/(\text{g}\cdot\text{K})](100.0-15.0)\text{ K}}{(750 \text{ W/m}^2)(1.5 \text{ m}^2)} = \boxed{320 \text{ s}}.$$

77. (a) Strategy Use Eqs. (14-4) and (14-9).

Solution Find the heat given up by the steam.
$$Q = -mc_w\Delta T + mL_v = (4.0 \text{ g})\{-[4.186 \text{ J}/(\text{g}\cdot\text{K})](45.0-100.0)\text{ K} + 2256 \text{ J/g}\} = \boxed{9.9 \text{ kJ}}$$

(b) Strategy Use Eq. (14-4).

Solution Compute the mass of the tissue.
$$m = \frac{Q}{c\Delta T} = \frac{9945 \text{ J}}{[3.5 \text{ J}/(\text{g}\cdot\text{K})](45.0-37.0)\text{ K}} = \boxed{360 \text{ g}}$$

81. Strategy The heat loss is proportional to the temperature difference.

Solution Compute the increase in heat loss.
$$\mathscr{P}_2 = \frac{\Delta T_2}{\Delta T_1}\mathscr{P}_1 = \frac{-8.0°\text{C}-(-18°\text{C})}{22°\text{C}-(-18°\text{C})}\mathscr{P}_1 = 0.25\mathscr{P}_1$$

Therefore, $\mathscr{P}_1 = 4.0\mathscr{P}_2$; thus, the heat loss would be $\boxed{4.0 \text{ times higher}}$ without the insulation.

85. Strategy Use Fourier's law of heat conduction, Eq. (14-10).

Solution Determine the rate of heat flow.
$$\mathscr{P} = \kappa A \frac{\Delta T}{d} = [67.5 \text{ W}/(\text{m}\cdot\text{K})]\pi(0.0130 \text{ m})^2 \frac{327 \text{ K}-37 \text{ K}}{1.00 \text{ m}} = \boxed{10.4 \text{ W}}$$

89. Strategy Gravitational potential energy is converted into internal energy. Use Eq. (14-9) and $U = mgh$.

Solution Find the mass of the ice melted by friction.
$$Q = m_m L_f = 0.75U = 0.75mgh, \text{ so } m_m = \frac{0.75mgh}{L_f} = \frac{0.75(75 \text{ kg})(9.80 \text{ m/s}^2)(2.43 \text{ m})}{333,700 \text{ J/kg}} = \boxed{4.0 \text{ g}}.$$

93. Strategy $W = Fd$ (work) and $P = Fv$ (power) so $W = Pd/v$. Use Eq. (14-4).

Solution Find the distance that the cheetah can run before it overheats.
$$0.700W = Q$$
$$\frac{0.700Pd}{v} = mc\Delta T$$
$$d = \frac{vmc\Delta T}{0.700P} = \frac{(110\times10^3 \text{ m/h})\left(\frac{1 \text{ h}}{3600 \text{ s}}\right)(50.0 \text{ kg})[3500 \text{ J}/(\text{kg}\cdot°\text{C})](41.0°\text{C}-38.0°\text{C})}{0.700(160,000 \text{ W})} = \boxed{140 \text{ m}}$$

95. Strategy Heat flows from the copper block to the water and iron pot. Use Eq. (14-4).

Solution Find the final temperature of the system.

$$0 = Q_w + Q_{Cu} + Q_{Fe}$$
$$Q_w = -Q_{Cu} - Q_{Fe}$$
$$m_w c_w (T_f - T_i) = -m_{Cu} c_{Cu} (T_f - T_{Cu}) - m_{Fe} c_{Fe} (T_f - T_i)$$
$$T_f (m_w c_w + m_{Cu} c_{Cu} + m_{Fe} c_{Fe}) = m_{Cu} c_{Cu} T_{Cu} + (m_{Fe} c_{Fe} + m_w c_w) T_i$$

Solve for T_f.

$$T_f = \frac{m_{Cu} c_{Cu} T_{Cu} + (m_{Fe} c_{Fe} + m_w c_w) T_i}{m_w c_w + m_{Cu} c_{Cu} + m_{Fe} c_{Fe}}$$
$$= \frac{(2.0 \text{ kg})[385 \text{ J/(kg}\cdot\text{K)}](100.0°\text{C}) + \{(2.0 \text{ kg})[440 \text{ J/(kg}\cdot\text{K)}] + (1.0 \text{ kg})[4186 \text{ J/(kg}\cdot\text{K)}]\}(25.0°\text{C})}{(1.0 \text{ kg})[4186 \text{ J/(kg}\cdot\text{K)}] + (2.0 \text{ kg})[385 \text{ J/(kg}\cdot\text{K)}] + (2.0 \text{ kg})[440 \text{ J/(kg}\cdot\text{K)}]} = \boxed{35°\text{C}}$$

97. (a) Strategy The work done by each animal is proportional to the heat generated. Use Eq. (14-4). Form a proportion.

Solution Compare the temperature changes.

$$\frac{Q_c}{Q_d} = \frac{2.00 mc\Delta T_c}{mc\Delta T_d} = \frac{0.700}{0.0500} = 14.0, \text{ so } \Delta T_c = 7.00 \Delta T_d.$$

The temperature change of the cheetah is $\boxed{7.00 \text{ times higher}}$ than that of the dog.

(b) Strategy Use the result from part (a).

Solution Find the final temperature of the dog.

$$T_c - T_i = 7.00(T_d - T_i)$$
$$T_c + 6.00 T_i = 7.00 T_d$$
$$T_d = \frac{T_c + 6.00 T_i}{7.00} = \frac{40.0°\text{C} + 6.00(35.0°\text{C})}{7.00} = \boxed{35.7°\text{C}}$$

$\boxed{\text{The dog is a much better regulator of temperature and, as a result, has more endurance.}}$

101. Strategy Multiply the energy required to melt the urethane by the molar mass and divide by the total mass to find the latent heat.

Solution Find the latent heat of fusion of urethane.

$$\frac{(17.10 \text{ kJ})[3(12.011) + 7(1.00794) + 2(15.9994) + 14.00674] \text{ g/mol}}{1.00 \times 10^2 \text{ g}} = \boxed{15.2 \text{ kJ/mol}}$$

Chapter 15

THERMODYNAMICS

Conceptual Questions

1. Yes, but it wouldn't be a very good heat pump. Like an electric heater, the heat output would be equal to the work input, with no heat being taken from the cold reservoir.

5. Energy is always conserved, so there would be just as much energy as before when the fossil fuels are exhausted. There wouldn't be as much high quality or useful energy though, so it would be better to call it something like a "high quality energy crisis."

9. No, entropy changes don't require a flow of heat. For example, when a gas expands freely into a vacuum, its entropy increases but there is no heat flow. Beating an egg is another example of a process that increases entropy with no flow of heat.

13. This is not a violation. Although the salt crystals are in a more ordered state, the surroundings of the bucket are in a more disordered state. The gaseous state of the evaporated sea water is less ordered than the water.

Problems

1. **Strategy** The work done by Ming is equal to the magnitude of the force of friction $f = \mu N$ times the total "rubbing" distance. Use the first law of thermodynamics.

   Solution Find the change in internal energy.
   $$\Delta U = Q + W = 0 + \mu N d = 0.45(5.0 \text{ N})[8(0.16 \text{ m})] = \boxed{2.9 \text{ J}}$$

3. **Strategy** Use the first law of thermodynamics. $\Delta U > 0$ and $W > 0$.

   Solution Find the heat flow.
   $$Q = \Delta U - W = 400 \text{ J} - 500 \text{ J} = -100 \text{ J}$$
   $\boxed{100 \text{ J of heat flows out of the system.}}$

5. **Strategy** No work is done during the constant volume process, but work is done during the constant pressure process. Use Eq. (15-3).

   Solution Compute the total work done by the gas.
   $$W = P_i \Delta V = (2.000 \text{ atm})(1.013 \times 10^5 \text{ Pa/atm})(2.000 \text{ L} - 1.000 \text{ L})(10^{-3} \text{ m}^3/\text{L}) = \boxed{202.6 \text{ J}}$$

9. **(a) Strategy** Oxygen gas is diatomic. Use Eqs. (15-7) and (15-9).

Solution Find the heat absorbed by the gas.

$$Q = nC_p\Delta T = n(C_v + R)\Delta T = n\left(\frac{5}{2}R + R\right)\Delta T = \frac{7}{2}nR\Delta T = \frac{7}{2}(1.00 \text{ mol})[8.314 \text{ J/(mol}\cdot\text{K)}](25.0°C - 10.0°C)$$

$$= \boxed{436 \text{ J}}$$

(b) Strategy Use the ideal gas law.

Solution Find the change in volume of the gas.

$$\Delta V = \frac{nR\Delta T}{P} = \frac{(1.00 \text{ mol})[8.314 \text{ J/(mol}\cdot\text{K)}](25.0°C - 10.0°C)}{(1.00 \text{ atm})(1.013\times10^5 \text{ Pa/atm})}\left(10^3 \text{ } \frac{\text{L}}{\text{m}^3}\right) = \boxed{1.23 \text{ L}}$$

(c) Strategy Use Eq. (15-2).

Solution Find the work done by the gas.

$$W = P\Delta V = (1.00 \text{ atm})(1.013\times10^5 \text{ Pa/atm})(1.23\times10^{-3} \text{ m}^3) = \boxed{125 \text{ J}}$$

(d) Strategy Use the first law of thermodynamics.

Solution Calculate the change in internal energy of the gas.

$$\Delta U = Q + W = 436.5 \text{ J} - 124.7 \text{ J} = \boxed{312 \text{ J}}$$

11. **(a) Strategy** For A–C (constant temperature), $W = nRT \ln V_i/V_f$, and for C–D (constant pressure), $W = -P_i\Delta V$. Use the ideal gas law to find T.

Solution

$$W_{\text{total}} = nRT \ln \frac{V_i}{V_f} - P\Delta V = nR\left(\frac{P_A V_A}{nR}\right)\ln \frac{V_A}{V_C} - P_i\Delta V$$

$$= \left[(2.000 \text{ atm})(4.000 \text{ L})\ln \frac{4.000 \text{ L}}{8.000 \text{ L}} - (1.000 \text{ atm})(16.000 \text{ L} - 8.000 \text{ L})\right](1.013\times10^5 \text{ Pa/atm})(10^{-3} \text{ m}^3/\text{L})$$

$$= \boxed{-1372 \text{ J}}$$

(b) Strategy For constant temperature, $\Delta U = 0$. For constant pressure,

$$\Delta U = Q + W = nC_p\Delta T - P_i\Delta V = \frac{5}{2}nR\left(\frac{P_i\Delta V}{nR}\right) - P_i\Delta V = \frac{3}{2}P_i\Delta V.$$

Solution

$$\Delta U = \frac{3}{2}(1.000 \text{ atm})(16.000 \text{ L} - 8.000 \text{ L})(1.013\times10^5 \text{ Pa/atm})(10^{-3} \text{ m}^3/\text{L}) = \boxed{1216 \text{ J}}.$$

The total heat flow is $Q = \Delta U - W = 1216 \text{ J} + 1372 \text{ J} = \boxed{2588 \text{ J}}$.

13. (a) Strategy The net work done in one cycle is equal to the area inside the graph.

Solution Compute the net work done per cycle.
$$W = (4.00\ \text{atm} - 1.00\ \text{atm})(1.013 \times 10^5\ \text{Pa/atm})(0.800\ \text{m}^3 - 0.200\ \text{m}^3) = \boxed{182\ \text{kJ}}$$

(b) Strategy and Solution The net heat flow into the engine is equal to the work done per cycle, so $Q_{\text{net}} = \boxed{182\ \text{kJ}}$.

17. (a) Strategy Use Eq. (15-12).

Solution Find the net work done by the engine.
$$W_{\text{net}} = eQ_{\text{H}} = 0.21(1.00\ \text{kJ}) = \boxed{210\ \text{J}}$$

(b) Strategy The net work done by an engine during one cycle is equal to the net heat flow into the engine during the cycle.

Solution Find the heat released by the engine.
$$W_{\text{net}} = Q_{\text{H}} - Q_{\text{C}}, \text{ so } Q_{\text{C}} = Q_{\text{H}} - W_{\text{net}} = 1.00 \times 10^3\ \text{J} - 210\ \text{J} = \boxed{790\ \text{J}}.$$

21. Strategy Use Eq. (15-12). The net work done by an engine during one cycle is equal to the net heat flow into the engine during the cycle.

Solution Find the efficiency of the engine.
$$e = \frac{W_{\text{net}}}{Q_{\text{in}}} = \frac{W_{\text{net}}}{W_{\text{net}} + Q_{\text{out}}} = \frac{1}{1 + \frac{Q_{\text{out}}}{W_{\text{net}}}} = \frac{1}{1 + \frac{0.450\ \text{kJ}}{0.100\ \text{kJ}}} = \boxed{0.182}$$

25. Strategy Use Eq. (15-17).

Solution Find the temperature of the cold reservoir.
$$e_r = 1 - \frac{T_{\text{C}}}{T_{\text{H}}}, \text{ so } T_{\text{C}} = T_{\text{H}}(1 - e_r) = (622\ \text{K})(1 - 0.725) = \boxed{171\ \text{K}}.$$

27. Strategy The minimum amount of heat is discharged when the steam engine is reversible. Use Eqs. (15-12) and (15-17), and $W_{\text{net}} = Q_{\text{H}} - Q_{\text{C}}$.

Solution Compute the efficiency of a reversible engine.
$$e_r = 1 - \frac{T_{\text{C}}}{T_{\text{H}}} = 1 - \frac{273.15\ \text{K} + 27\ \text{K}}{273.15\ \text{K} + 127\ \text{K}} = 0.250$$
Compute the minimum amount of heat discharged.
$$Q_{\text{C}} = Q_{\text{H}} - W_{\text{net}} = \frac{W_{\text{net}}}{e_r} - W_{\text{net}} = W_{\text{net}}\left(\frac{1}{e_r} - 1\right) = (8.34\ \text{kJ})\left(\frac{1}{0.250} - 1\right) = \boxed{25.0\ \text{kJ}}$$

29. Strategy Assume constant rates and reversibility. Use Eq. (15-17) and conservation of energy. $P = W_{net}/\Delta t$.

Solution Compute the efficiency.
$$e_r = 1 - \frac{T_C}{T_H} = 1 - \frac{273.15 \text{ K} + 2.0 \text{ K}}{273.15 \text{ K} + 40.0 \text{ K}} = 0.1213$$
Find the power used.
$$Q_C = Q_H - W_{net} = \frac{W_{net}}{e} - W_{net} = W_{net}\left(\frac{1}{e} - 1\right), \text{ so } \frac{Q_C}{\Delta t} = \frac{W_{net}}{\Delta t}\left(\frac{1}{e} - 1\right), \text{ and } P = \frac{Q_C/\Delta t}{\frac{1}{e} - 1} = \frac{0.10 \times 10^3 \text{ W}}{\frac{1}{0.1213} - 1} = \boxed{14 \text{ W}}.$$

33. (a) Strategy Use Eq. (15-17).

Solution Find the efficiency of the reversible engine.
$$e_r = 1 - \frac{T_C}{T_H} = 1 - \frac{273.15 \text{ K} + 300.0 \text{ K}}{273.15 \text{ K} + 600.0 \text{ K}} = \boxed{0.3436}$$

(b) Strategy Use Eq. (15-18).

Solution Find the amount of heat exhausted to the cold reservoir.
$$\frac{Q_C}{Q_H} = \frac{T_C}{T_H}, \text{ so } Q_C = \frac{T_C}{T_H}Q_H = \frac{273.15 \text{ K} + 300.0 \text{ K}}{273.15 \text{ K} + 600.0 \text{ K}}(420.0 \text{ kJ}) = \boxed{275.7 \text{ kJ}}.$$

37. Strategy The maximum possible efficiency occurs if the engine is reversible. Use Eq. (15-17).

Solution Find the maximum possible efficiency.
$$e_r = 1 - \frac{T_C}{T_H} = 1 - \frac{273.15 \text{ K} + 10.0 \text{ K}}{273.15 \text{ K} + 15.0 \text{ K}} = \boxed{0.0174}$$

41. Strategy The maximum rate at which the river can carry away heat is $Q_C/\Delta t = mc\Delta T/\Delta t$. Use energy conservation and the definition of efficiency of an engine.

Solution Find the maximum possible power the plant can produce.
$$\frac{W_{net}}{\Delta t} = \frac{Q_H - Q_C}{\Delta t} = \frac{W_{net}}{e\Delta t} - \frac{mc\Delta T}{\Delta t}, \text{ so}$$
$$\frac{W_{net}}{\Delta t} = \left(\frac{1}{e} - 1\right)^{-1}\frac{m}{\Delta t}c\Delta T = \left(\frac{1}{0.300} - 1\right)^{-1}(5.0 \times 10^6 \text{ kg/s})[4186 \text{ J/(kg} \cdot \text{K)}](0.50 \text{ K}) = \boxed{4.5 \text{ GW}}.$$

45. Strategy Use $W_{net} = Q_H - Q_C$, Eq. (15-17), and the definition of efficiency of an engine.

Solution
$$Q_C = Q_H - W_{net} = \frac{W_{net}}{e_r} - W_{net} = W_{net}\left(\frac{1}{e_r} - 1\right) = W_{net}\left(\frac{1}{1 - T_C/T_H} - 1\right) = W_{net}\left(\frac{T_H}{T_H - T_C} - \frac{T_H - T_C}{T_H - T_C}\right) = \frac{T_C}{T_H - T_C}W_{net}$$

47. Strategy and Solution The mass is the same for each case. For equal masses, water has more entropy than ice, and warmer water has more entropy than cooler water, so the order is $\boxed{\text{(b), (a), (c), (d)}}$.

49. Strategy The temperature is constant and the heat entering the system is $Q = mL_v$. Use Eq. (15-20).

Solution Find the change in the entropy of the water.

$$\Delta S = \frac{Q}{T} = \frac{mL_v}{T} = \frac{(1.00 \text{ kg})(2256 \text{ kJ/kg})}{273.15 \text{ K} + 100.0 \text{ K}} = \boxed{+6.05 \text{ kJ/K}}$$

Gas is more disordered than liquid, so the entropy increases.

51. Strategy Use Eq. (15-20).

Solution

(a) Compute the change in entropy of the block.

$$\Delta S_C = \frac{Q}{T_C} = \frac{1.0 \text{ J}}{273.15 \text{ K} + 20.0 \text{ K}} = \boxed{3.4 \times 10^{-3} \text{ J/K}}$$

(b) Compute the change in entropy for the water.

$$\Delta S_H = \frac{Q}{T_H} = \frac{-1.0 \text{ J}}{273.15 \text{ K} + 80.0 \text{ K}} = \boxed{-2.8 \times 10^{-3} \text{ J/K}}$$

(c) Calculate the change in entropy of the universe.

$$\Delta S_{\text{tot}} = \Delta S_H + \Delta S_C = \frac{Q}{T_H} + \frac{Q}{T_C} = -2.8 \times 10^{-3} \text{ J/K} + 3.4 \times 10^{-3} \text{ J/K} = \boxed{6 \times 10^{-4} \text{ J/K}}$$

53. Strategy The rate at which the entropy of the universe is changing is equal to the total change in entropy per unit time. Use Eq. (15-20).

Solution

$$\frac{\Delta S}{\Delta t} = \frac{Q}{\Delta t}\left(\frac{1}{T_C} - \frac{1}{T_H}\right) = (220.0 \text{ W})\left(\frac{1}{273.15 \text{ K} - 15.0 \text{ K}} - \frac{1}{273.15 \text{ K} + 20.0 \text{ K}}\right) = \boxed{0.102 \text{ J/(K} \cdot \text{s)}}$$

57. Strategy and Solution $\boxed{\text{The engine will not work.}}$ There is no energy available to do the work necessary to extract the water's internal energy.

61. (a) Strategy The work done per cycle is equal to the area contained within the curve.

Solution Compute the work done per cycle.

$$W = \frac{1}{2}(5.00 \text{ atm} - 1.00 \text{ atm})(1.013 \times 10^5 \text{ Pa/atm})(2.00 \text{ m}^3 - 0.500 \text{ m}^3) = \boxed{304 \text{ kJ}}$$

(b) Strategy Use the ideal gas law to compute the temperatures at the upper left and lower right points on the curve.

Solution Compute the temperatures.

$$PV = nRT, \text{ so } \frac{P_2 V_2}{P_1 V_1} = \frac{T_2}{T_1}.$$

$$T_{ul} = \frac{P_2 V_2 T_1}{P_1 V_1} = \frac{(5.00)(0.500)(470.0 \text{ K})}{(1.00)(0.500)} = 2350 \text{ K}$$

$$T_{lr} = \frac{P_2 V_2 T_1}{P_1 V_1} = \frac{(1.00)(2.00)(470.0 \text{ K})}{(1.00)(0.500)} = 1880 \text{ K}$$

The maximum temperature is $\boxed{2350 \text{ K}}$.

(c) Strategy Use the ideal gas law at the lower-left corner of the diagram.

Solution Find the number of moles of gas used in the engine.

$$n = \frac{PV}{RT} = \frac{(1.00 \text{ atm})(1.013 \times 10^5 \text{ Pa/atm})(0.500 \text{ m}^3)}{[8.314 \text{ J/(mol} \cdot \text{K)}](470.0 \text{ K})} = \boxed{13.0 \text{ mol}}$$

65. Strategy The energy of the mixed state, U, must equal the sum of the original (unmixed) states, U_1 and U_2. The energy for a monatomic ideal gas is related to the temperature by $U = \frac{3}{2} NkT$.

Solution Find the final temperature T of the mixture.

$$U = U_1 + U_2$$

$$\frac{3}{2}(N_1 + N_2)kT = \frac{3}{2}N_1 kT_1 + \frac{3}{2}N_2 kT_2$$

$$\frac{3}{2}(n_1 + n_2)RT = \frac{3}{2}n_1 RT_1 + \frac{3}{2}n_2 RT_2$$

$$T = \frac{n_1 T_1 + n_2 T_2}{n_1 + n_2} = \frac{(4.0 \text{ mol})(20.0°C) + (3.0 \text{ mol})(30.0°C)}{4.0 \text{ mol} + 3.0 \text{ mol}} = \boxed{24°C}$$

69. Strategy The amount of heat that flows into the water is given by $Q = mc\Delta T$. Assuming the heat flows at an average temperature gives $T = (20.0°C + 50.0°C)/2 = 35.0°C$. Use Eq. (15-20).

Solution Estimate the entropy change.

$$\Delta S = \frac{Q}{T} = \frac{mc\Delta T}{T} \approx \frac{(0.85 \text{ kg})[4186 \text{ J/(kg} \cdot \text{K)}](50.0 - 20.0) \text{ K}}{273.15 \text{ K} + 35.0 \text{ K}} = \boxed{350 \text{ J/K}}$$

73. **Strategy** Use Eqs. (14-9) and (15-12), and $W_{net} = Q_H - Q_C$.

Solution Find the time required to freeze the water.

$$W_{net} = Q_H - Q_C, \text{ so } Q_C = Q_H - W_{net} = \frac{W_{net}}{e} - W_{net} = W_{net}\left(\frac{1}{e} - 1\right).$$

Therefore, $\dfrac{1}{e} - 1 = \dfrac{Q_C}{W_{net}} = \dfrac{\text{heat removed}}{\text{net work input}} = \dfrac{\text{rate of heat removed}}{\text{power input}} = \dfrac{\Delta Q / \Delta t}{P}$, so

$$\Delta t = \frac{\Delta Q}{P\left(\frac{1}{e} - 1\right)} = \frac{mL_f}{P\left(\frac{1}{e} - 1\right)} = \frac{(1.0 \text{ kg})(333{,}700 \text{ J/kg})}{(186 \text{ W})\left(\frac{1}{0.333} - 1\right)}\left(\frac{1 \text{ min}}{60 \text{ s}}\right) = \boxed{15 \text{ min}}.$$

77. **(a) Strategy** Use Eq. (15-17).

Solution Compute the efficiency.

$$e_r = 1 - \frac{T_C}{T_H} = 1 - \frac{273.15 \text{ K} + 22 \text{ K} - 15 \text{ K}}{273.15 \text{ K} + 22 \text{ K}} = \boxed{0.051}$$

(b) Strategy The power supplied to the town is equal to the efficiency times the rate at which heat is supplied by the lake. Use Eq. (14-4) and the relationship between mass, density, and volume.

Solution Find the volume of water used each second.

$$P = e\frac{\Delta Q}{\Delta t} = \frac{emc\Delta T}{\Delta t} = \frac{e\rho Vc\Delta T}{\Delta t}, \text{ so}$$

$$V = \frac{P\Delta t}{e\rho c\Delta T} = \frac{(1.0\times10^8 \text{ W})(1.0 \text{ s})}{0.051(1.00\times10^3 \text{ kg/m}^3)[4186 \text{ J/(kg}\cdot\text{K)}](15 \text{ K})} = \boxed{31 \text{ m}^3}.$$

(c) Strategy The incident power of the Sun must be greater than the power required to run the engine. Power is equal to intensity times area.

Solution Compare the power supplied to the power required by the town.

$$P_{Sun} = IA = (200 \text{ W/m}^2)(8.0\times10^7 \text{ m}^2) = 1.6\times10^{10} \text{ W}$$

$$P_{engine} = \frac{P_{town}}{e} = \frac{1.0\times10^8 \text{ W}}{0.051} = 2.0\times10^9 \text{ W}$$

1.6×10^{10} W $> 2.0\times10^9$ W, so $P_{sun} > P_{engine}$, and $\boxed{\text{yes}}$, the lake can supply enough heat to meet the town's needs.

REVIEW AND SYNTHESIS: CHAPTERS 13–15

Review Exercises

1. **Strategy** Assume no heat is lost to the air. The potential energy of the water is converted into heating of the water. The internal energy of the water increases by an amount equal to the initial potential energy.

   Solution Find the change in internal energy.
   $$\Delta U = mgh = (1.00 \text{ m}^3)(1.00 \times 10^3 \text{ kg/m}^3)(9.80 \text{ m/s}^2)(11.0 \text{ m}) = \boxed{108 \text{ kJ}}$$

5. **Strategy** Set the sum of the heat flows equal to zero. Use Eqs. (14-4) and (14-9).

   Solution

   (a) Find the mass of ice required.
   $$0 = Q_w + Q_{ice} = m_w c_w \Delta T_w + m_{ice} L_f + m_{ice} c_{ice} \Delta T_{ice} = m_w c_w \Delta T_w + m_{ice}(L_f + c_{ice}\Delta T_{ice}), \text{ so}$$
   $$m_{ice} = -\frac{m_w c_w \Delta T_w}{L_f + c_{ice}\Delta T_{ice}} = -\frac{(0.250 \text{ kg})[4.186 \text{ kJ/(kg·K)}](-25.0 \text{ K})}{333.7 \text{ kJ/kg} + [2.1 \text{ kJ/(kg·K)}](10.0 \text{ K})} = \boxed{74 \text{ g}}.$$

   (b) Find the final temperature of the water, T, which includes the melted ice.
   $$0 = Q_w + Q_{ice}$$
   $$0 = m_w c_w \Delta T_w + m_{ice} L_f + m_{ice} c_{ice} \Delta T_1 + m_{ice} c_w \Delta T_2$$
   $$0 = m_w c_w (T - T_w) + m_{ice} L_f + m_{ice} c_{ice} \Delta T_{ice} + m_{ice} c_w (T - 273.15 \text{ K})$$
   $$0 = (m_w + m_{ice})c_w T + m_{ice}(L_f + c_{ice}\Delta T_{ice}) - c_w[m_w T_w + m_{ice}(273.15 \text{ K})]$$
   $$T = \frac{c_w[m_w T_w + m_{ice}(273.15 \text{ K})] - m_{ice}(L_f + c_{ice}\Delta T_{ice})}{(m_w + m_{ice})c_w}$$
   $$T = \frac{\begin{array}{c}[4.186 \text{ kJ/(kg·K)}][(0.250 \text{ kg})(273.15 \text{ K} + 25.0 \text{ K}) + (0.037 \text{ kg})(273.15 \text{ K})] \\ - (0.037 \text{ kg})\{333.7 \text{ kJ/kg} + [2.1 \text{ kJ/(kg·K)}](10.0 \text{ K})\}\end{array}}{(0.250 \text{ kg} + 0.037 \text{ kg})[4.186 \text{ kJ/(kg·K)}]} - 273.15 \text{ K} = \boxed{11°\text{C}}$$

9. (a) **Strategy** The maximum power emission is inversely proportional to the absolute temperature. Use Wien's law.

   Solution Compute the surface temperature of the star.
   $$T = \frac{2.898 \times 10^{-3} \text{ m·K}}{\lambda_{max}} = \frac{2.898 \times 10^{-3} \text{ m·K}}{700.0 \times 10^{-9} \text{ m}} = \boxed{4140 \text{ K}}$$

   (b) **Strategy** Use Stefan's law of blackbody radiation.

   Solution Compute the power radiated.
   $$\mathscr{P} = \sigma A T^4 = [5.670 \times 10^{-8} \text{ W/(m}^2 \text{·K}^4)][4\pi(7.20 \times 10^8 \text{ m})^2](4140 \text{ K})^4 = \boxed{1.09 \times 10^{26} \text{ W}}$$

   (c) **Strategy** Intensity is power radiated per unit area.

   Solution Compute the intensity measured by the Earth-based observer.
   $$I = \frac{\mathscr{P}}{A} = \frac{1.085 \times 10^{26} \text{ W}}{4\pi(9.78 \text{ ly})^2(9.461 \times 10^{15} \text{ m/ly})^2} = \boxed{1.01 \times 10^{-9} \text{ W/m}^2}$$

13. Strategy The temperature is constant and the heat entering the system is $Q = mL_f$.

Solution Find the change in entropy of the ice.
$$\Delta S = \frac{Q}{T} = \frac{mL_f}{T} = \frac{(2.00 \text{ kg})(333.7 \text{ kJ/kg})}{273.15 \text{ K} + 0.0 \text{ K}} = \boxed{2.44 \text{ kJ/K}}$$

17. Strategy The heat loss is proportional to the temperature difference.

Solution Compute the reduction in heat loss.
$$\mathcal{P}_2 = \frac{\Delta T_2}{\Delta T_1} \mathcal{P}_1 = \frac{81°C - 36°C}{81°C - 21°C} \mathcal{P}_1 = 0.75 \mathcal{P}_1$$

The heat loss was $\boxed{\text{reduced to 75\% of the original}}$.

21. (a) Strategy and Solution

The boiling temperature of water varies with pressure. If the pressure is high, the water molecules are pushed close together, making it harder for them to form a gas. (Gas molecules are farther apart from each other than are liquid molecules.) A higher pressure raises the temperature at which the coolant fluid will boil.

(b) Strategy and Solution

If you were to remove the cap on your radiator without first bringing the radiator pressure down to atmospheric pressure, the fluid would suddenly boil, sending out a jet of hot steam that could burn you.

25. (a) Strategy Use the ideal gas law, $PV = nRT$. Draw a qualitative diagram.

Solution First, the temperature is constant, so $P \propto V^{-1}$. Since the volume is reduced to one-eighth of its initial size, the pressure increases by a factor of eight. Next, the volume is constant, while the temperature and pressure increases. Then, the temperature is again constant. Finally, the volume is constant as the temperature and pressure decreases. The P-V diagram is shown.

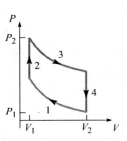

(b) Strategy Refer to the diagram in part (a). Calculate the quantities for each step in the cycle. Note that the gas is diatomic.

Solution Step 1, isothermal process:
The work done on the gas is $W = nRT \ln \dfrac{V_i}{V_f} = (2.00 \text{ mol})[8.314 \text{ J/(mol·K)}](325 \text{ K}) \ln 8 = 11.2 \text{ kJ}$.

The change in the internal energy of the gas is 0 for an isothermal process.
The heat transferred is $Q = -W = -11.2 \text{ kJ}$.

Step 2, isochoric process:
Without a displacement, work cannot be done, so $W = 0$.
The change in the internal energy of the gas is equal to the heat that enters the system, so the change in internal energy and the heat transferred are
$$\Delta U = Q = nC_v \Delta T = n \left(\frac{5}{2} R \right) \Delta T = \frac{5}{2}(2.00 \text{ mol})[8.314 \text{ J/(mol·K)}](985 \text{ K} - 325 \text{ K}) = 27.4 \text{ kJ}.$$

Step 3, isothermal process:

The work done on the gas is $W = nRT \ln \dfrac{V_i}{V_f} = (2.00 \text{ mol})[8.314 \text{ J/(mol·K)}](985 \text{ K}) \ln \dfrac{1}{8} = -34.1 \text{ kJ}$.

The change in the internal energy of the gas is 0 for an isothermal process.
The heat transferred is $Q = -W = 34.1 \text{ kJ}$.

Step 4, isochoric process:
Without a displacement, work cannot be done, so $W = 0$.
The change in the internal energy of the gas is equal to the heat that enters the system, so the change in internal energy and the heat transferred are

$$\Delta U = Q = nC_v \Delta T = n\left(\frac{5}{2}R\right)\Delta T = \frac{5}{2}(2.00 \text{ mol})[8.314 \text{ J/(mol·K)}](325 \text{ K} - 985 \text{ K}) = -27.4 \text{ kJ}.$$

The results of the processes and the totals are shown in the table. (Note that the totals for work and heat differ slightly from the sums of the values for each step due to round-off error.)

Process	W (kJ)	ΔU (kJ)	Q (kJ)
Step 1	11.2	0	−11.2
Step 2	0	27.4	27.4
Step 3	−34.1	0	34.1
Step 4	0	−27.4	−27.4
Total	−22.8	0	22.8

(c) **Strategy** The efficiency is equal to the ratio of the net work done by the gas to the heat transferred into the gas.

Solution The work done by the gas is negative the work done on the gas.
$W_{net} = -(-22.8 \text{ kJ}) = 22.8 \text{ kJ}$ and the heat transferred into the gas is $Q_{in} = 27.4 \text{ kJ} + 34.1 \text{ kJ} = 61.5 \text{ kJ}$.

The efficiency of the engine is $e = \dfrac{W_{net}}{Q_{in}} = \dfrac{22.8 \text{ kJ}}{61.5 \text{ kJ}} = \boxed{0.371 \text{ or } 37.1\%}$.

(d) **Strategy** Use Eq. (15-17).

Solution Compute the efficiency of a Carnot engine operating at the same extreme temperatures.

$$e_r = 1 - \frac{T_C}{T_H} = 1 - \frac{325 \text{ K}}{985 \text{ K}} = \boxed{0.670 \text{ or } 67.0\%}$$

29. (a) Strategy Use conservation of energy. Ignore air resistance.

Solution Find the escape speed.
$$K_i + U_i = K_f + U_f$$
$$\frac{1}{2}mv^2 - \frac{GMm}{R_E} = 0 + 0$$
$$v = \sqrt{\frac{2GM}{R_E}} = \sqrt{\frac{2(6.674\times10^{-11} \text{ N}\cdot\text{m}^2/\text{kg}^2)(5.974\times10^{24} \text{ kg})}{6.37\times10^6 \text{ m}}} = \boxed{11.2 \text{ km/s}}$$

(b) Strategy Use Eq. (13-22).

Solution Calculate the average speed.
$$v = \sqrt{\frac{3kT}{m}} = \sqrt{\frac{3(1.381\times10^{-23} \text{ J/K})(273.15 \text{ K})}{(2.00 \text{ u})(1.6605\times10^{-27} \text{ kg/u})]}} = \boxed{1850 \text{ m/s}}$$

(b) Strategy Use Eq. (13-22).

Solution Calculate the average speed.
$$v = \sqrt{\frac{3kT}{m}} = \sqrt{\frac{3(1.381\times10^{-23} \text{ J/K})(273.15 \text{ K})}{(32.0 \text{ u})(1.6605\times10^{-27} \text{ kg/u})]}} = \boxed{461 \text{ m/s}}$$

(d) Strategy and Solution

The atoms in the high end of the distribution are much faster than the average. Some of the hydrogen atoms have speeds greater than the escape speed, thus they can escape. This is not the case for oxygen, which is much more massive and, thus, much slower.

MCAT Review

1. Strategy and Solution According to the second law of thermodynamics, heat never flows spontaneously from a colder body to a hotter body, therefore, heat will not flow from bar A to bar B. The correct answer is $\boxed{C}$.

2. Strategy Assume that the specific heat capacity of seawater is approximately the same at 0°C and 5°C.

Solution Find the approximate temperature T.
$0 = Q_0 + Q_5 = mc(T - 0°C) + mc(T - 5°C)$, so $2T = 5°C$ or $T = 2.50°C$.
The correct answer is $\boxed{B}$.

3. Strategy Use the latent heat of fusion for water.

Solution The heat gained by the ice when melting is $Q = mL_f = (0.0180 \text{ kg})(333.7 \text{ kJ/kg}) = 6.01 \text{ kJ}$.
The correct answer is $\boxed{C}$.

4. Strategy and Solution Since $e = 1 - Q_C/Q_H = 1 - T_C/T_H$, decreasing the exhaust temperature will increase the steam engine's efficiency. The correct answer is $\boxed{B}$.

5. **Strategy and Solution** Since refrigerators remove heat by transferring it to a liquid that vaporizes, refrigerators are primarily dependent upon the heat of vaporization of the refrigerant liquid. The correct answer is $\boxed{\text{A}}$.

6. **Strategy and Solution** Steam is generally at a higher temperature than water and the specific heat of steam is lower than that of water, so water would be more effective than steam for changing steam to water. Circulating water brings more mass of water in contact with the condenser than stationary water, so it can carry away heat at a faster rate, therefore, it would be more effective for changing steam to water. The correct answer is $\boxed{\text{D}}$.

7. **Strategy and Solution** Since it is not possible to convert all of the input heat into output work, the amount of useful work that can be generated from a source of heat can only be less than the amount of heat. The correct answer is $\boxed{\text{A}}$.

8. **Strategy and Solution** The internal energy of the steam is converted into mechanical energy as it expands and moves the piston of the steam engine to the right, therefore, the correct answer is $\boxed{\text{C}}$.

9. **Strategy and Solution** The refrigerant must be able to vaporize (boil) at temperatures lower than the freezing point of water so that it can carry away heat (as a gas) from the contents of the refrigerator (which contain water) to cool and possibly freeze the contents. The correct answer is $\boxed{\text{B}}$.

10. **Strategy** The heat transferred to the water by the heaters was $Q_w = m_w c_w \Delta T_w$. The heat required for the oil is $Q_o = m_o c_o \Delta T_o$.

    Solution Form a proportion and use the temperature changes of the oil and water and the specific heat and the specific gravity of the oil to obtain a ratio of heat required for the oil to that transferred to the water.

    $$\frac{Q_o}{Q_w} = \frac{m_o c_o \Delta T_o}{m_w c_w \Delta T_w} = \frac{(0.7 m_w)(0.60 c_w)(60 - 20)}{m_w c_w (100 - 20)} = 0.21$$

    So, 21% of the amount of heat transferred to the water is required to heat the oil to 60°C. Assuming the heaters work at the same rate for both the water and the oil, the time required to raise the temperature of the oil from 20°C to 60°C is $0.21(15\text{ h}) = 3.2\text{ h}$. The correct answer is $\boxed{\text{A}}$.

11. **Strategy and Solution** The high pressure would increase the pressure on the plug, making it more difficult to lift. The pressure difference between the air in the tank and the air outside of the tank would increase the fluid velocity when the tank is drained, thus, decreasing the time required to drain the tank. The time required to heat the oil would be the least likely affected, since the oil is fairly incompressible. The correct answer is $\boxed{\text{A}}$.

Chapter 16

ELECTRIC FORCES AND FIELDS

Conceptual Questions

1. This proposition would not work because even with small net charges some objects would be observed to repel each other, which does not happen with gravity. To account for the weight of an object one may say, for example, that the Earth is slightly positively charged and the object negatively charged. A slightly positively charged object should then be repelled from the Earth and fall upward. Furthermore, increasing the charge on an object would increase the force from the Earth, but the weight is not observed to change by increasing an object's charge.

5. (a) Since the sphere is positively charged, it must have lost electrons to the charged rod, so its mass will be smaller. We know that electrons were transferred from the sphere to the rod, and that positive ions were not transferred from the rod to the sphere, because electrons in a metal are much more mobile than ions in any solid.

   (b) The rod must have been positively charged since it attracted electrons off of the sphere. Also, after the two objects are touched they must be at the same potential. Since the sphere ends up positively charged, the rod must end up positively charged as well.

9. No, the net charge—sum of all the charges with signs—is the same before and after.

13. (a) The foils are positively charged and will repel each other even after the rod is removed.

    (b) As another positively charged rod is brought close to the conducting sphere, the foils become further positively charged by induction. They will move farther apart, but will return to their previous position if the rod is removed.

    (c) If a negatively charged rod is now brought near the sphere, the foils will become less positively charged and will move closer together.

17. Electric flux is the total "amount" of electric field "flowing" through a surface. It is analogous to the volume flow rate of water through a pipe. For water, the product of the perpendicular velocity and the surface area gives the volume of water flowing through the surface per unit time. Electric flux originates from positive charges just as the flow of water originates from a faucet. Both are therefore given the name "sources". Electric flux flows toward negative charges just as water flows toward drains. Both are therefore given the name "sinks".

Problems

1. **Strategy** There are 10 protons in each water molecule. Multiply the elementary charge by Avogadro's number and the number of protons per molecule.

   Solution Find the total positive charge.
   $$10(1.0 \text{ mol})(6.022 \times 10^{23} \text{ mol}^{-1})(1.602 \times 10^{-19} \text{ C}) = \boxed{9.6 \times 10^5 \text{ C}}$$

3. **(a) Strategy and Solution** Since electrons have negative charge, and since the balloon acquired a negative net charge, electrons were added to the balloon.

    (b) Strategy Divide the net charge by the charge of an electron.

    Solution Compute the number of electrons transferred.
    $$\frac{-0.60\times10^{-9}\ \text{C}}{-1.602\times10^{-19}\ \text{C}} = \boxed{3.7\times10^{9}}$$

5. **Strategy and Solution**

    (a) When the rod is brought near sphere A, negative charge flows from sphere B to sphere A. The spheres are then moved apart and the rod is removed, so A is left with a net $\boxed{\text{negative charge}}$.

    (b) Sphere B has $\boxed{\text{an equal magnitude of positive charge}}$, since the two spheres were initially uncharged.

7. **Strategy** Each time a pair of spheres makes contact, their net charge is shared equally, with the exception of the time when C is grounded.

    Solution After A and B make contact and are separated, each sphere has a charge of $Q/2$. After B and C make contact and are separated, each sphere has zero charge because sphere C was grounded. After A and C make contact and are separated, each sphere has a charge of $(Q/2+0)/2 = Q/4$. The charges on spheres A and C are $\boxed{Q/4}$. The charge on sphere B is $\boxed{0}$.

9. **Strategy** Use Coulomb's law, Eq. (16-2).

    Solution Find the distance between the charges.
    $$F = \frac{k|q_1||q_2|}{r^2}, \text{ so } r = \sqrt{\frac{k|q_1||q_2|}{F}} = \sqrt{\frac{(8.988\times10^{9}\ \text{N}\cdot\text{m}^2/\text{C}^2)(1\ \text{C})^2}{10\ \text{N}}} = \boxed{30\ \text{km}}.$$

11. **Strategy** Divide the magnitude of the Coulomb force by the magnitude of the gravitational force.

    Solution Compute the ratio.
    $$\frac{F_q}{F_g} = \frac{\frac{kq^2}{r^2}}{\frac{Gm_p m_e}{r^2}} = \frac{kq^2}{Gm_p m_e} = \frac{(8.988\times10^{9}\ \text{N}\cdot\text{m}^2/\text{C}^2)(1.602\times10^{-19}\ \text{C})^2}{(6.674\times10^{-11}\ \text{N}\cdot\text{m}^2/\text{kg}^2)(1.673\times10^{-27}\ \text{kg})(9.109\times10^{-31}\ \text{kg})} = \boxed{2.268\times10^{39}}$$

13. **Strategy** The force is attractive. Use Coulomb's law, Eq. (16-2).

    Solution

    (a) Find the electric force on the positive charge.
    $$F = -\frac{k|q_1||q_2|}{r^2} = -\frac{(8.988\times10^{9}\ \text{N}\cdot\text{m}^2/\text{C}^2)(2.0\times10^{-9}\ \text{C})(3.0\times10^{-9}\ \text{C})}{(0.030\ \text{m})^2} = -6.0\times10^{-5}\ \text{N}$$
    So, $\vec{F} = \boxed{6.0\times10^{-5}\ \text{N toward the }-3.0\text{-nC charge}}$.

    (b) The force is equal in magnitude and opposite in direction to that found in part (a).
    So, $\vec{F} = \boxed{6.0\times10^{-5}\ \text{N toward the 2.0-nC charge}}$.

17. **Strategy** The force is attractive. Use Coulomb's law, Eq. (16-2).

    Solution Find the electric force on the potassium ion.
    $$F = -\frac{k|q_1||q_2|}{r^2} = -\frac{ke^2}{r^2} = -\frac{(8.988\times10^9 \ \text{N}\cdot\text{m}^2/\text{C}^2)(1.602\times10^{-19} \ \text{C})^2}{(9.0\times10^{-9} \ \text{m})^2} = -2.8\times10^{-12} \ \text{N}$$
    So, $\vec{F} = \boxed{2.8\times10^{-12} \ \text{N toward the Cl}^- \text{ ion}}$.

19. **Strategy** Use Coulomb's law, Eq. (16-2). The force on the 1.0-μC charge due to the −0.60-μC charge is to the left and that due to the 0.80-μC charge is along the line between the charges and away from the 0.80-μC charge.

    Solution Calculate the components of the force.

    $$F_x = -\frac{(8.988\times10^9 \ \text{N}\cdot\text{m}^2/\text{C}^2)(0.60\times10^{-6} \ \text{C})(1.0\times10^{-6} \ \text{C})}{\left(\sqrt{(0.100 \ \text{m})^2 - (0.080 \ \text{m})^2}\right)^2}$$
    $$+ \frac{(8.988\times10^9 \ \text{N}\cdot\text{m}^2/\text{C}^2)(0.80\times10^{-6} \ \text{C})(1.0\times10^{-6} \ \text{C})}{(0.100 \ \text{m})^2}\left(\frac{\sqrt{(0.100 \ \text{m})^2 - (0.080 \ \text{m})^2}}{0.100 \ \text{m}}\right) = -1.1 \ \text{N}$$
    $$F_y = -\frac{(8.988\times10^9 \ \text{N}\cdot\text{m}^2/\text{C}^2)(0.80\times10^{-6} \ \text{C})(1.0\times10^{-6} \ \text{C})}{(0.100 \ \text{m})^2}\left(\frac{0.080 \ \text{m}}{0.100 \ \text{m}}\right) = -0.58 \ \text{N}$$
    Calculate the magnitude of the force.
    $$F = \sqrt{F_x^2 + F_y^2} = \sqrt{(-1.067 \ \text{m})^2 + (-0.575 \ \text{m})^2} = 1.2 \ \text{N}$$
    Calculate the direction.
    $$\theta = \tan^{-1}\frac{F_y}{F_x} = \tan^{-1}\frac{-0.575}{-1.067} = 28°$$
    So, $\vec{F} = \boxed{1.2 \ \text{N at } 28° \text{ below the negative } x\text{-axis}}$.

21. **Strategy** The force is repulsive, so the charges have the same sign. Since we are concerned only with the magnitude of the charge on each sphere, we assume they are both positive for simplicity. Use Coulomb's law, Eq. (16-2).

    Solution Let the total charge be Q. Then, $Q = q_1 + q_2$. Find the charge on each sphere.
    $$F = \frac{kq_1q_2}{r^2} = \frac{k(Q-q_2)q_2}{r^2} = \frac{kQq_2 - kq_2^2}{r^2}, \text{ so } q_2^2 - Qq_2 + \frac{r^2F}{k} = 0. \text{ Solve for } q_2.$$
    $$q_2 = \frac{Q \pm \sqrt{Q^2 - \frac{4r^2F}{k}}}{2} = \frac{7.50\times10^{-6} \ \text{C} \pm \sqrt{(7.50\times10^{-6} \ \text{C})^2 - \frac{4(0.0600 \ \text{m})^2(20.0 \ \text{N})}{8.988\times10^9 \ \text{N}\cdot\text{m}^2/\text{C}^2}}}{2} = 6.21\times10^{-6} \ \text{C or } 1.29\times10^{-6} \ \text{C}$$
    Note that $7.50\times10^{-6} \ \text{C} - 6.21\times10^{-6} \ \text{C} = 1.29\times10^{-6} \ \text{C}$.
    Thus, the magnitudes of the charges are $\boxed{6.21 \ \mu\text{C and } 1.29 \ \mu\text{C}}$.

25. **Strategy** Use Eq. (16-4b).

    Solution Compute the force on the sphere.
    $$\vec{F} = q\vec{E} = (-6.0\times10^{-7} \ \text{C})(1.2\times10^6 \ \text{N/C west}) = \boxed{0.72 \ \text{N to the east}}$$

27. **Strategy** Use Newton's second law and Eq. (16-4b).

Solution $\vec{F} = q\vec{E} = e\vec{E}$ for a proton. Find the acceleration.

$$m\vec{a} = e\vec{E}, \text{ so } \vec{a} = \frac{e\vec{E}}{m} = \frac{(1.602\times10^{-19}\ \text{C})(33\times10^3\ \text{N/C up})}{1.673\times10^{-27}\ \text{kg}} = \boxed{3.2\times10^{12}\ \text{m/s}^2\ \text{up}}.$$

$\vec{E}\uparrow \quad \oplus_p \quad \uparrow\vec{a}$

29. **Strategy** The electric field at the midpoint is directed away from the positive charge and toward the negative charge. The magnitude of the field is the sum of the magnitudes of the fields due to each charge. Use Eq. (16-5).

    Solution Find the electric field midway between the two charges.

$$E = \frac{k|Q_1|}{(d/2)^2} + \frac{k|Q_2|}{(d/2)^2} = \frac{4k}{d^2}(|Q_1|+|Q_2|) = \frac{4(8.988\times10^9\ \text{N}\cdot\text{m}^2/\text{C}^2)}{(0.080\ \text{m})^2}(15\times10^{-6}\ \text{C}+12\times10^{-6}\ \text{C}) = 1.5\times10^8\ \text{N/C}$$

So, $\vec{E} = \boxed{1.5\times10^8\ \text{N/C directed toward the } -15\text{-}\mu\text{C charge}}$.

33. **Strategy** The electric field at $x = 2d$ is the vector sum of the electric fields due to each positive charge. Use Eq. (16-5).

    Solution Find the electric field.

$$E_x = \frac{k|q|}{(2d)^2} - \frac{k|2q|}{d^2} = \frac{k|q|}{4d^2} - \frac{2k|q|}{d^2} = -\frac{7k|q|}{4d^2}$$

The electric field is $\boxed{\dfrac{7k|q|}{4d^2} \text{ to the left}}$.

35. **Strategy and Solution** The electric field due to each charge is directed to the left for $x < 0$, and to the right for $x > 3d$; therefore, the electric field cannot be zero in these regions. In the region between the charges (on the x-axis), the electric fields due to the charges are in opposite directions; thus, the electric field is zero in the region $\boxed{0 < x < 3d}$.

37. **Strategy** Electric field lines begin on positive charges and end on negative charges. The magnitude of the negative charge is twice that of the positive charges (which have equal magnitude). The same number of field lines begins on each of the positive charges and all end on the negative charge. Field lines never cross. Use the principles of superposition and symmetry.

    Solution The electric field lines for the system of three charges:

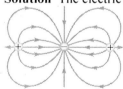

39. **Strategy** Let the x-direction be to the right and the y-direction be up. Due to symmetry, the x-components of the fields due to the two charges add to zero at point B; the vector sum of the y-components of the two charges is equal to twice that due to either one.

    Solution Find the electric field at point B.

$$E = 2E_y = 2\frac{k|q|}{r^2} = \frac{2(8.988\times10^9\ \text{N}\cdot\text{m}^2/\text{C}^2)(7.00\times10^{-6}\ \text{C})}{(0.300/2\ \text{m})^2 + (0.300\ \text{m})^2} = 1.12\times10^6\ \text{N/C}$$

The electric field is $\boxed{1.12\times10^6\ \text{N/C up}}$.

41. Strategy Let the x-direction be to the right and the y-direction be up. Label the charge on the left 1 and the charge on the right 2. Let d equal the side length of the square. The electric field due to the charge on the left is directed upward at point A, so it only has a y-component. Use Eq. (16-5).

Solution Find the electric field at point A.

$$E_x = E_{1x} + E_{2x} = 0 - \frac{k|q|}{2d^2}\cos 45° = \frac{\sqrt{2}k|q|}{4d^2}$$

$$E_y = E_{1y} + E_{2y} = \frac{k|q|}{d^2} + \frac{k|q|}{2d^2}\sin 45° = \frac{k|q|}{d^2} + \frac{\sqrt{2}k|q|}{4d^2} = \frac{k|q|(4+\sqrt{2})}{4d^2}$$

Compute the magnitude.

$$E = \sqrt{\left(\frac{\sqrt{2}k|q|}{4d^2}\right)^2 + \left[\frac{k|q|(4+\sqrt{2})}{4d^2}\right]^2} = \frac{k|q|}{4d^2}\sqrt{2+(4+\sqrt{2})^2}$$

$$= \frac{(8.988\times10^9 \text{ N}\cdot\text{m}^2/\text{C}^2)(7.00\times10^{-6}\text{ C})\sqrt{2+(4+\sqrt{2})^2}}{4(0.300 \text{ m})^2} = 9.78\times10^5 \text{ N/C}$$

Compute the direction.

$$\theta = \tan^{-1}\frac{E_y}{E_x} = \tan^{-1}\frac{4+\sqrt{2}}{\sqrt{2}} = 14.6° \text{ CCW from a vertical axis through the left side of the square}$$

The electric field is $\boxed{9.78\times10^5 \text{ N/C at } 14.6° \text{ CCW from a vertical axis through the left side of the square}}$.

45. Strategy $E_x = 0$ due to symmetry. If r is the distance to $x = 4.0$ m, $y = 3.0$ m from the two known charges, then $\sin\theta = y/r$. The field due to the known charges is upward, so that due to the unknown charge Q must be downward, thus $Q < 0$. Use Eq. (16-5) and the principle of superposition.

Solution Find the unknown charge.

$$E_y = \frac{2kq}{r^2}\left(\frac{y}{r}\right) - \frac{k|Q|}{y^2} = 0, \text{ so}$$

$$|Q| = \frac{2qy^3}{r^3} = \frac{2q(3.0 \text{ m})^3}{\left(\sqrt{(4.0 \text{ m})^2+(3.0 \text{ m})^2}\right)^3} = 0.43q.$$

$Q < 0$, so $Q = \boxed{-0.43q}$.

47. Strategy Use Eq. (16-5) and the principles of superposition and symmetry. If r is the distance to $(x, y) = (0.50$ m, 0.50 m$)$ from each charge, then $\cos\theta = x/r$ and $\sin\theta = y/r$.

Solution Find the magnitude of the electric field.

$$E = \sqrt{E_x^2 + E_y^2} = \sqrt{\left[\frac{kq_1}{r^2}\left(\frac{x}{r}\right) - \frac{kq_2}{r^2}\left(\frac{x}{r}\right)\right]^2 + \left[\frac{kq_1}{r^2}\left(\frac{y}{r}\right) + \frac{kq_2}{r^2}\left(\frac{y}{r}\right)\right]^2}$$

$$= \frac{k}{r^3}\sqrt{[(q_1-q_2)x]^2 + [(q_1+q_2)y]^2}$$

$$= \frac{8.988\times10^9 \text{ N}\cdot\text{m}^2/\text{C}^2}{\left(\sqrt{(0.50 \text{ m})^2+(0.50 \text{ m})^2}\right)^3}\sqrt{\begin{array}{c}[(20.0\times10^{-9}\text{ C}-10.0\times10^{-9}\text{ C})(0.50 \text{ m})]^2 \\ +[(20.0\times10^{-9}\text{ C}+10.0\times10^{-9}\text{ C})(0.50 \text{ m})]^2\end{array}} = \boxed{400 \text{ N/C}}$$

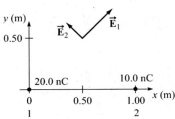

49. (a) **Strategy** Use Eq. (16-4b).

    Solution Find the force on the electron.

    $$\vec{F} = -e\vec{E} = -(1.602\times10^{-19} \text{ C})(500.0 \text{ N/C up}) = \boxed{8.010\times10^{-17} \text{ N down}}$$

    (b) **Strategy** Use the work-kinetic energy theorem. The work done on the electron is equal to the force on the electron times the deflection.

    Solution Find the increase in the kinetic energy of the electron.

    $$\Delta K = W = Fd = (8.010\times10^{-17} \text{ N})(0.00300 \text{ m}) = \boxed{2.40\times10^{-19} \text{ J}}$$

51. (a) **Strategy** Compare the electrical and gravitational forces.

    Solution The gravitational force is $mg = (0.00230 \text{ kg})(9.80 \text{ m/s}^2) = 2.25\times10^{-2}$ N. The electrical force is

    $$qE = (10.0\times10^{-6} \text{ C})(6.50\times10^3 \text{ N/C}) = 6.50\times10^{-2} \text{ N}.$$

    $$\boxed{\text{The gravitational force is about 1/3 of the electrical force, so the gravitational force can't be neglected.}}$$

    (b) **Strategy** Add the forces and find the total acceleration using Newton's second law. Then, use the formula for the range of a projectile.

    Solution Find the downward acceleration.

    $$a = \frac{F_g + F_e}{m}$$

    Find Δx.

    $$\Delta x = R = \frac{v_i^2 \sin 2\theta}{a} = \frac{mv_i^2 \sin 2\theta}{F_g + F_e} = \frac{(0.00230 \text{ kg})(8.50 \text{ m/s})^2 \sin[2(55.0°)]}{2.25\times10^{-2} \text{ N} + 6.50\times10^{-2} \text{ N}} = \boxed{1.78 \text{ m}}$$

53. **Strategy** Find the necessary acceleration in terms of the average electric field.

    Solution The acceleration is related to the average electric field by $a = qE/m$.

    $$v_{fx}^2 - v_{ix}^2 = v_{fx}^2 - 0 = 2a_x\Delta x = \frac{2qE\Delta x}{m}, \text{ so } E = \frac{mv_{fx}^2}{2q\Delta x} = \frac{(1.673\times10^{-27} \text{ kg})(1.0\times10^7 \text{ m/s})^2}{2(1.602\times10^{-19} \text{ C})(4.0 \text{ m})} = \boxed{1.3\times10^5 \text{ N/C}}.$$

57. **Strategy** The charge on the inner surface is induced by the net charge contained within the shell. The charge on the outer surface is equal in magnitude and opposite in sign to the charge on the inner surface plus the net charge.

    Solution

    (a) The 6 μC of charge within the shell induces a $\boxed{-6 \text{ μC}}$ charge on the inner surface of the shell.

    (b) The shell has a net charge of 6 μC, so the charge on the outer surface is 6 μC + 6 μC = $\boxed{12 \text{ μC}}$.

61. (a) **Strategy** Since the electric field points toward Earth, the charge is negative. Just outside of a conducting sphere, the field is nearly uniform since the curved surface is approximately flat for a small area A. Thus, the expression for the electric field between two oppositely-charged plates can be used.

    Solution Calculate the total charge.

$E = \dfrac{Q}{\epsilon_0 A}$, so $Q = \epsilon_0 AE = [8.854 \times 10^{-12} \text{ C}^2/(\text{N} \cdot \text{m}^2)]4\pi(6.371 \times 10^6 \text{ m})^2(-150 \text{ N/C}) = \boxed{-6.8 \times 10^5 \text{ C}}$.

Calculate the charge per unit area.

$\dfrac{Q}{A} = \epsilon_0 E = [8.854 \times 10^{-12} \text{ C}^2/(\text{N} \cdot \text{m}^2)](-150 \text{ N/C}) = \boxed{-1.3 \text{ nC/m}^2}$

(b) Strategy Since the Earth has negative charge, the charge density of the air must be positive.

Solution Calculate the charge density of the air.

$\Delta E = \dfrac{\Delta Q}{\epsilon_0 A} = \dfrac{\Delta Q}{\epsilon_0}\left(\dfrac{h}{V}\right) = \dfrac{1}{\epsilon_0}\dfrac{\Delta Q}{V}h$, so

$\dfrac{\Delta Q}{V} = \dfrac{\epsilon_0 \Delta E}{h} = \dfrac{[8.854 \times 10^{-12} \text{ C}^2/(\text{N} \cdot \text{m}^2)](150 \text{ N/C} - 120 \text{ N/C})}{250 \text{ m}} = \boxed{1 \times 10^{-12} \text{ C/m}^3}$.

65. Strategy Use the definition of electric flux, Eq. (16-8), Gauss's law, Eq. (16-9), and Coulomb's law.

Solution

(a) The expression for the electric flux is $\Phi_E = E_\perp A = E(4\pi r^2) = \boxed{4\pi r^2 E}$.

(b) Use Gauss's law.

$\Phi_E = cq = 4\pi r^2 E$, so $E = \dfrac{cq}{4\pi r^2}$, and by Coulomb's law, $E = \dfrac{q}{4\pi \epsilon_0 r^2}$.

Solve for c.

$\dfrac{cq}{4\pi r^2} = \dfrac{q}{4\pi \epsilon_0 r^2}$, so $c = \dfrac{1}{\epsilon_0}$.

69. Strategy Use the properties of electric fields and the rules for sketching field lines, and Gauss's law, Eq. (16-9).

Solution

(a) The electric field lines due to the (finite) sheet:

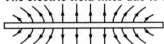

(b) The electric field lines due to an infinitely large sheet:

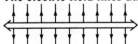

(c) The electric field lines in (b) are uniform, so

the field strength is independent of the distance from the sheet.

(d) The electric field lines in (a) are nearly uniform close to the sheet and far from the edges, so the answer is

yes.

(e) The Gaussian surface is a "pill box." It is a cylinder with its top and bottom circular surfaces parallel to the surface of the sheet, which bisects the cylinder. The electric field lines are approximately parallel to the side of the cylinder, so $\Phi_{E \text{ side}} = E_\perp A_{\text{side}} = 0$, or $E_\perp = 0$.

$$\Phi_{E\ net} = E_{top} A_{top} + E_{bottom} A_{bottom} = \frac{q}{\epsilon_0}$$

$\vec{E}_{top} = -\vec{E}_{bottom}$, and the outward normal of A_{top} is opposite to that for A_{bottom} and the areas are equal. Find the magnitude.

$$EA + (-E)(-A) = 2EA = \frac{q}{\epsilon_0}, \text{ so } E = \frac{1}{2\epsilon_0}\left(\frac{q}{A}\right) = \frac{\sigma}{2\epsilon_0}.$$

73. **Strategy** Use the definition of electric flux, Eq. (16-8), and Gauss's law, Eq. (16-9). The appropriate closed surface for the shell is a sphere. Due to symmetry, the electric field lines must be normal to the closed surface. Since the charge is positive, the electric field is directed radially away from the center of the spherical shell.

Solution Find the electric field outside of the shell.

$$\Phi_E = 4\pi kq = EA\cos\theta = EA = E(4\pi r^2), \text{ so } kq = Er^2 \text{ and } E = \frac{kq}{r^2}.$$

The electric field is $\boxed{\dfrac{kq}{r^2}\ (r > R) \text{ directed radially away from the center of the shell}}$.

77. **Strategy** Use Eq. (16-4b).

Solution Find the force on each electron.

$$\vec{F} = q\vec{E} = -e\vec{E} = -(1.602\times10^{-19}\ \text{C})(2.00\times10^5\ \text{N/C downward}) = \boxed{3.20\times10^{-14}\ \text{N upward}}$$

81. **Strategy** Use Coulomb's law and the principle of superposition.

Solution Find q.

$$0 = E_x(1.0\ \text{m}, 0) = \frac{kq_0}{d^2} + \frac{kq}{(d/2)^2} = q_0 + 4q, \text{ so}$$

$$q = -\frac{1}{4}q_0 = -\frac{1}{4}(6.0\ \text{nC}) = \boxed{-1.5\ \text{nC}}.$$

85. **Strategy** The net electric field at P is the vector sum of the electric fields at that location due to both of the charges. Let the left-hand charge by 1 and the right-hand charge be 2. Also, let $a = 0.0340$ m and $b = 0.0140$ m. Use Eq. (16-5) and the principle of superposition.

Solution Find the components of the electric field.

$$\begin{aligned} E_x &= E_{1x} + E_{2x} = 0 + \frac{k|q_2|}{a^2+b^2}\left(\frac{a}{\sqrt{a^2+b^2}}\right) = \frac{k|q_2|a}{(a^2+b^2)^{3/2}} \\ &= \frac{(8.988\times10^9\ \text{N}\cdot\text{m}^2/\text{C}^2)(47.0\times10^{-9}\ \text{C})(0.0340\ \text{m})}{[(0.0340\ \text{m})^2+(0.0140\ \text{m})^2]^{3/2}} \\ &= \boxed{2.89\times10^5\ \text{N/C}} \end{aligned}$$

$$\begin{aligned} E_y &= E_{1y} + E_{2y} = \frac{k|q_1|}{b^2} - \frac{k|q_2|}{a^2+b^2}\left(\frac{b}{\sqrt{a^2+b^2}}\right) = k\left[\frac{|q_1|}{b^2} - \frac{|q_2|b}{(a^2+b^2)^{3/2}}\right] \\ &= (8.988\times10^9\ \text{N}\cdot\text{m}^2/\text{C}^2)\left\{\frac{63.0\times10^{-9}\ \text{C}}{(0.0140\ \text{m})^2} - \frac{(47.0\times10^{-9}\ \text{C})(0.0140\ \text{m})}{[(0.0340\ \text{m})^2+(0.0140\ \text{m})^2]^{3/2}}\right\} \\ &= \boxed{2.77\times10^6\ \text{N/C}} \end{aligned}$$

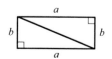

89. Strategy Use the properties of electric fields and the rules for sketching field lines.

Solution Since the semicircle is positively charged, the field lines point toward the center of curvature. Let the semicircle be oriented such that its ends are on the x-axis and its midpoint is on the negative y-axis. The x-components of $\vec{E}$ all cancel due to symmetry, and the y-components all add and point in the positive y-direction. So, the electric field at the center points away from the midpoint of the semicircle.

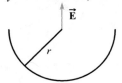

93. Strategy Let θ_+/θ_- be the angle between r_+/r_- and the x-axis. By symmetry, $E_{x,\,\text{net}} = 0$ on the x-axis, and the y-components of $\vec{E}$ due to each charge are directed downward. Use the binomial approximation $(1\pm x)^n \approx 1\pm nx$ for $x \ll 1$, and Coulomb's law.

Solution

(a) Write an expression for the magnitude of the electric field of the dipole on the positive x-axis.

$$E_y(x > 0,\, 0) = -\frac{kq}{r_+^2}\sin\theta_+ - \frac{kq}{r_-^2}\sin\theta_- = -\frac{kq}{r_+^2}\left(\frac{\frac{d}{2}}{r_+}\right) - \frac{kq}{r_-^2}\left(\frac{\frac{d}{2}}{r_-}\right) = -\frac{kq}{x^2 + \frac{d^2}{4}}\left(\frac{\frac{d}{2}}{\sqrt{x^2 + \frac{d^2}{4}}}\right) - \frac{kq}{x^2 + \frac{d^2}{4}}\left(\frac{\frac{d}{2}}{\sqrt{x^2 + \frac{d^2}{4}}}\right)$$

$$= -\frac{kqd}{\left(x^2 + \frac{d^2}{4}\right)^{3/2}} = -\frac{kqd}{x^3\left(1 + \frac{d^2}{4x^2}\right)^{3/2}}$$

If $x \gg d$, then

$$E_y(x \gg d,\, 0) \approx -\frac{kqd}{x^3}\left(1 - \frac{3d^2}{8x^2}\right) \approx -\frac{kqd}{x^3} \qquad [3d^2/(8x^2) \ll 1]$$

So, $|E_y| = E = \boxed{\dfrac{kqd}{x^3}}$.

(b) E_{+y} and E_{-y} are both directed downward, so $\boxed{\vec{E}(x, 0) \text{ is directed in the negative } y\text{-direction for all } x.}$

Chapter 17

ELECTRIC POTENTIAL

Conceptual Questions

1. **(a)** The electric field does positive work on $-q$ as it moves closer to $+Q$.

   (b) The potential increases as $-q$ moves closer to $+Q$.

   (c) The potential energy of $-q$ decreases.

   (d) If the fixed charge instead has a value $-Q$, the electric field does negative work, the potential decreases, and the potential energy increases.

5. Zero work is required to move a charge between two points at the same potential. An external force may need to be applied to move the charge but the work done to start the charge in motion will be negated by the work done to stop it.

9. If the electric field is zero throughout a region of space, the electric potential must be constant throughout that region.

13. It doesn't matter which points we choose because the potential on each plate is constant over the whole plate.

17. We can't say anything about the electric field if all we know is the potential at a single point. The electric field tells us how the potential *changes* if we move from one point to another.

21. In this case the capacitor plates are isolated, so it is the charge Q that remains constant. Again the capacitance decreases by a factor of 3, so ΔV must increase by a factor of 3 to keep $Q = C\Delta V$ the same. Thus, the electric field increases by a factor of 3 and the energy stored in the capacitor increases by a factor of 3 as well.

Problems

1. **Strategy** Use Eq. (17-1).

   Solution Compute the electric potential energy.
   $$U_E = k\frac{q_1 q_2}{r} = \frac{(8.988\times10^9 \text{ N}\cdot\text{m}^2/\text{C}^2)(5.0\times10^{-6} \text{ C})(-2.0\times10^{-6} \text{ C})}{5.0 \text{ m}} = \boxed{-18 \text{ mJ}}$$

5. **Strategy** The work done on the charges is equal to their potential energy. Let the upper charge by 1, the lower left-hand charge be 2, and the right-hand charge be 3. Also, let $a = 0.16$ m and $b = 0.12$ m. Use Eq. (17-2).

   Solution Compute the work done on the charges.
   $$W = U_E = k\left(\frac{q_1 q_2}{b} + \frac{q_1 q_3}{\sqrt{a^2+b^2}} + \frac{q_2 q_3}{a}\right)$$
   $$= (8.988\times10^9 \text{ N}\cdot\text{m}^2/\text{C}^2)\left[\frac{(5.5\times10^{-6} \text{ C})(-6.5\times10^{-6} \text{ C})}{0.12 \text{ m}} + \frac{(5.5\times10^{-6} \text{ C})(2.5\times10^{-6} \text{ C})}{\sqrt{(0.16 \text{ m})^2 + (0.12 \text{ m})^2}}\right.$$
   $$\left. + \frac{(-6.5\times10^{-6} \text{ C})(2.5\times10^{-6} \text{ C})}{0.16 \text{ m}}\right] = \boxed{-3.0 \text{ J}}$$

7. **Strategy** Let $q_1 = -q_2 = q = 10.0$ nC, $d = 4.00$ cm, and $q_3 = -4.2$ nC. Use Eq. (17-2).

   Solution Find the total electric potential energy of the three charges at point a.

   $$U_E = k\left(\frac{q_1 q_2}{r_{12}} + \frac{q_1 q_3}{r_{13}} + \frac{q_2 q_3}{r_{23}}\right) = k\left[-\frac{q^2}{2d} + \frac{qq_3}{d} - \frac{qq_3}{3d}\right] = \frac{kq}{d}\left[-\frac{q}{2} + q_3\left(1 - \frac{1}{3}\right)\right]$$

   $$= \frac{(8.988\times10^9 \text{ N}\cdot\text{m}^2/\text{C}^2)(10.0\times10^{-9} \text{ C})}{0.0400 \text{ m}}\left[-\frac{10.0\times10^{-9} \text{ C}}{2} + \frac{2(-4.2\times10^{-9} \text{ C})}{3}\right] = \boxed{-17.5 \text{ μJ}}$$

9. **Strategy** Let $q_1 = -q_2 = q = 10.0$ nC, $d = 4.00$ cm, and $q_3 = -4.2$ nC. Use Eq. (17-2).

   Solution Find the total electric potential energy of the three charges at point c.

   $$U_E = k\left(\frac{q_1 q_2}{r_{12}} + \frac{q_1 q_3}{r_{13}} + \frac{q_2 q_3}{r_{23}}\right) = k\left(-\frac{q^2}{2d} + \frac{qq_3}{2d} - \frac{qq_3}{2d}\right) = -\frac{kq^2}{2d}$$

   $$= -\frac{(8.988\times10^9 \text{ N}\cdot\text{m}^2/\text{C}^2)(10.0\times10^{-9} \text{ C})^2}{2(0.0400 \text{ m})} = \boxed{-11.2 \text{ μJ}}$$

11. **Strategy** Use Eqs. (6-8) and (17-1).

    Solution Compute the work done by the electric field.

    $$W_{\text{field}} = -\Delta U = U_i - U_f = U_{12} - (U_{12} + U_{13} + U_{23}) = -U_{13} - U_{23} = -\frac{kq_1 q_3}{r_{13}} - \frac{kq_2 q_3}{r_{23}}$$

    $$= -(8.988\times10^9 \text{ N}\cdot\text{m}^2/\text{C}^2)(2.00\times10^{-9} \text{ C})\left(\frac{8.00\times10^{-9} \text{ C}}{0.0400 \text{ m}} + \frac{-8.00\times10^{-9} \text{ C}}{0.0400 \text{ m} + 0.1200 \text{ m}}\right) = \boxed{-2.70 \text{ μJ}}$$

13. **Strategy** Use Eqs. (6-8) and (17-1).

    Solution Compute the work done by the electric field.

    $$W_{\text{field}} = -\Delta U = U_i - U_f = U_{12} + U_{13i} + U_{23i} - (U_{12} + U_{13f} + U_{23f}) = U_{13i} - U_{13f} + U_{23i} - U_{23f}$$

    $$= k\left[q_1 q_3\left(\frac{1}{r_{13i}} - \frac{1}{r_{13f}}\right) + q_2 q_3\left(\frac{1}{r_{23i}} - \frac{1}{r_{23f}}\right)\right]$$

    $$= (8.988\times10^9 \text{ N}\cdot\text{m}^2/\text{C}^2)\left[(8.00\times10^{-9} \text{ C})(2.00\times10^{-9} \text{ C})\left(\frac{1}{0.0400 \text{ m}} - \frac{1}{0.0800 \text{ m}}\right)\right.$$

    $$\left. + (-8.00\times10^{-9} \text{ C})(2.00\times10^{-9} \text{ C})\left(\frac{1}{0.0400 \text{ m} + 0.1200 \text{ m}} - \frac{1}{0.0400 \text{ m}}\right)\right] = \boxed{4.49 \text{ μJ}}$$

17. **Strategy** Use the principle of superposition and Eq. (17-9).

    Solution Sum the electric fields at the center due to each charge.
    $$\vec{E} = \vec{E}_a + \vec{E}_b + \vec{E}_c + \vec{E}_d = \vec{E}_a + \vec{E}_b - \vec{E}_a - \vec{E}_b = \boxed{0}$$
    Do the same for the potential at the center.

    $$V = \Sigma \frac{kQ_i}{r_i} = \frac{4kQ}{r} = \frac{4(8.988\times10^9 \text{ N}\cdot\text{m}^2/\text{C}^2)(9.0\times10^{-6} \text{ C})}{\dfrac{\sqrt{(0.020 \text{ m})^2 + (0.020 \text{ m})^2}}{2}} = \boxed{2.3\times10^7 \text{ V}}$$

21. Strategy and Solution

(a) Since V is positive, q is $\boxed{\text{positive}}$.

(b) $V \propto \dfrac{1}{r}$, so since the potential is doubled, the distance is halved or $\boxed{10.0 \text{ cm}}$.

25. Strategy Use Eq. (17-9).

Solution Find the electric potential at the third corner, B.

$$V = \Sigma \frac{kQ_i}{r_i} = \frac{k}{r}(Q_A + Q_B) = \frac{8.988 \times 10^9 \text{ N} \cdot \text{m}^2/\text{C}^2}{1.0 \text{ m}}(2.0 \times 10^{-9} \text{ C} - 1.0 \times 10^{-9} \text{ C}) = \boxed{9.0 \text{ V}}$$

29. (a) Strategy Let $d = 4.00$ cm, $r = 12.0$ cm, and $q = 8.00$ nC. Use Eq. (17-9). Let $V = 0$ at infinity.

Solution Find the potentials at points b and c.

$$V_b = \frac{kq_1}{r_1} + \frac{kq_2}{r_2} = \frac{kq}{2d} + \frac{k(-q)}{d} = -\frac{kq}{2d} = -\frac{(8.988 \times 10^9 \text{ N} \cdot \text{m}^2/\text{C}^2)(8.00 \times 10^{-9} \text{ C})}{2(0.0400 \text{ m})} = \boxed{-899 \text{ V}}$$

$$V_c = \frac{kq}{r} + \frac{k(-q)}{r} = \boxed{0}$$

(b) Strategy The potential difference is the change in electric potential energy per unit charge. Use $\Delta U_E = q\Delta V$.

Solution Compute the change in electric potential energy.

$$\Delta U_E = q\Delta V = (2.00 \times 10^{-9} \text{ C})(0 + 899 \text{ V}) = \boxed{1.80 \text{ μJ}}$$

33. Strategy $\Delta V = Ed$ for a uniform electric field.

Solution Find the distance between the equipotential surfaces.

$$d = \frac{\Delta V}{E} = \frac{1.0 \text{ V}}{100.0 \text{ N/C}} = \boxed{1.0 \text{ cm}}$$

35. Strategy Equipotential surfaces are perpendicular to electric field lines at all points. For equipotential surfaces drawn such that the potential difference between adjacent surfaces is constant, the surfaces are closer together where the field is stronger. The electric field always points in the direction of maximum potential decrease.

Solution Outside the cylinder, $\vec{\text{E}}$ is radially directed away from the axis of the cylinder. The equipotential surfaces are perpendicular to $\vec{\text{E}}$ at any point, so they are $\boxed{\text{cylinders}}$.

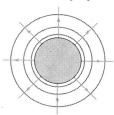

37. **(a) Strategy** Equipotential surfaces are perpendicular to electric field lines at all points. For equipotential surfaces drawn such that the potential difference between adjacent surfaces is constant, the surfaces are closer together where the field is stronger. The electric field always points in the direction of maximum potential decrease.

   Solution The electric field lines are radial. They begin on the point charge and end on the inner surface of the shell. Then they begin again on the surface and extend to infinity.

   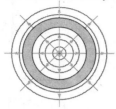

   (b) Strategy Use Eqs. (16-5) and (17-8), and the principle of superposition.

   Solution For $r < r_1$, E is that due to the point charge, $E = kq/r^2$. For $r_1 < r < r_2$, $E = 0$, since this is inside a conductor. For $r > r_2$, E once again is that due to the point charge, kq/r^2. For $r < r_1$, $V = kq/r$ (point charge). For $r_1 < r < r_2$, $V = kq/r_1$, since V is continuous, and it is constant in a conductor. For $r > r_2$,

   $$V = \frac{kq}{r_1} + \left(\frac{kq}{r} - \frac{kq}{r_2} \right)$$ (to preserve continuity). The graphs of the electric field magnitude and potential:

   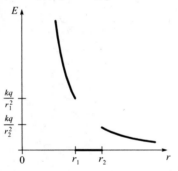 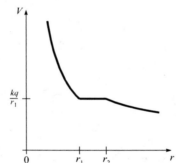

39. **Strategy** Since the electric field is uniform, we can use Eq. (17-10). Use Newton's second law.

   Solution Find the magnitude of the charge on the drop.
   $\Sigma F = qE - mg = 0$, so

   $$|q| = \left| \frac{mg}{E} \right| = \left| \frac{mg}{-\Delta V/d} \right| = \frac{mgd}{\Delta V} = \frac{(1.0 \times 10^{-15} \text{ kg})(9.80 \text{ m/s}^2)(0.16 \text{ m})}{9.76 \times 10^3 \text{ V}} = \boxed{1.6 \times 10^{-19} \text{ C} = e}.$$

41. **Strategy** Use conservation of energy and Eq. (17-7).

   Solution Find the potential difference.

   $$\Delta U = -e\Delta V = -\Delta K = -\frac{1}{2}mv^2, \text{ so } \Delta V = \frac{mv^2}{2e} = \frac{(9.109 \times 10^{-31} \text{ kg})(7.26 \times 10^6 \text{ m/s})^2}{2(1.602 \times 10^{-19} \text{ C})} = \boxed{150 \text{ V}}.$$

43. **Strategy** According to Example 17.8, the speed of the electrons at the anode is proportional to the square root of the potential difference. Use a proportion.

    Solution Find the speed of the electrons.

    $v \propto \sqrt{\Delta V}$, so $\dfrac{v_2}{v_1} = \sqrt{\dfrac{\Delta V_2}{\Delta V_1}}$ and $v_2 = v_1 \sqrt{\dfrac{\Delta V_2}{\Delta V_1}} = (6.5 \times 10^7 \text{ m/s}) \sqrt{\dfrac{6.0 \text{ kV}}{12 \text{ kV}}} = \boxed{4.6 \times 10^7 \text{ m/s}}$.

45. **Strategy and Solution**

    (a) Electrons travel opposite the direction of the electric field, so $\vec{E}$ is directed $\boxed{\text{upward}}$.

    (b) For a uniform electric field, $\Sigma F_y = eE = \dfrac{e\Delta V}{d} = ma_y$, so $a_y = \dfrac{e\Delta V}{md}$. Thus, $\Delta t = \dfrac{v_y}{a_y} = \boxed{\dfrac{v_y md}{e\Delta V}}$.

    (c) Since the electron gains kinetic energy, its potential energy $\boxed{\text{decreases}}$.

47. **Strategy** Use conservation of energy and Eq. (17-7).

    Solution Find the final kinetic energy.
    $\Delta K = K_f - K_i = -\Delta U = -q\Delta V = -2e\Delta V$, so

    $K_f = K_i - 2e\Delta V = 1.20 \times 10^{-16} \text{ J} - 2(1.602 \times 10^{-19} \text{ C})(-0.50 \times 10^3 \text{ V}) = \boxed{2.8 \times 10^{-16} \text{ J}}$.

49. **Strategy** The electron must have enough kinetic energy at point A to overcome the potential decrease between A and C. Use conservation of energy and Eq. (17-7).

    Solution Find the required kinetic energy.
    $K_A = \Delta U = -e\Delta V = -(1.602 \times 10^{-19} \text{ C})(-60.0 \text{ V} - 100.0 \text{ V}) = \boxed{2.56 \times 10^{-17} \text{ J}}$

51. **Strategy** Use the definition of capacitance, Eq. (17-14).

    Solution Find the magnitude of the charge on each plate.
    $Q = C\Delta V = (2.0 \text{ } \mu\text{F})(9.0 \text{ V}) = \boxed{18 \text{ } \mu\text{C}}$

53. **Strategy** Use the definition of capacitance, Eq. (17-14).

    Solution

    $Q = C\Delta V = (10.2 \times 10^{-6} \text{ F})(-60.0 \text{ V}) = -6.12 \times 10^{-4} \text{ C}$

    $\boxed{612 \text{ } \mu\text{C}}$ of charge must be removed from each plate.

55. **Strategy and Solution**

    (a) Since $\vec{E}$ does not depend upon the separation of the plates ($E = \sigma/\epsilon_0$), it $\boxed{\text{stays the same}}$.

    (b) Since $\Delta V \propto d$, ΔV $\boxed{\text{increases}}$ if d increases.

57. Strategy and Solution

    (a) Since $\Delta V \propto d$, ΔV $\boxed{\text{decreases}}$ if d decreases.

    (b) Since $\bar{E}$ does not depend upon the separation of the plates $(E = \sigma/\epsilon_0)$, it $\boxed{\text{stays the same}}$.

    (c) Since $\Delta V \propto d$, $C \propto d^{-1}$, and $Q = C\Delta V$, the charge on the plates $\boxed{\text{stays the same}}$.

61. Strategy Use the definition of capacitance, Eq. (17-14).

    Solution Find the capacitance of the spheres.

$$Q = C\Delta V, \text{ so } C = \frac{Q}{\Delta V} = \frac{3.2 \times 10^{-14} \text{ C}}{0.0040 \text{ V}} = \boxed{8.0 \text{ pF}}.$$

65. (a) Strategy Use Eq. (16-6).

    Solution The electric field between the plates is

$$E = \frac{Q}{\epsilon_0 A} = \frac{4.0 \times 10^{-11} \text{ C}}{[8.854 \times 10^{-12} \text{ C}^2/(\text{N} \cdot \text{m}^2)](0.062 \text{ m})(0.022 \text{ m})} = \boxed{3.3 \times 10^3 \text{ V/m}}.$$

    (b) Strategy Use the definition of the dielectric constant, Eq. (17-17).

    Solution Find the electric field between the plates of the capacitor with the dielectric.

$$\kappa = \frac{E_0}{E}, \text{ so } E = \frac{E_0}{\kappa} = \frac{3.3 \times 10^3 \text{ V/m}}{5.5} = \boxed{6.0 \times 10^2 \text{ V/m}}.$$

69. Strategy The spark flies between the spheres when the electric field between them exceeds the dielectric strength. The magnitude of the electric field is given by $\Delta V/d$, where d is the distance between the spheres.

    Solution Find d.

$$E = \frac{\Delta V}{d}, \text{ so } d = \frac{\Delta V}{E} = \frac{900 \text{ V}}{3.0 \times 10^6 \text{ V/m}} = \boxed{0.30 \text{ mm}}.$$

71. Strategy Use Eq. (17-16).

    Solution Compute the capacitance of the capacitor.

$$C = \kappa \frac{\epsilon_0 A}{d} = \frac{2.5[8.854 \times 10^{-12} \text{ C}^2/(\text{N} \cdot \text{m}^2)](0.30 \text{ m})(0.40 \text{ m})}{0.030 \times 10^{-3} \text{ m}} = \boxed{89 \text{ nF}}$$

73. **(a) Strategy** Use Eq. (17-18c).

Solution Compute the capacitance.

$$U = \frac{Q^2}{2C}, \text{ so } C = \frac{Q^2}{2U} = \frac{(8.0\times10^{-2}\text{ C})^2}{2(450\text{ J})} = \boxed{7.1\ \mu\text{F}}.$$

(b) Strategy Use Eq. (17-18a).

Solution Compute the potential difference.

$$U = \frac{1}{2}Q\Delta V, \text{ so } \Delta V = \frac{2U}{Q} = \frac{2(450\text{ J})}{8.0\times10^{-2}\text{ C}} = \boxed{1.1\times10^4\text{ V}}.$$

77. **(a) Strategy** Use Eq. (17-15).

Solution Find the capacitance for the thundercloud.

$$C = \frac{\epsilon_0 A}{d} = \frac{[8.854\times10^{-12}\text{ C}^2/(\text{N}\cdot\text{m}^2)](4500\text{ m})(2500\text{ m})}{550\text{ m}} = \boxed{0.18\ \mu\text{F}}$$

(b) Strategy Use Eq. (17-18c).

Solution Find the energy stored in the capacitor.

$$U = \frac{Q^2}{2C} = \frac{(18\text{ C})^2}{2(0.1811\times10^{-6}\text{ F})} = \boxed{8.9\times10^8\text{ J}}$$

79. **(a) Strategy** Use the definition of capacitance, Eq. (17-14), and Eq. (17-15).

Solution Find the charge on the capacitor.

$$Q = C\Delta V = \frac{\epsilon_0 A}{d}\Delta V = \frac{[8.854\times10^{-12}\text{ C}^2/(\text{N}\cdot\text{m}^2)](0.100\text{ m})^2(150\text{ V})}{0.75\times10^{-3}\text{ m}} = \boxed{18\text{ nC}}$$

(b) Strategy Use Eq. (17-18a).

Solution Compute the energy stored in the capacitor.

$$U = \frac{1}{2}Q\Delta V = \frac{1}{2}(17.7\times10^{-9}\text{ C})(150\text{ V}) = \boxed{1.3\ \mu\text{J}}$$

81. **(a) Strategy** $U = P\Delta t$ where $P = 10.0\text{ kW}$ and $\Delta t = 2.0\text{ ms}$. Use Eq. (17-18b).

Solution Find the initial potential difference.

$$U = \frac{1}{2}C(\Delta V)^2, \text{ so } \Delta V = \sqrt{\frac{2U}{C}} = \sqrt{\frac{2(10.0\text{ kW})(2.0\text{ ms})}{100.0\times10^{-6}\text{ F}}} = \boxed{630\text{ V}}.$$

(b) Strategy Use Eq. (17-18c).

Solution Find the initial charge.

$$U = \frac{Q^2}{2C}, \text{ so } Q = \sqrt{2CU} = \sqrt{2(100.0\times10^{-6}\text{ F})(10.0\text{ kW})(2.0\text{ ms})} = \boxed{0.063\text{ C}}.$$

85. **(a) Strategy** Use the definition of capacitance, Eq. (17-14).

    Solution Compute the charge that passes through the body tissues.
    $$Q = C\Delta V = (15 \times 10^{-6} \text{ F})(9.0 \times 10^{3} \text{ V}) = \boxed{0.14 \text{ C}}$$

    (b) Strategy Use Eq. (17-18b) and the definition of average power.

    Solution Find the average power delivered to the tissues.
    $$P_{\text{av}} = \frac{\Delta E}{\Delta t} = \frac{U}{\Delta t} = \frac{C(\Delta V)^2}{2\Delta t} = \frac{(15 \times 10^{-6} \text{ F})(9.0 \times 10^{3} \text{ V})^2}{2(2.0 \times 10^{-3} \text{ s})} = \boxed{0.30 \text{ MW}}$$

89. **(a) Strategy** Let $q_L = -q_R = 10.0$ nC. Use Eqs. (17-5) and (17-9).

    Solution Find the potential energy of the point charge at each location.
    $$U_a = qV_a = q\left(\frac{kq_L}{r_L} + \frac{kq_R}{r_R} \right) = kqq_L\left(\frac{1}{r_L} - \frac{1}{r_R} \right)$$
    $$= (8.988 \times 10^{9} \text{ N} \cdot \text{m}^2/\text{C}^2)(-4.2 \times 10^{-9} \text{ C})(10.0 \times 10^{-9} \text{ C})\left(\frac{1}{0.0400 \text{ m}} - \frac{1}{0.1200 \text{ m}} \right) = \boxed{-6.3 \text{ μJ}}$$
    $$U_b = qV_b = (8.988 \times 10^{9} \text{ N} \cdot \text{m}^2/\text{C}^2)(-4.2 \times 10^{-9} \text{ C})(10.0 \times 10^{-9} \text{ C})\left(\frac{1}{0.0400 \text{ m}} - \frac{1}{0.0400 \text{ m}} \right) = \boxed{0}$$
    $$U_c = qV_c = (8.988 \times 10^{9} \text{ N} \cdot \text{m}^2/\text{C}^2)(-4.2 \times 10^{-9} \text{ C})(10.0 \times 10^{-9} \text{ C})\left(\frac{1}{0.0800 \text{ m}} - \frac{1}{0.0800 \text{ m}} \right) = \boxed{0}$$

    (b) Strategy The work done by the external force is negative the work done by the field. Use Eq. (6-8).

    Solution Find the work required to move the point charge.
    $$W = -W_{\text{field}} = \Delta U = U_a - U_b = -6.3 \text{ μJ} - 0 = \boxed{-6.3 \text{ μJ}}$$

93. **(a) Strategy** Electric field lines begin on positive charges and end on negative charges. The same number of field lines begins on the plate and ends on the negative point charge. Field lines never cross. Use the principles of superposition and symmetry.

    Solution The electric field lines for the cylinder and sheet:

    (b) Strategy Equipotential surfaces are perpendicular to electric field lines at all points. For equipotential surfaces drawn such that the potential difference between adjacent surfaces is constant, the surfaces are closer together where the field is stronger. The electric field always points in the direction of maximum potential decrease.

    Solution The equipotential surfaces for the cylinder and sheet:

97. **Strategy** The negatively charged particle will accelerate toward the positively charged plate while it is between the plates of the capacitor.

Solution The particle is between the plates for a time given by $\Delta t = \Delta x / v_x$. During this time, the particle travels a vertical distance $\Delta y = 0.00100$ m. Find the acceleration of the particle.

$$\Delta y = \frac{1}{2} a (\Delta t)^2 = \frac{1}{2} a \left(\frac{\Delta x}{v_x} \right)^2 = \frac{a(\Delta x)^2}{2 v_x^2}, \text{ so } a = \frac{2 v_x^2 \Delta y}{(\Delta x)^2}.$$

The magnitude of the electrical force on the particle is $NeE = Ne\dfrac{\Delta V}{d}$, where d is the plate separation and N is the number of excess electrons on the particle. According to Newton's second law, the acceleration of the particle is $a = \dfrac{Ne\frac{\Delta V}{d}}{m} = \dfrac{Ne\Delta V}{md}$. We set the two expressions for the acceleration of the particle equal and solve for N.

$$\frac{Ne\Delta V}{md} = \frac{2 v_x^2 \Delta y}{(\Delta x)^2}, \text{ so } N = \frac{2 m d v_x^2 \Delta y}{e \Delta V (\Delta x)^2} = \frac{2(5.00 \times 10^{-19} \text{ kg})(0.00200 \text{ m})(35.0 \text{ m/s})^2 (0.00100 \text{ m})}{(1.602 \times 10^{-19} \text{ C})(3.00 \text{ V})(0.0100 \text{ m})^2} = \boxed{51}.$$

101. **Strategy** The energy in the capacitor is converted into heat in the water. Use Eqs. (14-4) and (17-18b).

Solution Find the temperature change of the water.

$$Q = mc\Delta T = U = \frac{1}{2} C (\Delta V)^2, \text{ so } \Delta T = \frac{C(\Delta V)^2}{2mc} = \frac{(200.0 \times 10^{-6} \text{ F})(12.0 \text{ V})^2}{2(1.00 \text{ cm}^3)(1.00 \text{ g/cm}^3)[4.186 \text{ J/(g} \cdot \text{K)}]} = \boxed{3.44 \text{ mK}}.$$

105. (a) **Strategy** For a parallel plate capacitor, $E = \sigma / \epsilon_0$ and $\Delta V = Ed$.

Solution Find the potential difference between the plates.

$$\Delta V = Ed = \frac{\sigma d}{\epsilon_0} = \frac{(4.0 \times 10^{-6} \text{ C/m}^2)(0.0060 \text{ m})}{8.854 \times 10^{-12} \text{ C}^2/(\text{N} \cdot \text{m}^2)} = \boxed{2.7 \text{ kV}}$$

(b) **Strategy** Use conservation of energy and the fact that $\Delta U = q\Delta V$.

Solution Find the kinetic energy of each point charge just before it hits the positive plate.

$$\Delta K = K_f - K_i = -\Delta U = -q\Delta V, \text{ so } K_f = K_i - q\Delta V = 0 - (-2.5 \times 10^{-9} \text{ C})(2711 \text{ V}) = \boxed{6.8 \text{ } \mu\text{J}}.$$

109. **Strategy** The energy stored in the capacitor is directly proportional to the capacitance. Use Eq. (17-16) and the fact that ΔV is constant. Form a proportion.

Solution Determine what happens to the energy stored in the capacitor.

$$\frac{\Delta U}{U_0} = \frac{\Delta C}{C_0} = \frac{\kappa \epsilon_0 A \left(\frac{1}{d_f} - \frac{1}{d_i} \right)}{\frac{\kappa \epsilon_0 A}{d_i}} = \frac{d_i}{d_f} - 1 = \frac{1}{1.25} - 1 = -0.200, \text{ so } \boxed{\text{the energy is reduced by 20.0\%.}}$$

113. (a) **Strategy** Use Eq. (17-10).

   Solution Compute the minimum thickness of the titanium dioxide.

   $$d = \frac{\Delta V}{E} = \frac{5.00 \text{ V}}{4.00 \times 10^6 \text{ V/m}} = \boxed{1.25 \text{ μm}}$$

(b) **Strategy** Use Eq. (17-16).

   Solution Find the area of the plates.

   $$C = \kappa \frac{\epsilon_0 A}{d}, \text{ so } A = \frac{dC}{\kappa \epsilon_0} = \frac{(1.25 \times 10^{-6} \text{ m})(1.0 \text{ F})}{90.0[8.854 \times 10^{-12} \text{ C}^2/(\text{N} \cdot \text{m}^2)]} = \boxed{1600 \text{ m}^2}.$$

Chapter 18

ELECTRIC CURRENT AND CIRCUITS

Conceptual Questions

1.

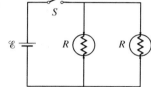

5. This statement is not exactly true. The current flowing through a branch in a circuit is inversely proportional to the resistance of the branch. Thus, more current follows the path of least resistance than follows any other path, but every path has some current.

9. Electric stoves and clothes dryers require relatively large amounts of power to operate. Supplying them with 240 V instead of 120 V decreases the magnitude of required current to supply the power. This reduces the rate ($P = I^2 R$) at which energy is dissipated in the wiring.

13. Batteries convert chemical energy into electrical energy. As a battery is used, its supply of chemical energy is depleted. Recharging the battery does not actually put additional charges back into the battery, but instead converts electrical energy into chemical energy.

17. An electrician working on live wiring wears insulated shoes to avoid being grounded and therefore to reduce the chance of starting a flow of current through the body if a live wire is touched. Similarly, an electrician using two hands would risk completing a circuit so that current flows from one hand to the other through the body and near the heart.

21. **(a)** It increases.

   (b) It decreases.

   (c) It increases.

Problems

1. **Strategy** Use the definition of electric current.

   Solution Compute the total charge.

   $I = \dfrac{\Delta q}{\Delta t}$, so $\Delta q = I \Delta t = (3.0 \text{ A})(4.0 \text{ h})(3600 \text{ s/h}) = \boxed{4.3 \times 10^4 \text{ C}}$.

3. **(a) Strategy and Solution** The electrons flow from the filament to the anode; since they are negatively charged, the current flows $\boxed{\text{from the anode to the filament}}$.

   (b) Strategy Use the definition of electric current.

   Solution Compute the current in the tube.

   $I = \dfrac{\Delta q}{\Delta t} = \Delta q \times f = (1.602 \times 10^{-19} \text{ C})(6.0 \times 10^{12} \text{ s}^{-1}) = \boxed{0.96 \text{ μA}}$

5. **Strategy** Use the definition of electric current and the elementary charge of an electron.

   Solution Find the number of electrons per second that hit the screen.
   $$I = \frac{\Delta q}{\Delta t} = \frac{Ne}{\Delta t}, \text{ so } \frac{N}{\Delta t} = \frac{I}{e} = \frac{320 \times 10^{-6} \text{ A}}{1.602 \times 10^{-19} \frac{C}{\text{electron}}} = \boxed{2.0 \times 10^{15} \text{ electrons/s}}.$$

7. **Strategy** Since the oppositely charged ions move in opposite directions, they both contribute to the current in the same direction. Use the definition of electric current.

   Solution Compute the current in the solution.
   $$I = \frac{\Delta q}{\Delta t} = \frac{Ne}{\Delta t} = \frac{[2(3.8 \times 10^{16}) + 6.2 \times 10^{16}](1.602 \times 10^{-19} \text{ C})}{1.0 \text{ s}} = \boxed{22.1 \text{ mA}}$$

9. **Strategy** The total energy stored in a battery is equal to the total work the battery is able to do. Use Eq. (18-2).

   Solution Compute the energy stored in the battery.
   $$W = \mathscr{E}q = (1.20 \text{ V})(675 \text{ C}) = \boxed{810 \text{ J}}$$

11. **(a) Strategy** Use the definition of electric current.

   Solution Compute the amount of charge pumped by the battery.
   $$\Delta q = I\Delta t = (220.0 \text{ A})(1.20 \text{ s}) = \boxed{264 \text{ C}}$$

   (b) Strategy The electrical energy supplied is equal to the work done by the battery. Use Eq. (18-2).

   Solution Compute the amount of electrical energy supplied by the battery.
   $$W = \mathscr{E}q = (12.0 \text{ V})(264 \text{ C}) = \boxed{3.17 \text{ kJ}}$$

13. **Strategy** Use Eq. (18-3).

   Solution Form a proportion.
   $$\frac{I_1}{I_2} = 1 = \frac{neA_1v_1}{neA_2v_2} = \frac{ne\left(\frac{1}{4}\pi d_1^2\right)v_1}{ne\left(\frac{1}{4}\pi d_2^2\right)v_2} = \frac{d_1^2 v_1}{d_2^2 v_2}, \text{ so } v_1 = \left(\frac{d_2}{d_1}\right)^2 v_2 = \left(\frac{2}{1}\right)^2 v_2 = 4v_2.$$

   The relationship between the drift speeds is $\boxed{v_1 = 4v_2}$.

15. **Strategy** Use Eq. (18-3) and $\Delta x = v_D \Delta t$.

   Solution Find the drift speed of the conduction electrons in the wire.
   $$I = neAv_D, \text{ so } v_D = \frac{I}{neA} = \frac{I}{ne(\pi r^2)} = \frac{I}{\pi ner^2}.$$
   Find the time to travel 1.00 m along the wire.
   $$\Delta t = \frac{\Delta x}{v_D} = \frac{\Delta x}{\frac{I}{\pi ner^2}} = \frac{\pi ner^2 \Delta x}{I} = \frac{\pi(8.47 \times 10^{28} \text{ m}^{-3})(1.602 \times 10^{-19} \text{ C})(0.00100 \text{ m}/2)^2(1.00 \text{ m})}{10.0 \text{ A}}\left(\frac{1 \text{ min}}{60 \text{ s}}\right)$$
   $$= \boxed{17.8 \text{ min}}$$

17. Strategy Let h be the thickness of the strip so that the cross-sectional area is $A = hw$, where w is the width. Use Eq. (18-3).

Solution Find the thickness of the strip.

$$I = neAv_D = ne(hw)v_D, \text{ so } h = \frac{I}{newv_D} = \frac{130 \times 10^{-6}\,\text{A}}{(8.8 \times 10^{22}\,\text{m}^{-3})(1.602 \times 10^{-19}\,\text{C})(260 \times 10^{-6}\,\text{m})(0.44\,\text{m/s})} = \boxed{81\,\mu\text{m}}.$$

19. Strategy Use Eq. (18-3) and $n = 1.3\rho N_A/M$, the number of electrons per unit volume.

Solution Find the drift speed of the conduction electrons.

$$v_D = \frac{I}{neA} = \frac{IM}{1.3\rho N_A eA} = \frac{(2.0\,\text{A})(64\,\text{g/mol})}{1.3(9.0 \times 10^6\,\text{g/m}^3)(6.022 \times 10^{23}\,\text{mol}^{-1})(1.602 \times 10^{-19}\,\text{C})(1.00 \times 10^{-6}\,\text{m}^2)}$$
$$= \boxed{0.11\,\text{mm/s}}$$

21. Strategy Use the definition of resistance.

Solution Compute the current through the resistor.

$$R = \frac{\Delta V}{I}, \text{ so } I = \frac{\Delta V}{R} = \frac{16\,\text{V}}{12\,\Omega} = \boxed{1.3\,\text{A}}.$$

25. (a) Strategy Use the definition of resistance.

Solution Compute the required potential difference between the electrician's hands.

$$\Delta V = IR = (50\,\text{mA})(1\,\text{k}\Omega) = \boxed{50\,\text{V}}$$

(b) Strategy and Solution An electrician working on a "live" circuit keeps one hand behind his or her back $\boxed{\text{to avoid becoming part of the circuit}}$.

29. Strategy As found in Example 18.4, $R/R_0 = 1 + \alpha\Delta T$. Find T using this and the definition of resistance.

Solution Estimate the temperature of the tungsten filament.

$$1 + \alpha\Delta T = \frac{R}{R_0}, \text{ so } T = \frac{1}{\alpha}\left(\frac{R}{R_0} - 1\right) + T_0 = \frac{1}{4.50 \times 10^{-3}\,°\text{C}^{-1}}\left(\frac{\frac{2.90\,\text{V}}{0.300\,\text{A}}}{1.10\,\Omega} - 1\right) + 20.0°\text{C} = \boxed{1750°\text{C}}.$$

33. Strategy Use the definition of resistance, the relationship between voltage and uniform electric field, and Eq. (18-8).

Solution $V = IR = EL$ and $R = \rho L/A$. Find E.

$$V = EL = IR = I\rho\frac{L}{A}, \text{ so } \boxed{E = \rho\frac{I}{A}, \text{ where } \rho \text{ is the resistivity}}.$$

37. (a) Strategy Sum the individual emfs with those with their left terminal at the higher potential being positive.

Solution Compute the equivalent emf.

$$\mathscr{E}_{eq} = 3.0 \text{ V} + 3.0 \text{ V} + 2.5 \text{ V} - 1.5 \text{ V} = \boxed{7.0 \text{ V}}$$

(b) Strategy Use the definition of resistance.

Solution Find the value of the resistor.

$$R = \frac{\Delta V}{I} = \frac{\mathscr{E}_{eq}}{I} = \frac{7.0 \text{ V}}{0.40 \text{ A}} = \boxed{18 \text{ }\Omega}$$

39. (a) Strategy $C_{eq} = \Sigma C_i$ for the capacitors, which are in parallel.

Solution Compute the equivalent capacitance.

$$C_{eq} = 4.0 \text{ }\mu\text{F} + 2.0 \text{ }\mu\text{F} + 3.0 \text{ }\mu\text{F} + 9.0 \text{ }\mu\text{F} + 5.0 \text{ }\mu\text{F} = \boxed{23.0 \text{ }\mu\text{F}}$$

(b) Strategy Use Eq. (17-14).

Solution Compute the charge on the equivalent capacitor.

$$Q = C\Delta V = C_{eq}\mathscr{E} = (23.0\times10^{-6} \text{ F})(16.0 \text{ V}) = \boxed{368 \text{ }\mu\text{C}}.$$

(c) Strategy Use Eq. (17-14).

Solution Compute the charge on the capacitor.

$$Q = C\Delta V = C\mathscr{E} = (3.0\times10^{-6} \text{ F})(16.0 \text{ V}) = \boxed{48 \text{ }\mu\text{C}}.$$

41. (a) Strategy Use Eqs. (18-13) and (18-17).

Solution Compute the resistance between points A and B.

$$R_{eq} = \left(\frac{1}{2.0 \text{ }\Omega} + \frac{1}{1.0 \text{ }\Omega + 1.0 \text{ }\Omega}\right)^{-1} + 4.0 \text{ }\Omega = \boxed{5.0 \text{ }\Omega}$$

(b) Strategy Label the currents on a diagram. Use Kirchhoff's rules.

Solution The current through the emf is $I = \mathscr{E}/R_{eq} = I_1 + I_2$, where the currents labeled 1 and 2 are shown in the diagram. Applying the loop rule, we have $I_2(2R_2) - I_1R_1 = 0$, so $I_2 = \frac{R_1}{2R_2}I_1 = \frac{2.0 \text{ }\Omega}{2(1.0 \text{ }\Omega)}I_1 = 1.0I_1$.

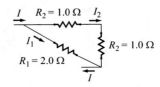

Solve for the current through the 2.0-Ω resistor, I_1.

$$I = I_1 + I_2 = I_1 + 1.0I_1 = 2.0I_1, \text{ so } I_1 = \frac{I}{2.0} = \frac{\mathscr{E}}{2.0R_{eq}} = \frac{20 \text{ V}}{2.0(5.0 \text{ }\Omega)} = \boxed{2.0 \text{ A}}.$$

45. Strategy Use the concept of equivalent resistance. The equivalent resistance of two identical resistances R in parallel is half or $R/2$.

Solution

(a) The two 2.0-Ω resistors are in series, so their equivalent resistance is $2.0\ \Omega + 2.0\ \Omega = 4.0\ \Omega$. These two resistors are in parallel with the rightmost 4.0-Ω resistor. Because the resistances of each branch of this parallel circuit are equal, the current is split evenly. Let the current through each branch be called I_3. We must determine I_3. Now, the equivalent resistance of this parallel circuit is $2.0\ \Omega$, and this is in series with the rightmost 3.0-Ω resistor and the rightmost 1.0-Ω, so the equivalent series resistance is $6.0\ \Omega$. This resistance is in parallel with the 6.0-Ω resistor, so the current is again split evenly. Let it be called I_2; then, $I_3 = I_2/2$. The equivalent resistance of this parallel circuit is 3.0-Ω, and this is in series with the middle 1.0-Ω resistor, so the equivalent series resistance is $4.0\ \Omega$. This equivalent resistance is in parallel with the leftmost 4.0-Ω resistor, so the current is again split evenly. Let it be called I_1; then, $I_3 = I_2/2 = I_1/4$. The equivalent resistance of this parallel circuit is 2.0-Ω, and this is in series with the leftmost 1.0-Ω and 3.0-Ω resistors, so the equivalent resistance of the entire circuit is $6.0\ \Omega$. If the current through the emf is I; then, $I_3 = I_2/2 = I_1/4 = I/8$. The current though the emf is given by $I = \mathscr{E}/R_{eq}$. Compute the current through one of the 2.0-Ω resistors.

$$I_3 = \frac{I}{8} = \frac{\mathscr{E}}{8R_{eq}} = \frac{24\text{ V}}{8(6.0\ \Omega)} = \boxed{0.50\text{ A}}$$

(b) The current through the 6.0-Ω resistor is I_2, which is one-fourth of the current through the emf.

$$I_2 = \frac{I}{4} = \frac{\mathscr{E}}{4R_{eq}} = \frac{24\text{ V}}{4(6.0\ \Omega)} = \boxed{1.0\text{ A}}$$

(c) The current through the leftmost 4.0-Ω resistor is I_1, which is half of the current through the emf.

$$I_1 = \frac{I}{2} = \frac{\mathscr{E}}{2R_{eq}} = \frac{24\text{ V}}{2(6.0\ \Omega)} = \boxed{2.0\text{ A}}$$

49. (a) Strategy Use Eqs. (18-15) and (18-18).

Solution Find the equivalent capacitance.

$$C_{eq} = \left(\frac{1}{12\ \mu\text{F}} + \frac{1}{12\ \mu\text{F} + 12\ \mu\text{F}}\right)^{-1} = \boxed{8.0\ \mu\text{F}}$$

(b) Strategy Since the capacitor at the left side of the diagram (1) is in series with the parallel combination of the other two capacitors (2), the charge Q on the capacitor 1 is the same as that on capacitor 2. (Think of the parallel combination as one capacitor with capacitance $C_2 = 12\ \mu\text{F} + 12\ \mu\text{F} = 24\ \mu\text{F}$.) Use the definition of capacitance.

Solution Find the potential difference across C_1. Let this potential difference be V_1 and the potential difference across C_2 be V_2. Then, $\mathscr{E} = V_1 + V_2$. Form a proportion.

$$\frac{V_2}{V_1} = \frac{\mathscr{E} - V_1}{V_1} = \frac{\mathscr{E}}{V_1} - 1 = \frac{Q/C_2}{Q/C_1} = \frac{C_1}{C_2}, \text{ so } V_1 = \frac{\mathscr{E}}{1 + \frac{C_1}{C_2}} = \frac{25\text{ V}}{1 + \frac{12\ \mu\text{F}}{24\ \mu\text{F}}} = \boxed{17\text{ V}}.$$

(c) Strategy The charge on the capacitor at the far right of the circuit (1) is half of the charge on the capacitor at the left of the circuit (2).

Solution Find the charge on the capacitor.

$$Q_2 = C_2 V_2 = 2Q_1, \text{ so } Q_1 = \frac{1}{2} C_2 V_2 = \frac{1}{2}(12 \times 10^{-6} \text{ F})(17 \text{ V}) = \boxed{1.0 \times 10^{-4} \text{ C}}.$$

53. Strategy Use Kirchhoff's rules. Let I_1 be the top branch, I_2 be the middle branch, and I_3 be the bottom branch. Assume that each current flows right to left.

Solution Find the current in each branch of the circuit.

(1) $I_1 = -I_2 - I_3$ (2) $0 = 25.00 \text{ V} + (5.6 \text{ }\Omega)I_2 - (122 \text{ }\Omega)I_1$

(3) $0 = 25.00 \text{ V} + 5.00 \text{ V} + (75 \text{ }\Omega)I_3 - (122 \text{ }\Omega)I_1$ (4) $0 = 5.00 \text{ V} + (75 \text{ }\Omega)I_3 - (5.6 \text{ }\Omega)I_2$

Substitute (1) into (2).

(5) $0 = 25.00 \text{ V} + (122 \text{ }\Omega + 5.6 \text{ }\Omega)I_2 + (122 \text{ }\Omega)I_3$

Multiply (4) by 5 and subtract from (5).

$$0 = [122 \text{ }\Omega + 5.6 \text{ }\Omega + 5(5.6 \text{ }\Omega)]I_2 + [122 \text{ }\Omega - 5(75 \text{ }\Omega)]I_3, \text{ so } I_2 = \frac{5(75 \text{ }\Omega) - 122 \text{ }\Omega}{122 \text{ }\Omega + 5.6 \text{ }\Omega + 5(5.6 \text{ }\Omega)} I_3 = 1.6 I_3 \ (1.626 I_3).$$

Substitute the result above into (4).

$$0 = 5.00 \text{ V} + (75 \text{ }\Omega)I_3 - (5.6 \text{ }\Omega)(1.626 I_3), \text{ so } I_3 = \frac{5.00 \text{ V}}{1.626(5.6 \text{ }\Omega) - 75 \text{ }\Omega} = -0.076 \text{ A}.$$

So, $I_2 = 1.626(-0.076 \text{ A}) = -0.12 \text{ A}$ and $I_1 = -(-0.12 \text{ A}) - (-0.076 \text{ A}) = 0.20 \text{ A}$.

Branch	I (A)	Direction
AB	0.20	right to left
FC	0.12	left to right
ED	0.076	left to right

55. Strategy Use Kirchhoff's rules. Let the current on the left be I, the one in the middle be I_1, and the one on the right be I_2. I_1 flows downward.

Solution Find the unknown emf and the unknown resistor.

$I_1 = I + I_2 = 1.00 \text{ A} + 10.00 \text{ A} = 11.00 \text{ A}$

Loop *ABCFA*:

$0 = -\mathscr{E} - (6.00 \text{ }\Omega)(1.00 \text{ A}) - (4.00 \text{ }\Omega)(11.00 \text{ A}) + 125 \text{ V}, \text{ so } \mathscr{E} = \boxed{75 \text{ V}}.$

Loop *ABCDEFA*:

$0 = -\mathscr{E} - (6.00 \text{ }\Omega)(1.00 \text{ A}) + (10.00 \text{ A})R = -75 \text{ V} - 6.00 \text{ V} + (10.00 \text{ A})R, \text{ so } R = \dfrac{81 \text{ V}}{10.00 \text{ A}} = \boxed{8.1 \text{ }\Omega}.$

57. Strategy Use Eq. (18-20).

Solution Compute the power dissipated by the resistor.

$P = \mathscr{E}I = (2.00 \text{ V})(2.0 \text{ A}) = \boxed{4.0 \text{ W}}$

61. **Strategy and Solution** $\boxed{\text{Yes}}$; the power rating can be determined by $P = IV = (5.0 \text{ A})(120 \text{ V}) = \boxed{600 \text{ W}}$.

65. **(a) Strategy** Use Eqs. (8-13) and (8-17) to find the equivalent resistance. Then draw the diagram.

   Solution Compute the equivalent resistance.
   $$R_{eq} = 20.0 \ \Omega + 50.0 \ \Omega + \left(\frac{1}{70.0 \ \Omega + 20.0 \ \Omega} + \frac{1}{40.0 \ \Omega + 20.0 \ \Omega} \right)^{-1} = 106.0 \ \Omega$$
   The simplest equivalent circuit contains the emf and one 106.0-Ω resistor.

   106.0 Ω

   120 V

   (b) Strategy Use the definition of resistance.

   Solution Compute the current that flows from the battery.
   $$I = \frac{\mathcal{E}}{R_{eq}} = \frac{120 \text{ V}}{106.0 \ \Omega} = \boxed{1.1 \text{ A}}$$

   (c) Strategy Compute the resistance between A and B. Then use the definition of resistance to find the potential difference.

   Solution Compute the resistance. It is in series with the two resistors not between A and B.
   $106.0 \ \Omega - 20.0 \ \Omega - 50.0 \ \Omega = 36.0 \ \Omega$
   Compute the potential difference.
   $V = IR = (1.13 \text{ A})(36.0 \ \Omega) = \boxed{41 \text{ V}}$

   (d) Strategy The current that flows through the battery is shared by each branch between points A and B. Draw a diagram with equivalent resistances. Use Kirchhoff's rules.

   Solution The diagram is shown.
   $I = I_1 + I_2$ and $I_1 (60.0 \ \Omega) - I_2 (90.0 \ \Omega) = 0$, so $I_2 = \frac{60.0}{90.0} I_1$ and
   $$I = I_1 + \frac{60.0}{90.0} I_1 = \left(1 + \frac{60.0}{90.0} \right) I_1.$$

   So, the current through the upper branch is $I_1 = \left(1 + \frac{60.0}{90.0} \right)^{-1} I = \left(1 + \frac{60.0}{90.0} \right)^{-1} (1.13 \text{ A}) = \boxed{0.68 \text{ A}}$, and the

   current through the lower branch is $I_2 = \frac{60.0}{90.0} I_1 = \frac{60.0}{90.0}(0.68 \text{ A}) = \boxed{0.45 \text{ A}}$.

   (e) Strategy Use $P = I^2 R$.

   Solution Determine the power dissipated in the resistors.
   $P_{50} = (1.13 \text{ A})^2 (50.0 \ \Omega) = \boxed{64 \text{ W}}$, $P_{70} = (0.45 \text{ A})^2 (70.0 \ \Omega) = \boxed{14 \text{ W}}$, and
   $P_{40} = (0.68 \text{ A})^2 (40.0 \ \Omega) = \boxed{18 \text{ W}}$.

69. (a) Strategy Use Eq. (18-10).

Solution Compute the voltage across the terminals of the battery.
$$V = \mathcal{E} - Ir = 6.00 \text{ V} - (1.20 \text{ A})(0.600 \text{ } \Omega) = \boxed{5.28 \text{ V}}$$

(b) Strategy Use Eq. (18-19).

Solution Compute the power supplied by the battery.
$$P = IV = (1.20 \text{ A})(5.28 \text{ V}) = \boxed{6.34 \text{ W}}$$

73. Strategy Voltmeters are connected in parallel.

Solution Redraw the circuit to include the voltmeters.

(a)

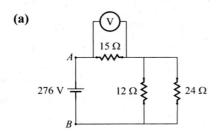

(b)

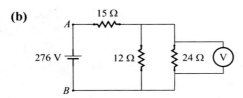

75. Strategy Ammeters are connected in series. Voltmeters are connected in parallel.

Solution

(a) Redraw the circuit to include the voltmeter.

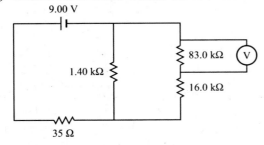

(b) Find the reading of the voltmeter. Assume the voltmeter to be ideal ($R = 0$). Use Kirchhoff's rules.

(1) $I_1 = I_2 + I_3$

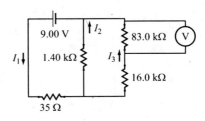

(2) $0 = 9.00\ \text{V} - I_1(35\ \Omega) - I_2(1.40\times10^3\ \Omega)$

$0 = I_2(1.40\times10^3\ \Omega) - I_3(99.0\times10^3\ \Omega)$

$I_3 = \dfrac{1.40\times10^3\ \Omega}{99.0\times10^3\ \Omega} I_2$

(3) $I_3 = \dfrac{1.40}{99.0} I_2$

Substitute (3) into (1).

$I_1 = I_2 + \dfrac{1.40}{99.0} I_2$

(4) $I_1 = \left(1 + \dfrac{1.40}{99.0}\right) I_2$

Substitute (4) into (2).

$0 = 9.00\ \text{V} - \left(1 + \dfrac{1.40}{99.0}\right) I_2(35\ \Omega) - I_2(1.40\times10^3\ \Omega)$, so $I_2 = \dfrac{9.00\ \text{V}}{\left(1 + \frac{1.40}{99.0}\right)(35\ \Omega) + 1.40\times10^3\ \Omega} = 6.27\ \text{mA}$.

Find I_3 using (3).

$I_3 = \dfrac{1.40}{99.0} I_2 = \dfrac{1.40}{99.0}(6.27\ \text{mA}) = 88.7\ \mu\text{A}$

The reading of the voltmeter is $V = IR = (88.7\ \mu\text{A})(83.0\ \Omega) = \boxed{7.36\ \text{V}}$.

(c) Find the reading of the voltmeter. Use Kirchhoff's rules.

(1) $I_1 = I_2 + I_3$

(2) $0 = 9.00\ \text{V} - I_1(35\ \Omega) - I_2(1.40\times10^3\ \Omega)$

$0 = I_2(1.40\times10^3\ \Omega)$

$-I_3\left[16.0\times10^3\ \Omega + \left(\dfrac{1}{83.0\times10^3\ \Omega} + \dfrac{1}{1.00\times10^6\ \Omega}\right)^{-1}\right]$

$I_3 = \dfrac{1.40\times10^3\ \Omega}{16.0\times10^3\ \Omega + \left(\frac{1}{83.0\times10^3\ \Omega} + \frac{1}{1.00\times10^6\ \Omega}\right)^{-1}} I_2$

(3) $I_3 = 0.01511 I_2$

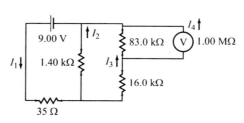

Substitute (3) into (1).

$I_1 = I_2 + 0.01511 I_2$

(4) $I_1 = 1.01511 I_2$

Substitute (4) into (2).

$0 = 9.00\ \text{V} - 1.01511 I_2(35\ \Omega) - I_2(1.40\times10^3\ \Omega)$, so $I_2 = \dfrac{9.00\ \text{V}}{1.01511(35\ \Omega) + 1.40\times10^3\ \Omega} = 6.27\ \text{mA}$.

Find I_3 using (3).

$I_3 = 0.01511 I_2 = 0.01511(6.27\ \text{mA}) = 94.7\ \mu\text{A}$

The reading of the voltmeter is $V = IR = (94.7\ \mu\text{A})(83.0\ \Omega) = \boxed{7.86\ \text{mV}}$.

77. **Strategy** The resistances are in series, so $V = IR_{\text{eq}}$.

Solution Find the required resistance of the series resistor.

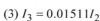

$(0.120\times10^{-3}\ \text{A})(R_S + 34.0\ \Omega) = 100.0\ \text{V}$, so $R_S = \dfrac{100.0\ \text{V}}{0.120\times10^{-3}\ \text{A}} - 34.0\ \Omega = \boxed{833\ \text{k}\Omega}$.

79. **Strategy** The resistances are in series, so $V = IR_{eq}$.

    Solution Find the required resistances of the series resistors.

    (a) $(2.0 \times 10^{-3} \text{ A})(R_S + 75 \text{ } \Omega) = 50.0 \text{ V}$, so $R_S = \dfrac{50.0 \text{ V}}{2.0 \times 10^{-3} \text{ A}} - 75 \text{ } \Omega = \boxed{25 \text{ k}\Omega}$.

    (b) $R_S = \dfrac{500.0 \text{ V}}{2.0 \times 10^{-3} \text{ A}} - 75 \text{ } \Omega = \boxed{250 \text{ k}\Omega}$

81. **Strategy** The resistances are in series, so $V = IR_{eq}$.

    Solution The two equations to solve simultaneously are
    $I(9850 \text{ } \Omega + R_G) = 25.0 \text{ V}$ and $I(3850 \text{ } \Omega + R_G) = 10.0 \text{ V}$.
    Find R_G.

    $$\frac{I(9850 \text{ } \Omega + R_G)}{I(3850 \text{ } \Omega + R_G)} = \frac{25.0 \text{ V}}{10.0 \text{ V}} = 2.50$$
    $$9850 \text{ } \Omega + R_G = 2.50(3850 \text{ } \Omega + R_G)$$
    $$R_G = \frac{2.50(3850 \text{ } \Omega) - 9850 \text{ } \Omega}{-1.50} = \boxed{150 \text{ } \Omega}$$

    Find I.

    $$I(9850 \text{ } \Omega + 150 \text{ } \Omega) = 25.0 \text{ V}, \text{ so } I = \frac{25.0 \text{ V}}{10.0 \times 10^{3} \text{ } \Omega} = \boxed{2.50 \text{ mA}}.$$

85. (a) **Strategy** The energy dissipated by the resistor is equal to the energy initially stored in the capacitor. Use Eqs. (18-24), (18-25), and (17-18b), and the definition of resistance.

    Solution Find the time constant.
    $$I(t = \tau) = I_0 e^{-1} \approx 0.368 I_0 = 0.368(100.0 \text{ mA}) = 36.8 \text{ mA}, \text{ so } \tau \approx 12.8 \text{ ms} = RC.$$
    $$R = \frac{V_0}{I_0} = \frac{9.0 \text{ V}}{100.0 \times 10^{-3} \text{ A}} = \boxed{90 \text{ } \Omega}; \quad C = \frac{\tau}{R} = \frac{0.0128 \text{ s}}{90 \text{ } \Omega} = \boxed{140 \text{ } \mu\text{F}};$$
    $$U = \frac{1}{2} C V_0^2 = \frac{1}{2}(142 \times 10^{-6} \text{ F})(9.0 \text{ V})^2 = \boxed{5.8 \text{ mJ}}$$

    (b) **Strategy** The energy is directly proportional to the voltage across the capacitor squared. Solve Eq. (18-26) for t.

    Solution The energy is half its initial value when
    $$V_C^2 = \frac{1}{2} V_0^2 = \left(\frac{V_0}{\sqrt{2}}\right)^2 \text{ or } V_C = \frac{V_0}{\sqrt{2}}.$$
    $$V_C = V_0 e^{-t/\tau} = \frac{V_0}{\sqrt{2}}, \text{ so } -\frac{t}{\tau} = \ln\frac{1}{\sqrt{2}} \text{ or } t = \frac{1}{2}\tau \ln 2 = \frac{1}{2}(12.8 \text{ ms})(0.693) = \boxed{4.4 \text{ ms}}.$$

(c) **Strategy** Use Eq. (18-26).

Solution Substitute numerical values and graph the voltage across the capacitor.

$$V_C(t) = V_0 e^{-t/\tau} = (9.0 \text{ V})e^{-t/(12.8 \text{ ms})}$$

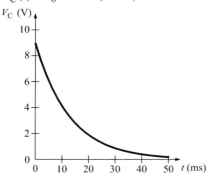

87. (a) **Strategy** Use Eq. (17-18b).

Solution Find the required initial potential difference.

$$U = \frac{1}{2}C(\Delta V)^2, \text{ so } \Delta V = \sqrt{\frac{2U}{C}} = \sqrt{\frac{2(20.0 \text{ J})}{100.0 \times 10^{-6} \text{ F}}} = \boxed{632 \text{ V}}.$$

(b) **Strategy** Use the definition of capacitance.

Solution Find the initial charge.

$$Q = C(\Delta V) = (100.0 \times 10^{-6} \text{ F})(632 \text{ V}) = \boxed{63.2 \text{ mC}}$$

(c) **Strategy** Solve for R using $I = I_0 e^{-t/\tau}$ where $\tau = RC$.

Solution Find the resistance of the lamp.

$$0.050I_0 = I_0 e^{-t/\tau}$$

$$\ln 0.050 = -\frac{t}{RC}$$

$$R = -\frac{t}{C \ln 0.050} = -\frac{0.0020 \text{ s}}{(100.0 \times 10^{-6} \text{ F}) \ln 0.050} = \boxed{6.7 \ \Omega}$$

89. **Strategy** Use Eqs. (18-24) and (18-25) and the definition of resistance.

Solution

(a) Initially ($t = 0$), the capacitor has nearly zero resistance. Find the currents and the voltages.

$$I_1 = I_2 = \frac{V}{R} = \frac{12 \text{ V}}{40.0 \times 10^3 \ \Omega} = \boxed{0.30 \text{ mA}} \text{ and } V_1 = V_2 = \boxed{12 \text{ V}}.$$

(b) Calculate the time constant.

$$\tau = (40.0 \times 10^3 \ \Omega)(5.0 \times 10^{-8} \text{ F}) = 2.0 \text{ ms}$$

Find the currents.

$$I_1 = I_2 = I_0 e^{-t/\tau} = (3.0 \times 10^{-4} \text{ A})e^{-(1.0 \text{ ms})/(2.0 \text{ ms})} = \boxed{0.18 \text{ mA}}$$

Find V_1 and V_2.

$V_1 = \boxed{12 \text{ V}}$ and $V_2 = I_2R = (1.82 \times 10^{-4} \text{ A})(40.0 \times 10^3 \text{ } \Omega) = \boxed{7.3 \text{ V}}$.

(c) Find the currents and the voltages.

$I_1 = I_2 = (3.0 \times 10^{-4} \text{ A})e^{-(5.0 \text{ ms})/(2.0 \text{ ms})} = \boxed{25 \text{ } \mu A}$, $V_1 = \boxed{12 \text{ V}}$, and

$V_2 = I_2R = (2.463 \times 10^{-5} \text{ A})(40.0 \times 10^3 \text{ } \Omega) = \boxed{0.99 \text{ V}}$.

93. (a) Strategy According to the figure, $I_0 \approx 0.070$ A. Use Eq. (18-25).

Solution Compute the current at $t = \tau$.

$I(t = \tau) = I_0e^{-1} = (0.070 \text{ A})e^{-1} = 0.026$ A

So, according to the figure, $\tau \approx 0.060$ s. The final charge is $Q = I_0 \Delta t = I_0 \tau \approx (0.070 \text{ A})(0.060 \text{ s}) = \boxed{4.2 \text{ mC}}$.

(b) Strategy Use the definition of capacitance.

Solution Find the capacitance.

$C = \dfrac{Q}{V} \approx \dfrac{0.0042 \text{ C}}{9.0 \text{ V}} = \boxed{470 \text{ } \mu F}$

(c) Strategy Use Eq. (18-24).

Solution Find the total resistance in the circuit.

$R = \dfrac{\tau}{C} \approx \dfrac{0.060 \text{ s}}{470 \times 10^{-6} \text{ F}} = \boxed{130 \text{ } \Omega}$

(d) Strategy Use Eqs. (17-18b) and (18-23) and the fact that $U = U_0/2$.

Solution Solve for t.

$$U = \frac{1}{2}CV^2 = \frac{1}{2}CV_0^2(1 - e^{-t/\tau})^2 = U_0(1 - e^{-t/\tau})^2$$

$$\pm\sqrt{\frac{U}{U_0}} = 1 - e^{-t/\tau}$$

$$e^{-t/\tau} = 1 \pm \sqrt{\frac{U}{U_0}}$$

$$-\frac{t}{\tau} = \ln\left(1 \pm \sqrt{\frac{U}{U_0}}\right)$$

$$t = -\tau \ln\left(1 \pm \sqrt{\frac{U}{U_0}}\right) \approx -(0.060 \text{ s})\ln\left(1 \pm \sqrt{\frac{1}{2}}\right) = -32 \text{ ms or } 74 \text{ ms}$$

t cannot be negative, so the answer is $\boxed{74 \text{ ms}}$.

95. (a) Strategy Use the definition of resistance.

Solution Compute the current that passes through Oscar.

$$I = \frac{V}{R} = \frac{100.0 \text{ V}}{2.0 \times 10^3 \ \Omega} = \boxed{50 \text{ mA}}$$

(b) Strategy I_1 passes through the 15 Ω resistor, and I_2 passes through Oscar. They are in parallel, so the voltage across each is the same, 100.0 V. Use Kirchhoff's rules.

Solution

(1) $1.00 \text{ A} = I_1 + I_2$

$V = I_1 R_1 = I_2 R_2$, so

(2) $(15 \ \Omega)I_1 = (2.0 \times 10^3 \ \Omega)I_2$

Solve (2) for I_1 and substitute it into (1).

$(15 \ \Omega)I_1 = (2.0 \times 10^3 \ \Omega)I_2$, so $I_1 = \dfrac{2.0 \times 10^3}{15} I_2$.

Substitute.

$$1.00 \text{ A} = \frac{2.0 \times 10^3}{15} I_2 + I_2, \text{ so } I_2 = \frac{1.00 \text{ A}}{\frac{2.0 \times 10^3}{15} + 1} = \boxed{7.4 \text{ mA}}.$$

97. (a) Strategy and Solution The circuit breaker should be placed at $\boxed{D}$. This will best protect the household against a short circuit in case any or all of the appliances overload the circuit or short out.

(b) Strategy Use Eq. (18-19).

Solution Compute the current.

$$I = \frac{P}{V} = \frac{1500 \text{ W} + 300 \text{ W} + 1200 \text{ W}}{120 \text{ V}} = 25 \text{ A}$$

> The devices cannot all be operated at the same time since the total current would be 25 A, which is greater than the rated 20.0 A.

101. Strategy Use the definition of resistance and Eqs. (18-13) and (18-17).

Solution

(a) The current through A_1 is the same as that through the emf.

$$I = \frac{V}{R_{eq}} = \frac{10.0 \text{ V}}{2.00 \ \Omega + \left(\frac{1}{2.00 \ \Omega} + \frac{1}{3.00 \ \Omega} + \frac{1}{6.00 \ \Omega}\right)^{-1} + 2.00 \ \Omega} = \boxed{2.00 \text{ A}}$$

(b) Since $\left(\dfrac{1}{3.00 \ \Omega} + \dfrac{1}{6.00 \ \Omega}\right)^{-1} = 2.00 \ \Omega$, the current is split evenly at the first junction to the right of A_1. So,

$$I = \frac{2.00 \text{ A}}{2} = \boxed{1.00 \text{ A}}.$$

105. Strategy Use Eq. (18-19).

Solution Compute the current drawn by the motor.

$$I = \frac{P}{V} = \frac{1.5 \text{ hp}}{120 \text{ V}}(745.7 \text{ W/hp}) = \boxed{9.3 \text{ A}}$$

107. Strategy Use Eq. (18-21b).

Solution Since the resistances are equal and in series, the voltage is dropped by half by each resistor in Circuit 1.

So, the power is $P_1 = \dfrac{V^2}{R} = \dfrac{(\mathcal{E}/2)^2}{R} = \dfrac{\mathcal{E}^2}{4R} = 5.0 \text{ W}.$

Since the bulbs are connected in parallel in Circuit 2, the voltage across each is $\mathcal{E}$ and, thus, the power dissipated

by each is $P_2 = \dfrac{V^2}{R} = \dfrac{\mathcal{E}^2}{R} = 4P_1 = 4(5.0 \text{ W}) = \boxed{20 \text{ W}}.$

109. Strategy Use Eqs. (18-21a) and (18-21b).

Solution

(a) $P = V^2/R$, so $R = V^2/P$. Compute the resistances.

$$R_{60} = \frac{(120 \text{ V})^2}{60.0 \text{ W}} = \boxed{240 \text{ }\Omega} \quad \text{and} \quad R_{100} = \frac{(120 \text{ V})^2}{100.0 \text{ W}} = \boxed{140 \text{ }\Omega}.$$

(b) Since $P = I^2 R$, and I is the same through each bulb (connected in series), the bulb with the larger resistance dissipates more power and, thus, shines brighter. So, the $\boxed{60.0\text{-W bulb}}$ shines brighter ($240 \text{ }\Omega > 140 \text{ }\Omega$).

(c) When the bulbs are connected in parallel, the voltage across each is the same; and since $P = V^2/R$, the bulb with the smaller resistance dissipates the most power, therefore, the $\boxed{100.0\text{-W bulb}}$ shines brighter ($140 \text{ }\Omega < 240 \text{ }\Omega$).

113. **(a) Strategy** Use Eq. (18-21b).

    Solution Find the resistance of the hair dryer.

    $$P = \frac{V^2}{R}, \text{ so } R = \frac{V^2}{P} = \frac{(120 \text{ V})^2}{1500 \text{ W}} = \boxed{9.6 \ \Omega}.$$

    (b) Strategy Use the definition of resistance.

    Solution Find the current through the hair dryer.

    $$I = \frac{V}{R} = \frac{120 \text{ V}}{9.6 \ \Omega} = \boxed{13 \text{ A}}$$

    (c) Strategy The total energy used by the hair dryer is equal to its power times the time of usage.

    Solution Find the cost to run the hair dryer for five minutes.

    $$\text{Cost} = E \times \text{rate} = P \Delta t \times \text{rate} = (1.5 \text{ kW})(5.00 \text{ min}) \frac{1 \text{ h}}{60 \text{ min}} \times \frac{10 \text{ cents}}{\text{kW} \cdot \text{h}} = \boxed{1.3 \text{ cents}}$$

    (d) Strategy Use Eq. (18-21b).

    Solution Find the power used.

    $$P = \frac{V^2}{R} = \frac{(240 \text{ V})^2}{9.6 \ \Omega} = \boxed{6.0 \text{ kW}}$$

    (e) Strategy Use the definition of resistance.

    Solution Find the current through the hair dryer.

    $$I = \frac{V}{R} = \frac{240 \text{ V}}{9.6 \ \Omega} = \boxed{25 \text{ A}}$$

117. **Strategy and Solution**

    (a) Olivia should use one of the 1.5 V batteries to oppose the 6.0 V battery; $6.0 \text{ V} + (-1.5 \text{ V}) = 4.5 \text{ V}$.

    (b) The current will flow in the wrong direction through the 1.5 V-battery. This current may be too large for the battery to handle, since the 1.5-V battery is not meant to be recharged.

121. **Strategy** The current is related to the drift speed by $I = neAv_{\text{D}}$. Form a proportion.

    Solution Compare the drift speeds.

    $$\frac{I_{\text{Al}}}{I_{\text{Au}}} = \frac{n_{\text{Al}} e A v_{\text{Al}}}{n_{\text{Au}} e A v_{\text{Au}}} = \frac{n_{\text{Al}} v_{\text{Al}}}{n_{\text{Au}} v_{\text{Au}}} = \frac{(3 n_{\text{Au}}) v_{\text{Al}}}{n_{\text{Au}} v_{\text{Au}}} = \frac{3 v_{\text{Al}}}{v_{\text{Au}}} = 1, \text{ so } \boxed{v_{\text{Au}} = 3 v_{\text{Al}}}.$$

125. **(a) Strategy** It is okay to treat the Earth-ionosphere system as a parallel plate capacitor, since
$\frac{d}{R} = \frac{5.0 \times 10^4 \text{ m}}{6.371 \times 10^6 \text{ m}} \approx 10^{-2}$; locally, the Earth is flat when compared with the distance between the "plates".
Use Eq. (17-15).

Solution Find the capacitance.
$$C = \frac{\epsilon_0 A}{d} = \frac{[8.854 \times 10^{-12} \text{ C}^2/(\text{N} \cdot \text{m}^2)] 4\pi (6.371 \times 10^6 \text{ m})^2}{5.0 \times 10^4 \text{ m}} = \boxed{0.090 \text{ F}}$$

(b) Strategy Use Eqs. (17-10) and (17-18b).

Solution Find the energy stored in the capacitor.
$$U = \frac{1}{2} CV^2 = \frac{1}{2} C(Ed)^2 = \frac{1}{2} (0.090 \text{ F})(150 \text{ V/m})^2 (5.0 \times 10^4 \text{ m})^2 = \boxed{2.5 \text{ TJ}}$$

(c) Strategy Use the definition of resistance and Eqs. (18-8) and (17-10).

Solution Compute the resistance.
$$R = \rho \frac{L}{A} = (3.0 \times 10^{14} \text{ } \Omega \cdot \text{m}) \frac{5.0 \times 10^4 \text{ m}}{4\pi (6.371 \times 10^6 \text{ m})^2} = \boxed{29 \text{ k}\Omega}$$

Compute the current.
$$I = \frac{V}{R} = \frac{Ed}{R} = \frac{\left(150 \frac{\text{V}}{\text{m}}\right)(5.0 \times 10^4 \text{ m})}{29{,}408 \text{ } \Omega} = \boxed{260 \text{ A}}.$$

(d) Strategy The system can be modeled by an *RC* circuit. The voltage across the capacitor while it is discharging is given by $V = V_0 e^{-t/(RC)}$. Since $Q = CV$, $Q = Q_0 e^{-t/(RC)}$ assuming *C* doesn't change.

Solution Solve for *t*.
$$Q = Q_0 e^{-t/(RC)}$$
$$e^{t/(RC)} = \frac{Q_0}{Q}$$
$$\frac{t}{RC} = \ln \frac{Q_0}{Q}$$
$$t = RC \ln \frac{Q_0}{Q} = \left[(29 \times 10^3 \text{ } \Omega)(0.090 \text{ F}) \ln \frac{1}{0.01} \right] \left(\frac{1 \text{ min}}{60 \text{ s}} \right) = \boxed{200 \text{ min}}$$

REVIEW AND SYNTHESIS: CHAPTERS 16–18

Review Exercises

1. **Strategy and Solution** Since the spheres are identical, the charge will be shared evenly by the two spheres, so the spheres will have $(18.0\ \mu C + 6.0\ \mu C)/2 = \boxed{12.0\ \mu C}$ of charge each.

5. **Strategy** The force on charge A due to charge B is equal and opposite to the horizontal component of the tension. The sign of charge A is negative, since the sign of charge B is positive and the force is attractive. Use Newton's second law and Eq. (16-2).

   Solution

   (a) Find the tension.

   $$\Sigma F_y = T\cos 7.20° - mg = 0,\ \text{so}\ T = \frac{mg}{\cos 7.20°}.$$

   The horizontal component of the tension is $T_x = \dfrac{mg}{\cos 7.20°}\sin 7.20° = mg\tan 7.20°$.

   Solve for the second charge.

   $$\frac{k|q_A||q_B|}{r^2} = mg\tan 7.20°,\ \text{so}$$

   $$|q_A| = \frac{r^2 mg\tan 7.20°}{k|q_B|} = \frac{(0.0500\ \text{m})^2(0.0900\ \text{kg})(9.80\ \text{m/s}^2)\tan 7.20°}{(8.988\times10^9\ \text{N}\cdot\text{m}^2/\text{C}^2)(130\times10^{-9}\ \text{C})} = 238\ \text{nC}.$$

   Thus, the charge on A is $\boxed{-238\ \text{nC}}$.

   (b) The tension in the thread is $T = \dfrac{mg}{\cos 7.20°} = \dfrac{(0.0900\ \text{kg})(9.80\ \text{m/s}^2)}{\cos 7.20°} = \boxed{0.889\ \text{N}}$.

9. (a) **Strategy** Use Eqs. (18-13) and (18-17).

   Solution Find the equivalent resistance.

   $$R_{eq} = 15.0\ \Omega + \left(\frac{1}{40.0\ \Omega} + \frac{1}{20.0\ \Omega} + \frac{1}{40.0\ \Omega}\right)^{-1} + 10.0\ \Omega = \boxed{35.0\ \Omega}$$

   (b) **Strategy** The current that flows through resistor R_1 is the current that flows through the emf.

   Solution Find the current.

   $$I = \frac{V}{R_{eq}} = \frac{24.0\ \text{V}}{35.0\ \Omega} = \boxed{0.686\ \text{A}}$$

   (c) **Strategy** Use Eq. (18-21b).

   Solution Find the power dissipated in the circuit.

   $$P = \frac{V^2}{R_{eq}} = \frac{(24.0\ \text{V})^2}{35.0\ \Omega} = \boxed{16.5\ \text{W}}$$

(d) Strategy R_2, R_3, and R_4 are in parallel, so the potential difference across each is the same. Use Kirchhoff's loop rule.

Solution Find the potential difference across R_3, V_3.

$V - IR_1 - V_3 - IR_5 = 0$, so $V_3 = V - IR_1 - IR_5 = 24.0\text{ V} - (0.686\text{ A})(15.0\ \Omega + 10.0\ \Omega) = \boxed{6.9\text{ V}}$.

(e) Strategy Use the definition of resistance.

Solution Find the current through R_3, I_3.

$$I_3 = \frac{V_3}{R_3} = \frac{6.86\text{ V}}{20.0\ \Omega} = \boxed{0.34\text{ A}}$$

(f) Strategy Use Eq. (18-21b).

Solution Find the power dissipated in R_3.

$$P = \frac{V_3^2}{R_3} = \frac{(6.9\text{ V})^2}{20.0\ \Omega} = \boxed{2.4\text{ W}}$$

13. (a) Strategy After the switch has been closed for a long time, the capacitor is fully charged and acts like a resistor with infinite resistance. Thus, all of the current passes through the 12-Ω resistor, thereby bypassing the capacitor. Therefore, the current through the 12-Ω resistor is equal to the current through the emf.

Solution The resistors are in series, so the equivalent resistance of the circuit is 27 Ω. Compute the current.

$$I = \frac{V}{R_{eq}} = \frac{12\text{ V}}{27\ \Omega} = \boxed{0.44\text{ A}}$$

(b) Strategy The capacitor and 12-Ω resistor are in parallel, so the voltage across each is the same.

Solution Find the voltage across the capacitor.

$$V_{cap} = V_R = IR = \frac{V}{R_{eq}}R = \frac{(12\text{V})(12\ \Omega)}{27\ \Omega} = \boxed{5.3\text{ V}}$$

17. Strategy Use the definition of resistance.

Solution No current flows through the upper branch of the circuit, since V_x is open. The voltage across R is $\mathscr{E} = 45.0$ V. So, the current is $I = \mathscr{E}/R$. The voltage across R_x is V_x, and the current is I. Find R_x.

$$R_x = \frac{V_x}{I} = \frac{V_x}{\mathscr{E}}R = \frac{30.0\text{ V}}{45.0\text{ V}}(100.0\ \Omega) = \boxed{66.7\ \Omega}$$

21. (a) **Strategy** Use Eq. (17-18b).

    Solution Find the capacitance.
    $$U = \frac{1}{2}C(\Delta V)^2, \text{ so } C = \frac{2U}{(\Delta V)^2} = \frac{2(32 \text{ J})}{(300 \text{ V})^2} = \boxed{710 \text{ } \mu\text{F}}.$$

    (b) **Strategy** Use Eq. (17-16).

    Solution Find the dielectric constant.
    $$C = \frac{\kappa A}{4\pi k d}, \text{ so } \kappa = \frac{4\pi k d C}{A} = \frac{4\pi(8.988\times10^9 \text{ N}\cdot\text{m}^2/\text{C}^2)(1.1\times10^{-6} \text{ m})(710\times10^{-6} \text{ F})}{9.0 \text{ m}^2} = \boxed{9.8}.$$

    (c) **Strategy** The average power produced is equal to the energy stored in the capacitor divided by the time it takes to discharge it.

    Solution Compute the average power.
    $$P_{av} = \frac{U}{\Delta t} = \frac{32 \text{ J}}{4.0\times10^{-3} \text{ s}} = \boxed{8.0 \text{ kW}}$$

    (d) **Strategy** The capacitance of a parallel plate capacitor is inversely proportional to the plate separation, and the energy stored in a capacitor is directly proportional to its capacitance. Thus, the energy stored in a capacitor is inversely proportional to the plate separation.

    Solution Form a proportion to find the new energy capacity of the capacitor.
    $$\frac{U_2}{U_1} = \frac{d_1}{d_2}, \text{ so } U_2 = \frac{d_1}{d_2}U_1 = \frac{d_1}{\frac{1}{2}d_1}U_1 = 2U_1 = 2(32 \text{ J}) = \boxed{64 \text{ J}}.$$

25. **Strategy** Use the relationships between power, resistance, voltage, charge, and time. Use Eqs. (14-4) and (14-9).

    Solution To find the volume change in the wax, we use the following.
    $$\Delta V = \pi r^2 \Delta x = \pi(0.0020 \text{ m})^2(0.010 \text{ m}) = 1.257\times10^{-7} \text{ m}^3 = 0.1257 \text{ cm}^3 = 0.1257 \text{ ml}$$
    Find the volume of wax that must melt.
    $$0.15 V_{melt} = \Delta V = 0.1257 \text{ ml}$$
    $$V_{melt} = \frac{0.1257 \text{ ml}}{0.15} = 8.38\times10^{-7} \text{ ml}$$
    Calculate the heat required to melt the wax.
    $$Q = mc\Delta T + m_{melt} L_f = (2.0 \text{ ml})\frac{0.90 \text{ g}}{\text{ml}}[0.80 \text{ J/(g}\cdot{}^\circ\text{C)}](70^\circ\text{C}) + (0.838 \text{ ml})\frac{0.90 \text{ g}}{\text{ml}}(60 \text{ J/g}) = 146 \text{ J}$$
    Find the time it takes for the valve to fully open.
    $$t = \frac{Q}{P} = \frac{QR}{V^2} = \frac{(146 \text{ J})(200 \text{ }\Omega)}{(24 \text{ V})^2} = \boxed{51 \text{ s}}$$

MCAT Review

1. **Strategy and Solution** At a given temperature, the resistance R of a wire to direct current is given by $R = \rho L / A$, where ρ is the resistivity, L is the length, and A is the cross-sectional area. Therefore, the correct answer is $\boxed{D}$.

2. **Strategy** There are ten electric immersion heaters that each use 5 kW of power. Use Eq. (18-19).

   Solution The total power requirement to run all ten heaters is 50 kW. Find the current.

   $$P = I\Delta V, \text{ so } I = \frac{P}{\Delta V} = \frac{50\times10^3 \text{ W}}{600 \text{ V}} = 83 \text{ A}.$$

   The correct answer is $\boxed{C}$.

3. **Strategy** Each heater draws 20 A, so five heaters draw 100 A. Use Eq. (18-19).

   Solution Find the total power usage of the heaters.
   $$P = I\Delta V = (100 \text{ A})(800 \text{ V}) = 80 \text{ kW}$$

   The correct answer is $\boxed{C}$.

4. **Strategy** Use Kirchhoff's rules and the definition of resistance.

   Solution Find the current flowing through R_L. According to the loop rule,

   $$I_L R_L - I_S R_S = 0, \text{ so } I_S = \frac{R_L}{R_S} I_L = \frac{1.0 \text{ }\Omega}{2.0 \text{ }\Omega} I_L = 0.50 I_L. \text{ According to the junction rule,}$$

   $$I = I_L + I_S = I_L + 0.50 I_L = 1.50 I_L, \text{ so } I_L = \frac{I}{1.50}. \text{ Thus, the voltage drop across } R_L \text{ is}$$

   $$V_L = I_L R_L = \frac{I R_L}{1.50} = \frac{(0.5 \text{ A})(1.0 \text{ }\Omega)}{1.50} = 0.33 \text{ V}. \text{ The correct answer is } \boxed{B}.$$

5. **Strategy** Use Kirchhoff's rules and Eq. (18-21a).

   Solution Find the current flowing through R_S. According to the loop rule,

   $$I_L R_L - I_S R_S = 0, \text{ so } I_L = \frac{R_S}{R_L} I_S = \frac{3.0 \text{ }\Omega}{1.0 \text{ }\Omega} I_S = 3.0 I_S. \text{ According to the junction rule,}$$

   $$I = I_L + I_S = 3.0 I_S + I_S = 4.0 I_S, \text{ so } I_S = 0.25 I. \text{ Thus, the power dissipated in } R_S \text{ is}$$

   $$P_S = I_S^2 R_S = (0.25 I)^2 R_S = [0.25(1.2 \text{ A})]^2 (3.0 \text{ }\Omega) = 0.27 \text{ W}. \text{ The correct answer is } \boxed{A}.$$

6. **Strategy and Solution** As current flows through R_L, power is dissipated at a constant rate by R_L as heat that enters the water, increasing its energy and raising its temperature (and the temperature of the system). Thus, the entropy of the system increases, as well. The correct answer is $\boxed{D}$.

7. **Strategy and Solution** Energy is stored in the battery as chemical energy. This energy is converted into electrical energy when the current flows. The electrical energy is dissipated as heat by the resistor. The correct answer is $\boxed{A}$.

8. **Strategy and Solution** As R_L increases with time, the amount of the current I passing through it decreases and the amount passing through R_S increases. The correct answer is $\boxed{C}$.

9. **Strategy** Use Eq. (14-4).

   Solution The water is heated at a rate of $Q/\Delta t = mc\Delta T/\Delta t = 1.0$ W. So, the time it takes for the temperature of the water to increase $1.0°C$ is $\Delta t = \dfrac{mc\Delta T}{1.0 \text{ W}} = \dfrac{(1.0 \text{ kg})[4.2\times10^3 \text{ J}/(\text{kg}\cdot°\text{C})](1.0°\text{C})}{1.0 \text{ W}} = 4200$ s. The correct answer is $\boxed{D}$.

10. **Strategy** Use Eq. (18-19).

    Solution Compute the current required.
    $P = I\Delta V$, so $I = \dfrac{P}{\Delta V} = \dfrac{1.2\times10^4 \text{ W}}{120 \text{ V}} = 100$ A. The correct answer is $\boxed{C}$.

11. **Strategy** Refer to the table to compute the initial and final resistances.

    Solution Compute the resistances.
    $R_i = (10^5 \text{ m})\dfrac{3.4\times10^{-1} \text{ } \Omega}{10^3 \text{ m}} = 34 \text{ } \Omega$ and $R_f = (10^5 \text{ m})\dfrac{3.8\times10^{-1} \text{ } \Omega}{10^3 \text{ m}} = 38 \text{ } \Omega$.
    The change in resistance is $R_f - R_i = 38 \text{ } \Omega - 34 \text{ } \Omega = 4 \text{ } \Omega$. The correct answer is $\boxed{C}$.

12. **Strategy** Use Eq. (18-21a).

    Solution Compute the power lost as heat.
    $P = I^2 R = (2 \text{ A})^2(3 \text{ } \Omega) = 12$ W, so the correct answer is $\boxed{C}$.

13. **Strategy and Solution** The ten residences require $10\times10^4 \text{ W} = 10^5$ W of power and 5×10^3 W of power is lost as heat, so the total power requirement is $10^5 \text{ W} + 5\times10^3 \text{ W} = 1.05\times10^5$ W. The correct answer is $\boxed{C}$.

Chapter 19

MAGNETIC FORCES AND FIELDS

Conceptual Questions

1. The magnetic field cannot be described as the magnetic force per unit charge because unlike the electric force, the magnetic force depends upon the velocity of the charge.

5. A constant magnetic field does zero work on a moving charge and therefore cannot change its kinetic energy—the speed of a particle with constant kinetic energy must also be constant.

9. It is impossible to completely eliminate the magnetic fields generated by power lines. Current in the lines produces magnetic fields that encircle them—the magnitude of these fields falls fairly rapidly but, in principle, has infinite extent.

13. The magnetic field around the speaker would deflect the electron beams, and thus, distort the image on the computer monitor.

17. (a) The metal bar was placed in an external magnetic field and the domains, on average, lined up with the field.

    (b) The metal is ferromagnetic as evidenced by the fact that it contains small magnetic domains in which the magnetic moments of the atoms in the metal are all lined up.

Problems

1. **Strategy** The magnetic field is strong where field lines are close together and weak where they are far apart.

   Solution

   (a) The field lines are farthest apart (lowest density) at point $\boxed{F}$, so the magnetic field strength is smallest there.

   (b) The field lines are closest together (highest density) at point $\boxed{A}$, so the magnetic field strength is largest there.

3. **Strategy** A bar magnet is a magnetic dipole. Field lines emerge from a bar magnet at its north pole and enter at its south pole. Magnetic field lines are closed loops. Use symmetry.

   Solution

   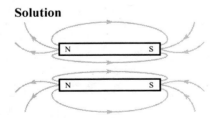

5. **Strategy** A bar magnet is a magnetic dipole. Field lines emerge from a bar magnet at its north pole and enter at its south pole. Magnetic field lines are closed loops. Use symmetry.

   Solution

   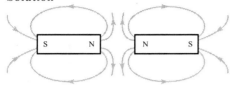

9. **Strategy** The magnetic force on a moving point charge is given by Eq. (19-5). The direction of the force is determined by the right-hand rule and the sign of the charge.

   Solution Find the magnetic force exerted on the proton.

   $$\vec{F} = q\vec{v} \times \vec{B} = e\vec{v} \times \vec{B} = (1.602 \times 10^{-19} \text{ C})[(6.0 \times 10^{6} \text{ m/s east}) \times (2.50 \text{ T north})] = \boxed{2.4 \times 10^{-12} \text{ N up}}$$

11. **Strategy** Determine the speed of the electron using its kinetic energy. Then determine the magnetic force on it using Eq. (19-5).

    Solution Find the speed.

    $$K = \frac{1}{2}mv^2, \text{ so } v = \sqrt{\frac{2K}{m}}.$$

    Calculate the force.

    $$\vec{F} = -e\vec{v} \times \vec{B} = -(1.602 \times 10^{-19} \text{ C})\left\{ \left[\sqrt{\frac{2(7.2 \times 10^{-18} \text{ J})}{9.109 \times 10^{-31} \text{ kg}}} \text{ east} \right] \times (0.800 \text{ T up}) \right\} = \boxed{5.1 \times 10^{-13} \text{ N north}}$$

13. **Strategy** The magnetic force on a moving point charge is given by Eq. (19-5). The direction of the force is determined by the right-hand rule and the sign of the charge.

    Solution Find the magnetic force on the electron at point *b*.
    $$\vec{F} = -e[(v \text{ at } 20.0° \text{ above left}) \times (B \text{ down})]$$

    $$= (1.602 \times 10^{-19} \text{ C})(8.0 \times 10^{5} \text{ m/s})(0.40 \text{ T})\sin 110.0° \text{ into the page} = \boxed{4.8 \times 10^{-14} \text{ N into the page}}$$

    (e) **Strategy and Solution** Since the force due to the magnetic field is always perpendicular to the velocity of the electrons, it does not increase the electrons' speed but only changes their direction.

17. **Strategy** Use Eqs. (19-1) and (19-5) to find the magnitude and direction of the force, respectively.

    Solution The charge of the muon is negative. According to the RHR and $\vec{\mathbf{F}}_B = q\vec{\mathbf{v}} \times \vec{\mathbf{B}}$, the magnetic force on the muon is into the page, or to the west.
    Compute the magnitude.

    $F_B = |q|(v \sin \theta)B$
    $$= (1.602 \times 10^{-19} \text{ C})(4.5 \times 10^7 \text{ m/s}) \sin(55° + 90°)(5.0 \times 10^{-5} \text{ T}) = 2.1 \times 10^{-16} \text{ N}$$

    Thus, the force on the muon is $\boxed{2.1 \times 10^{-16} \text{ N to the west}}$.

21. **Strategy** Since the magnetic field points south and the magnetic force is upward, the (negatively charged) electron's component of velocity perpendicular to the magnetic field points east.

    Solution Find the angle.

    $$evB \sin \theta = F, \text{ so } \theta = \sin^{-1} \frac{F}{evB} = \sin^{-1} \frac{1.6 \times 10^{-14} \text{ N}}{(1.602 \times 10^{-19} \text{ C})(2.0 \times 10^5 \text{ m/s})(1.4 \text{ T})} = 21°.$$

    $\boxed{\text{There are two possibilities: } 21° \text{ E of N and } 21° \text{ E of S.}}$

23. **Strategy** $\vec{\mathbf{B}}$ and $\vec{\mathbf{v}}$ are perpendicular. Use Eq. (19-6).

    Solution Find the magnitude of the magnetic field.
    $$F = evB, \text{ so } B = \frac{F}{ev} = \frac{1.0 \times 10^{-13} \text{ N}}{(1.602 \times 10^{-19} \text{ C})(8.0 \times 10^5 \text{ m/s})} = \boxed{0.78 \text{ T}}.$$

25. **Strategy** Solve Eq. (19-7) for the speed of the proton.

    Solution Find the speed.
    $$\frac{v^2}{r} = \frac{evB}{m}, \text{ so } v = \frac{reB}{m} = \frac{(0.820 \text{ m})(1.602 \times 10^{-19} \text{ C})(0.360 \text{ T})}{1.673 \times 10^{-27} \text{ kg}} = \boxed{2.83 \times 10^7 \text{ m/s}}.$$

29. **Strategy** Use conservation of energy and Equation (19-7).

    Solution Find the speed of the ion.
    $$\Delta K = \frac{1}{2}mv^2 = -\Delta U = eV, \text{ so } v = \sqrt{\frac{2eV}{m}}.$$
    Solve for the magnitude of the magnetic field.
    $$\frac{evB}{m} = \frac{v^2}{r}, \text{ so}$$
    $$B = \frac{mv}{er} = \frac{m}{er}\sqrt{\frac{2eV}{m}} = \frac{1}{r}\sqrt{\frac{2mV}{e}} = \frac{1}{0.21 \text{ m}}\sqrt{\frac{2(12.0 \text{ u})(1.6605 \times 10^{-27} \text{ kg/u})(5.0 \times 10^3 \text{ V})}{1.602 \times 10^{-19} \text{ C}}} = \boxed{0.17 \text{ T}}.$$

31. **Strategy** Since the ions are accelerated through the same potential difference and they have the same charge, they have the same kinetic energy. Let m_u be the mass of the unknown element. Use Eq. (19-7) and refer to the periodic table.

    Solution

    (a) Relate the speeds to the masses.

    $$\frac{1}{2}m_S v_S^2 = \frac{1}{2}m_u v_u^2, \text{ so } \frac{v_S}{v_u} = \sqrt{\frac{m_u}{m_S}}.$$

    $$\frac{v^2}{r} = \frac{|q|vB}{m}, \text{ so } r = \frac{mv}{|q|B}. \text{ Form a proportion.}$$

    $$\frac{r_S}{r_u} = \frac{m_S v_S}{m_u v_u} = \frac{m_S}{m_u}\sqrt{\frac{m_u}{m_S}} = \sqrt{\frac{m_S}{m_u}}, \text{ or } m_u = \left(\frac{r_u}{r_S}\right)^2 m_S = \left(\frac{r_S + 1.07 \text{ cm}}{r_S}\right)^2 m_S = \left(1 + \frac{1.07 \text{ cm}}{r_S}\right)^2 m_S.$$

    Similarly, we have $\dfrac{r_{Mn}}{r_S} = \sqrt{\dfrac{m_{Mn}}{m_S}}$, so $r_{Mn} = r_S\sqrt{\dfrac{m_{Mn}}{m_S}}$. Find r_S.

    $$r_{Mn} - r_S = r_S\left(\sqrt{\frac{m_{Mn}}{m_S}} - 1\right), \text{ so } r_S = \frac{r_{Mn} - r_S}{\sqrt{\dfrac{m_{Mn}}{m_S}} - 1} = \frac{3.20 \text{ cm}}{\sqrt{\dfrac{55 \text{ u}}{32 \text{ u}}} - 1} = 10.3 \text{ cm}.$$

    Therefore, the mass of the unknown element is $m_u = \left(1 + \dfrac{1.07 \text{ cm}}{10.3 \text{ cm}}\right)^2 (32 \text{ u}) = \boxed{39 \text{ u}}$.

    (b) According to the periodic table, the unknown element is $\boxed{\text{potassium}}$.

33. **Strategy** Use Eq. (19-6) and Newton's second law.

    Solution $\Sigma F_r = qvB = ma_r = mv^2/r$, so $v = qrB/m$. The period is $T = C/v = 2\pi r/v$. Substitute for v in the equation for T.

    $$T = \frac{2\pi r}{\frac{qrB}{m}} = \boxed{\frac{2\pi m}{qB}}, \text{ which is independent of the particle's speed.}$$

37. **Strategy** As found in Example 19.7, the Hall voltage is given by $V_H = BI/(net)$.

    Solution Find the density of the carriers.

    $$n = \frac{BI}{etV_H} = \frac{(0.43 \text{ T})(54 \text{ A})}{(1.602 \times 10^{-19} \text{ C})(0.00024 \text{ m})(7.2 \times 10^{-6} \text{ V})} = \boxed{8.4 \times 10^{28} \text{ m}^{-3}}$$

41. **Strategy** The blood speed is equal to the Hall drift speed $v_D = E_H/B = V_H/(wB)$, where the width w is the diameter of the artery. The flow rate is equal to the drift speed times the cross-sectional area of the artery. The magnetic force on a moving charge is given by Eq. (19-5). The direction of the force is determined by the right-hand rule and the sign of the charge.

    Solution

    (a) Compute the drift speed.

    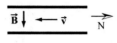

    $$v_D = \frac{V_H}{wB} = \frac{0.35 \times 10^{-3} \text{ V}}{(0.0040 \text{ m})(0.25 \text{ T})} = \boxed{0.35 \text{ m/s}}$$

    (b) Compute the flow rate.

    $$\text{flow rate} = v_D A = (0.35 \text{ m/s})\pi(0.0020 \text{ m})^2 = \boxed{4.4 \times 10^{-6} \text{ m}^3/\text{s}}$$

    (c) The positive ions are deflected east (south $\times$ down), so $\boxed{\text{the east lead}}$ is at the higher potential.

45. **Strategy** The magnitude of the force is given by $F = ILB \sin \theta$.

    Solution

    (a) Solve for B.

    $$B = \frac{F}{IL \sin \theta}$$

    B is a minimum when $\sin \theta = 1$ ($\theta = 90°$).

    $$B_{\min} = \frac{F}{IL(1)} = \frac{4.12 \text{ N}}{(33.0 \text{ A})(0.25 \text{ m})} = \boxed{0.50 \text{ T}}$$

    (b) $\boxed{\text{We do not know the directions of the current and the field; therefore, we set } \sin \theta = 1 \text{ and get the minimum field strength.}}$

47. (a) **Strategy** Use Eq. (19-12a). Refer to the figure.

    Solution The current flows from east to west. According to $\vec{F} = I\vec{L} \times \vec{B}$ and the RHR, the force on the rod is directed to the $\boxed{\text{north}}$.

    (b) **Strategy** The rod will accelerate northward. Use Newton's second law and Eq. (19-12b).

    Solution Find the acceleration of the rod.

    $$\Sigma F = ILB = ma, \text{ so } a = \frac{ILB}{m}.$$

    Find the speed of the rod.

    $$v_f^2 - v_i^2 = v^2 - 0 = 2a\Delta r = 2\frac{ILB}{m}\Delta r, \text{ so}$$

    $$v = \sqrt{\frac{2ILB\Delta r}{m}} = \sqrt{\frac{2(2.00 \text{ A})(0.500 \text{ m})(0.750 \text{ T})(8.00 \text{ m})}{0.0500 \text{ kg}}} = \boxed{15.5 \text{ m/s}}.$$

49. Strategy Use Eq. (19-12a).

Solution

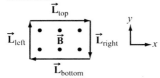

(a) Calculate the force on each wire segment.

$\vec{F}_{top} = I\vec{L}_{top} \times \vec{B} = (1.0 \text{ A})(0.300 \text{ m right}) \times (2.5 \text{ T out of the page}) = \boxed{0.75 \text{ N in the } -y\text{-direction}}$

$\vec{F}_{bottom} = I\vec{L}_{bottom} \times \vec{B} = -I\vec{L}_{top} \times \vec{B} = \boxed{0.75 \text{ N in the } +y\text{-direction}}$

$\vec{F}_{left} = I\vec{L}_{left} \times \vec{B} = (1.0 \text{ A})(0.200 \text{ m up}) \times (2.5 \text{ T out of the page}) = \boxed{0.50 \text{ N in the } +x\text{-direction}}$

$\vec{F}_{right} = I\vec{L}_{right} \times \vec{B} = -I\vec{L}_{left} \times \vec{B} = \boxed{0.50 \text{ N in the } -x\text{-direction}}$

(b) Compute the components of the net force.

$F_{net, x} = 0.50 \text{ N} - 0.50 \text{ N} = 0$ and $F_{net, y} = 0.75 \text{ N} - 0.75 \text{ N} = 0$, so $\vec{F}_{net} = \boxed{0}$.

51. Strategy Use Eq. (19-12a).

Solution

(a) $\vec{F} = I\vec{L} \times \vec{B}$, where $\vec{L} = L$ west and $\vec{B} = 0.48 \text{ mT}$ at 72° below the horizontal with the horizontal component due north. The diagram below shows the orientation of $\vec{L}$, $\vec{B}$, and $\vec{F}$ where $\vec{F}$ was determined by the right hand rule.

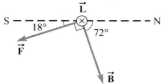

$\vec{F}$ is directed $\boxed{18° \text{ below the horizontal with the horizontal component due south}}$.

(b) $F = ILB$, so $I = \dfrac{F/L}{B} = \dfrac{0.020 \text{ N/m}}{0.48 \times 10^{-3} \text{ T}} = \boxed{42 \text{ A}}$.

53. (a) Strategy The torque on a current loop is $\tau = NIAB \sin\theta$. In this case (maximum torque) $\sin\theta = \sin 90° = 1$.

Solution Find B.

$\tau = NIAB$, so $B = \dfrac{\tau}{NIA} = \dfrac{0.0020 \text{ N} \cdot \text{m}}{100(0.075 \text{ A})\pi(0.020 \text{ m})^2} = \boxed{0.21 \text{ T}}$.

(b) Strategy Use Eq. (19-12a) and symmetry.

Solution The field points from the north pole to the south. Due to symmetry, $\vec{L}_{av}$ is parallel to the faces of the poles and into the page for the right half of the loop; $\vec{L}_{av}$ is perpendicular to $\vec{B}$. So, $\vec{L}_{av} \times \vec{B}$ is directed downward. By similar reasoning, the force on the right half of the loop is equal and opposite. Thus, the torque is $\boxed{\text{clockwise}}$.

57. Strategy Use Eq. (19-13a).

Solution

(a) $\tau = IAB$ and for constant I and B, $\tau \propto A$. Areas $A_{\text{one turn}} = (L/4)^2 = L^2/16$ and $A_{\text{two turns}} = (L/8)^2 = L^2/64$, so the fewer turns the larger the area and, thus, the greater the torque. Therefore, one turn gives the maximum torque.

(b) $\tau = IAB = I\left(\dfrac{L}{4}\right)^2 B = \dfrac{1}{16}L^2 IB$

61. Strategy Use the principle of superposition and the field due to a long straight current-carrying wire, Eq. (19-14).

Solution According to RHR 2, the field due to the bottom wire is out of the page and that due to the top wire is into the page. Since the bottom wire is closer to P, the net field is out of the page.

$$B = \frac{\mu_0 I}{2\pi}\left(\frac{1}{r_{\text{bottom}}} - \frac{1}{r_{\text{top}}}\right) = \frac{(4\pi \times 10^{-7}\ \text{T} \cdot \text{m/A})(10.0\ \text{A})}{2\pi}\left(\frac{1}{0.25\ \text{m}} - \frac{1}{0.25\ \text{m} + 0.0030\ \text{m}}\right) = 9 \times 10^{-8}\ \text{T}$$

So, $\vec{\mathbf{B}}(P) = \boxed{9 \times 10^{-8}\ \text{T out of the page}}$.

63. Strategy Use the principle of superposition and the field due to a long straight current-carrying wire, Eq. (19-14).

Solution The fields are equal in magnitude $(I_1 = I_2;\ r_1 = r_2)$ and opposite in direction at point P, so $\vec{\mathbf{B}}(P) = \boxed{0}$.

65. Strategy Use Eqs. (19-5) and (19-14).

Solution $\vec{\mathbf{B}}$ is into the page at the electron. $\vec{\mathbf{v}} \times \vec{\mathbf{B}}$ is to the left, so $-\vec{\mathbf{v}} \times \vec{\mathbf{B}}$ is to the right (parallel to the current).

$$\vec{\mathbf{F}} = q\vec{\mathbf{v}} \times \vec{\mathbf{B}} = ev\left(\frac{\mu_0 I}{2\pi r}\right) \text{ parallel to the current}$$

$$= \frac{(1.602 \times 10^{-19}\ \text{C})(1.0 \times 10^7\ \text{m/s})(4\pi \times 10^{-7}\ \text{T} \cdot \text{m/A})(50.0\ \text{A})}{2\pi(0.050\ \text{m})} \text{ parallel to the current}$$

$$= \boxed{3.2 \times 10^{-16}\ \text{N parallel to the current}}$$

67. Strategy The magnitude of the magnetic field due to a long strait wire is inversely proportional to the perpendicular distance from the axis of the wire. Use the principle of superposition.

Solution Form a proportion.

$$\frac{B_{\text{vert}}}{B_{\text{horiz}}} = \frac{r_{\text{horiz}}}{r_{\text{vert}}}$$

At A, the fields are equal in magnitude and opposite in direction, so $B = 0$.

At B, the fields are equal in magnitude and in the same direction, so $B = 2B_d = 2\left(\dfrac{\mu_0 I}{2\pi d}\right)$.

At C, $B = B_d - B_{2d} = B_d - \dfrac{B_d}{2} = \dfrac{B_d}{2}$. At D, $B = B_d + B_{2d} = B_d + \dfrac{B_d}{2} = \dfrac{3B_d}{2}$.

The ranking is: $\boxed{B,\ D,\ C,\ A}$.

69. **Strategy** The magnetic field at each point in question is equal to the vector sum of the two fields generated by the currents. Use Eqs. (19-12a) and (19-14).

    Solution Due to symmetry, the magnitudes of the fields due to each wire are the same at each location, only the directions differ. According to the RHR, the field at A and B due to the vertical wire (1) is directed out of the page and that due to the horizontal wire (2) is directed into the page at A and out of the page at B, so the direction of the total magnetic field is either into or out of the page at A and out of the page at B. Find the magnitudes of the fields due to each wire.

    $B_1 = \dfrac{\mu_0 I}{2\pi d}$ and $B_2 = \dfrac{\mu_0 I}{2\pi d}$, so $B_1 = B_2 = B$.

    Find the field at A.

    $\vec{B}_1 + \vec{B}_2 = B$ out of the page $+ B$ into the page $= -B$ into the page $+ B$ into the page $= \boxed{0}$

    Find the field at B.

    $\vec{B}_1 + \vec{B}_2 = B$ out of the page $+ B$ out of the page $= 2B$ out of the page

    $= \dfrac{\mu_0 I}{\pi d}$ out of the page $= \dfrac{(4\pi\times10^{-7}\ \text{T}\cdot\text{m/A})(5.70\times10^{-2}\ \text{A})}{\pi(0.0675\ \text{m})}$ out of the page

    $= \boxed{3.38\times10^{-7}\ \text{T out of the page}}$

73. **Strategy** Since the magnetic field in the center of the concentric circular wires is zero, the current in the smaller loop (1) must be opposite in direction (CCW) to that of the larger loop (2). Use Eq. (19-16).

    Solution Find the magnitude of the current in the smaller loop, I_1.

    $0 = B_2 - B_1 = \dfrac{\mu_0 I_2}{2r_2} - \dfrac{\mu_0 I_1}{2r_1} = \dfrac{\mu_0}{2}\left(\dfrac{I_2}{r_2} - \dfrac{I_1}{r_1}\right)$, so $I_1 = \dfrac{r_1}{r_2}I_2 = \dfrac{4.42}{6.20}(8.46\ \text{A}) = 6.03\ \text{A}$.

    Thus, the current in the smaller loop is $\boxed{6.03\ \text{A, CCW}}$.

75. **Strategy** The currents have the same magnitude and are equidistant from the point at which the net magnetic field's direction is evaluated. So, the magnitudes of the four fields are the same.

    Solution According to RHR 2 and symmetry, the field directions are the following:

	Current	Direction at the center of the square
1	top left	toward the top right wire
2	top right	toward the bottom right wire
3	bottom left	toward the bottom right wire
4	bottom right	toward the top right wire

    Since the magnitudes of the fields are equal, the vertical component of the net field is zero and the horizontal component is to the right.

    Each field is at a 45° angle to the horizontal (either above or below). The y-components cancel and the x-components add. The field for a long thin wire is $B = \mu_0 I/(2\pi r)$. Find the magnitude of the field.

    $B_x = \dfrac{\mu_0 I}{2\pi r}\cos 45° = \dfrac{\mu_0 I}{2\sqrt{2}\pi r}$. Thus, $B_{\text{net},x} = \dfrac{4\mu_0 I}{2\sqrt{2}\pi r} = \dfrac{\sqrt{2}\mu_0 I}{\pi r} = B_{\text{net}}$, where $r = \sqrt{\left(\dfrac{s}{2}\right)^2 + \left(\dfrac{s}{2}\right)^2} = \sqrt{\dfrac{s^2}{2}} = \dfrac{s}{\sqrt{2}}$.

    So, $B_{\text{net}} = \dfrac{2\mu_0 I}{\pi s} = \dfrac{2(4\pi\times10^{-7}\ \text{T}\cdot\text{m/A})(10.0\ \text{A})}{\pi(0.10\ \text{m})} = 80\ \mu\text{T}$, and $\vec{B} = \boxed{80\ \mu\text{T to the right}}$.

77. Strategy Use RHR 2, symmetry, and the expression for a long thin wire.

Solution By symmetry and RHR 2, the fields due to the two currents on the left are directed to the right at point R. By symmetry and RHR 2, the vertical components of the fields due to the two currents on the right cancel at point R, and the horizontal components are equal in magnitude and are directed to the right.

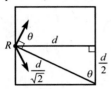

From the figure (which does not show the field vectors for the two currents on the left), we see that if $d = 0.10$ m, the distance from each of the two currents on the right to point P is $d/\sqrt{2}$. Find the angle θ.

$$\theta = \tan^{-1}\frac{d}{d/2} = \tan^{-1} 2 \approx 63.435°$$

The vertical components cancel; the horizontal components add. The field for a long thin wire is $B = \mu_0 I/(2\pi r)$. Find the magnitude of the field.

$$B_{\text{net},x} = 2B_{\text{left}} + 2B_{\text{right}} = 2\frac{\mu_0 I}{2\pi(d/2)} + 2\frac{\mu_0 I}{2\pi(d/\sqrt{2})}\cos 63.435° = \frac{2\mu_0 I}{\pi d} + \frac{\sqrt{2}\mu_0 I}{\pi d}\cos 63.435°$$

$$= \frac{\mu_0 I}{\pi d}(2 + \sqrt{2}\cos 63.435°) = \frac{(4\pi\times10^{-7}\text{ T}\cdot\text{m/A})(10.0\text{ A})(2 + \sqrt{2}\cos 63.435°)}{\pi(0.10\text{ m})} = 0.11\text{ mT}$$

Thus, $\vec{\mathbf{B}} = \boxed{0.11\text{ mT to the right}}$.

81. Strategy Use Ampère's law for each region.

Solution

(a) $r \le a$:

Let the closed circular path be concentric with the center of the cross-section and have radius $r = a$. Inside the path, $I = 0$, so $\Sigma B_{\parallel}\Delta l = 0$, or $B_{\parallel} = 0$. According to RHR 2, the field is such that $B_{\perp} = 0$, so $\vec{\mathbf{B}} = 0$.

$a \le r \le b$:

The amount of current flowing through the shell is proportional to its cross-sectional area. Let the closed circular path be as before except with radius r between a and b. The current enclosed I' is

$$\frac{I'}{I} = \frac{A'}{A} = \frac{\pi(r^2 - a^2)}{\pi(b^2 - a^2)}, \text{ so } I' = \frac{\pi(r^2 - a^2)}{\pi(b^2 - a^2)}I = \frac{r^2 - a^2}{b^2 - a^2}I.$$

So, $\Sigma B_{\parallel}\Delta l = \mu_0 I\dfrac{r^2 - a^2}{b^2 - a^2}$. Since $B_{\parallel} = B$, $B = \dfrac{\mu_0 I}{2\pi r}\left(\dfrac{r^2 - a^2}{b^2 - a^2}\right)$. According to RHR 2, the field lines are circles

and are directed CCW.

$r \ge b$:

Now, the closed path has radius $r \ge b$. So, $\Sigma B_{\parallel}\Delta l = \mu_0 I = B(2\pi r)$,

or $B = \dfrac{\mu_0 I}{2\pi r}$.

As before, the field is CCW.

(b) As found in part (a), $\vec{\mathbf{B}}(r > b) = \boxed{\dfrac{\mu_0 I}{2\pi r} \text{ CCW as seen from above}}$.

85. Strategy The maximum torque occurs when the field and the dipole moment are perpendicular. Use Eq. (19-13a).

Solution Compute the maximum torque.

$$\tau_{\text{max}} = NIAB = (9.3 \times 10^{-24} \text{ A} \cdot \text{m}^2)(1.0 \text{ T}) = \boxed{9.3 \times 10^{-24} \text{ N} \cdot \text{m}}$$

89. Strategy and Solution According to RHR 2, the magnetic field due to the current at the compass is directed downward. Since magnetic field lines end at south poles and emerge at north poles, magnets align with their north poles pointing in the direction of an external magnetic field. So, the north end of the needle points downward, or $\boxed{\text{south}}$ on the compass face.

93. Strategy Use Eqs. (19-12) and (19-14) and Newton's second law.

Solution

(a) According to Newton's second law, $\Sigma F_y = N - mg = N - \lambda Lg = 0$, so $N = \lambda Lg$ for each

wire, where λ is the mass per unit length. The force of static friction, $f_s = \mu_s N = \mu_s \lambda Lg$, opposes the magnetic force. The magnetic field due to each wire is perpendicular to the axis of the opposite wire, so the magnitude of the force on one wire due to the other is

$$F = ILB = IL\frac{\mu_0 I}{2\pi r} = \frac{\mu_0 LI^2}{2\pi r}.$$

Set this result and the magnitude of the force due to static friction equal and solve for the minimum current necessary to make the wires start to move.

$$\frac{\mu_0 LI^2}{2\pi r} = \mu_s \lambda Lg, \text{ so } I = \sqrt{\frac{2\pi r \mu_s \lambda g}{\mu_0}} = \sqrt{\frac{2\pi (0.0025 \text{ m})(0.035)(0.0250 \text{ kg/m})(9.80 \text{ m/s}^2)}{4\pi \times 10^{-7} \text{ T} \cdot \text{m/A}}} = \boxed{10 \text{ A}}.$$

(b) The currents flow in opposite directions. According to $\vec{F} = I\vec{L} \times \vec{B}$ and the RHR, the force due to one wire on the other is repulsive, so the wires move $\boxed{\text{farther apart}}$.

97. Strategy v is that of the velocity selector, E_H / B, and $E_H = V_H/d$ for a uniform field.

Solution Compute the average speed.

$$v = \frac{V_H}{dB} = \frac{88.0 \times 10^{-6} \text{ V}}{(3.80 \times 10^{-3} \text{ m})(0.115 \text{ T})} = \boxed{20.1 \text{ cm/s}}$$

101. Strategy The velocity of the electrons is to the right. Use Eq. (19-5).

Solution The magnetic field points upward, toward the south pole of the magnet. According to $\vec{F} = -e\vec{v} \times \vec{B}$ and RHR 1, the electrons are deflected (and thus the beam moves) $\boxed{\text{into the page}}$.

105. Strategy Use Eqs. (19-12a) and (19-14).

Solution

(a) The field due to the long wire is given by $B = \mu_0 I_2/(2\pi r)$. The direction of the field is given by RHR 2; it is out of the page. The force on a side is $\vec{F} = I_1\vec{L}\times\vec{B}$. RHR 1 gives the following directions for $\vec{L}\times\vec{B}$:

Side	Current direction	Field direction	Force direction
top	right	out of the page	attracted to long wire
bottom	left	out of the page	repelled by long wire
left	up	out of the page	right
right	down	out of the page	left

(b) Due to symmetry, the magnitudes of the forces on the left and right sides are equal. The force directions are opposite, so they cancel. The force due to the bottom side is up, and since it is closer to the long wire, its magnitude is larger than that due to the top side; thus, the direction of the net force is up (away from the long wire). Calculate the magnitude.

$$F = I_1 L_{\text{bottom}} B_{\text{bottom}} - I_1 L_{\text{top}} B_{\text{top}} = I_1 L B_{\text{bottom}} - I_1 L B_{\text{top}} = I_1 L \frac{\mu_0 I_2}{2\pi}\left(\frac{1}{r_{\text{bottom}}} - \frac{1}{r_{\text{top}}}\right)$$

$$= \frac{(4\pi\times 10^{-7}\ \text{T}\cdot\text{m/A})(0.0020\ \text{A})(8.0\ \text{A})(0.090\ \text{m})}{2\pi}\left(\frac{1}{0.020\ \text{m}} - \frac{1}{0.070\ \text{m}}\right) = 1.0\times 10^{-8}\ \text{N}$$

So, $\vec{F} = \boxed{1.0\times 10^{-8}\ \text{N away from the long wire}}$.

109. Strategy Use Eqs. (16-4b) and (19-5).

Solution Sum the electric and magnetic forces and use RHR 1.
$$\vec{F} = \vec{F}_{\text{E}} + \vec{F}_{\text{B}} = q\vec{E} + q\vec{v}\times\vec{B} = -e(E\ \text{east}) - e[(v\ \text{south})\times(B\ \text{east})] = eE\ \text{west} - e(vB\ \text{up})$$
$$= eE\ \text{west} + evB\ \text{down}$$
Calculate the magnitude.
$$F = \sqrt{e^2 E^2 + e^2 v^2 B^2} = e\sqrt{E^2 + v^2 B^2} = (1.602\times 10^{-19}\ \text{C})\sqrt{(3.0\times 10^4\ \text{V/m})^2 + (5.0\times 10^6\ \text{m/s})^2(0.080\ \text{T})^2}$$
$$= 6.4\times 10^{-14}\ \text{N}$$
Calculate the direction.
$$\theta = \tan^{-1}\frac{F_y}{F_x} = \tan^{-1}\frac{evB}{eE} = \tan^{-1}\frac{vB}{E} = \tan^{-1}\frac{(5.0\times 10^6\ \text{m/s})(0.080\ \text{T})}{3.0\times 10^4\ \text{V/m}} = 86°$$

So, $\vec{F} = \boxed{6.4\times 10^{-14}\ \text{N at 86° below west}}$.

113. (a) Strategy Apply Newton's second law to the circular part of the proton's motion. Use $v_\perp = v\sin\theta$.

Solution Find the radius of the helix.

$$\Sigma F_r = ev_\perp B = m\frac{v_\perp^2}{r}, \text{ so } r = \frac{mv_\perp}{eB} = \frac{mv\sin\theta}{eB} = \frac{(1.673\times10^{-27}\text{ kg})(4.0\times10^7\text{ m/s})\sin 25^\circ}{(1.602\times10^{-19}\text{ C})(1.0\times10^{-6}\text{ T})} = \boxed{180\text{ km}}.$$

(b) Strategy The time for one revolution is given by $T = 1/f = C/v_\perp = 2\pi r/(v\sin\theta)$.

Solution The distance the proton moves along a field line in time T is

$$d = v_\| T = v\cos\theta T = \frac{v\cos\theta(2\pi r)}{v\sin\theta} = \frac{2\pi r}{\tan\theta} = \frac{2\pi(1.77\times10^5\text{ m})}{\tan 25^\circ} = \boxed{2.4\times10^6\text{ m}}.$$

Chapter 20

ELECTROMAGNETIC INDUCTION

Conceptual Questions

1. As the loop rotates, the magnitude of the flux increases as it approaches the maximum positive value with the loop perpendicular to the field and then decreases to zero as the loop becomes parallel to the field. As it turns through the maximum position and the flux changes from increasing to decreasing, the induced current reverses direction, because the sign of the change it's opposing has flipped. As the loop continues to rotate, the flux reaches its maximum negative value when the loop is once again perpendicular to the field but facing the opposite way. As it rotates through this position and the flux changes from decreasing to increasing, the induced current will once again reverse direction. Thus, the induced current reverses its direction twice per rotation.

5. (a) In position 1, there is an eddy current circulating clockwise to oppose the increase in magnetic flux as the plate enters the magnetic field.

   (b) In position 2, there is an eddy current circulating counterclockwise to oppose the decrease in magnetic flux as the plate leaves the magnetic field.

   (c) From the right hand rule, the induced magnetic field due to the induced current produces a magnetic force on the metal plate that opposes the motion of the plate. This force slows the pendulum each time it enters or leaves the magnetic field and rapidly brings the plate to rest.

9. No; knowing the flux through the surface and the area of the surface would be sufficient to calculate the average component of the magnetic field perpendicular to the surface only.

13. (a) A transformer only works for alternating currents because its operation is based upon the principle of magnetic induction. A current can only be induced by a changing magnetic flux which requires a changing primary current—impossible to achieve with a direct current.

   (b) The back emf in the primary coil limits the size of the current. For dc emf, no back emf exists, and thus, the current is much too large (limited only by the small resistance of the coils).

17. The back emf decreases as the mixer's motor slows and the current to the mixer therefore increases. As the current increases, resistive heating causes the motor to heat up. The motor will "burn out" if the temperature gets too high.

Problems

1. **Strategy** Use Eqs. (20-2a), (19-5), and (19-12a).

   Solution

   (a) The motional emf is $\mathscr{E} = vBL$, so $I = \dfrac{\mathscr{E}}{R} = \boxed{\dfrac{vBL}{R}}$.

   (b) By the RHR and $\vec{\mathbf{F}} = -e\vec{\mathbf{v}} \times \vec{\mathbf{B}}$, the direction of the force on the electrons in the rod is down. So, the direction of the current is $\boxed{\text{CCW}}$.

   (c) By the RHR and $\vec{\mathbf{F}} = I\vec{\mathbf{L}} \times \vec{\mathbf{B}}$, the direction of the magnetic force on the rod is $\boxed{\text{left}}$.

(d) $F = ILB = \dfrac{vBL}{R} LB = \boxed{\dfrac{vB^2 L^2}{R}}$

3. **Strategy** Use the result of Problem 1d, $P = Fv$, and Eqs. (20-2a) and (18-21b).

   Solution

   (a) According to Problem 1d, the magnitude of the magnetic force on the rod is $vB^2 L^2/R$. The net force must be

   zero for constant velocity. So, $F_{\text{ext}} = F_B = \boxed{\dfrac{vB^2 L^2}{R}}$.

   (b) $\dfrac{\Delta W}{\Delta t} = P = Fv = \boxed{\dfrac{v^2 B^2 L^2}{R}}$

   (c) $\mathcal{E} = vBL$, so $P = \dfrac{V^2}{R} = \boxed{\dfrac{v^2 B^2 L^2}{R}}$.

   (d) $\boxed{\text{Energy is conserved since the rate at which the external force does work is equal to the power dissipated in the resistor.}}$

5. **(a) Strategy** Use Newton's second law, Eq. (19-12b), and Eq. (20-2a).

   Solution $\Sigma F_y = F_B - mg = 0$, so $F_B = mg$ when the rod is falling with constant velocity. The magnitude of

   the magnetic force is $F_B = ILB = \dfrac{\mathcal{E}}{R} LB = \dfrac{vBL}{R} LB = \dfrac{vL^2 B^2}{R}$. Set this equal to mg and solve for the terminal

   speed.

   $\dfrac{vL^2 B^2}{R} = mg$, so $v = \dfrac{mgR}{L^2 B^2} = \dfrac{(0.0150\ \text{kg})(9.80\ \text{m/s}^2)(8.00\ \Omega)}{(1.30\ \text{m})^2 (0.450\ \text{T})^2} = \boxed{3.44\ \text{m/s}}$.

   (b) Strategy Use the potential energy in a uniform gravitational field and Eqs. (18-21b) and (20-2a).

   Solution The change in gravitational energy per second is

   $$\dfrac{\Delta U}{\Delta t} = \dfrac{mg\Delta y}{\Delta t} = -mgv = -\dfrac{m^2 g^2 R}{L^2 B^2} = -\dfrac{(0.0150\ \text{kg})^2 (9.80\ \text{m/s}^2)^2 (8.00\ \Omega)}{(1.30\ \text{m})^2 (0.450\ \text{T})^2} = -0.505\ \text{W},$$

   so the magnitude of the change is 0.505 W. The power dissipated is

   $$P = \dfrac{\mathcal{E}^2}{R} = \dfrac{v^2 B^2 L^2}{R} = \left(\dfrac{mgR}{L^2 B^2} \right)^2 \dfrac{B^2 L^2}{R} = \dfrac{m^2 g^2 R}{L^2 B^2} = 0.505\ \text{W}.$$

   $\boxed{\text{The magnitude of the change in gravitational potential energy per second and the power dissipated in the resistor are the same, 0.505 W.}}$

9. **Strategy** Use Eq. (19-5).

   Solution

   (a) The force on the electrons is given by $\vec{\mathbf{F}} = -e\vec{\mathbf{v}} \times \vec{\mathbf{B}}$.
   According to the RHR, $\vec{\mathbf{v}} \times \vec{\mathbf{B}}$ is out of the page, so $\vec{\mathbf{F}}$ is into the page.

   (b) The answer is no; the electrons are pushed perpendicular to the length of the wire.

   (c) I along the *length* of the wire is zero, so there is no induced emf.

   (d) The velocity of the left side of wire 3 is equal in magnitude and opposite in direction to that of the right side. So, for the left side:

   > $\vec{\mathbf{F}}$ is out of the page; no, electrons are pushed perpendicular to the length of the wire; there is no induced emf. The situation for side 1 is identical to that of side 3.

13. **Strategy** The angle between $\vec{\mathbf{B}}$ and the normal to the area is $\theta = 90° - 65° = 25°$. Use Eq. (20-5).

    Solution Compute the magnetic flux through the surface of the desk.
    $$\Phi_{\mathrm{B}} = BA\cos\theta = (0.44 \times 10^{-3}\ \mathrm{T})(1.3\ \mathrm{m})(1.0\ \mathrm{m})\cos 25° = \boxed{5.2 \times 10^{-4}\ \mathrm{Wb}}$$

15. (a) **Strategy** According to Lenz's law, the direction of an induced current in a loop always opposes the *change* in magnetic flux that induces the current.

    Solution According to the RHR, $\vec{\mathbf{B}}$ is directed into the page at the loop. Since $B \propto r^{-1}$ for a long straight wire, as the loop moves closer to the wire, B increases. The induced current in the loop flows such that it generates a magnetic field opposite to $\vec{\mathbf{B}}$, or out of the page. To generate this field, the current must be CCW.

    (b) **Strategy** According to Faraday's law, the rate of change of the magnetic flux is equal to the induced emf.

    Solution Find the rate of change of the flux.
    $$\left|\frac{\Delta\Phi_{\mathrm{B}}}{\Delta t}\right| = |\mathscr{E}| = 3.5\ \mathrm{mV} = \boxed{3.5 \times 10^{-3}\ \mathrm{W/s}}.$$

17. **Strategy** According to Lenz's law, the direction of an induced current in a loop always opposes the *change* in magnetic flux that induces the current. According to Faraday's law, the rate of change of the magnetic flux is equal to the induced emf.

    Solution While I_1 is increasing, a changing magnetic flux is induced; the field is increasing and, by the RHR, it is directed to the left. By Lenz's law, the changing magnetic flux is opposed by an induced current in loop 2, so the answer is yes. To generate a magnetic flux to oppose that generated by loop 1, the current in loop 2 must flow opposite to the current in loop 1, or counterclockwise as viewed from the right.

21. **(a) Strategy** Use Eqs. (20-5) and (20-6a).

    Solution The initial flux is $\Phi_i = BA\cos 0° = BA$. The final flux is $\Phi_i = BA\cos 180° = -BA$.
    Compute the average induced emf.

    $$\mathcal{E}_{av} = -\frac{\Delta\Phi_B}{\Delta t} = -\frac{\Phi_f - \Phi_i}{\Delta t} = -\frac{-BA - BA}{\Delta t} = \frac{2BA}{\Delta t} = \frac{2(0.880\ \text{T})\pi(0.0340\ \text{m})^2}{0.222\ \text{s}} = \boxed{28.8\ \text{mV}}$$

    (b) Strategy Use the definition of resistance and Eq. (18-8).

    Solution Compute the average current that flows through the coil.

    $$R = \rho\frac{L}{A} = \frac{\mathcal{E}}{I}, \text{ so } I_{av} = \frac{\mathcal{E}_{av}A}{\rho L} = \frac{\mathcal{E}_{av}\frac{1}{4}\pi d^2}{\rho(2\pi r)} = \frac{\mathcal{E}_{av}d^2}{8\rho r} = \frac{(0.02879\ \text{V})(0.000900\ \text{m})^2}{8(1.67\times10^{-8}\ \Omega\cdot\text{m})(0.0340\ \text{m})} = \boxed{5.13\ \text{A}}.$$

25. **(a) Strategy** Use Faraday's law, Eq. (20-6a), and Eq. (20-5) for the flux.

    Solution In one rotation, the change in the flux is BA, and the frequency is $f = 1/\Delta t = \omega/(2\pi)$. Find the magnitude of the emf.

    $$|\mathcal{E}| = \left|\frac{\Delta\Phi_B}{\Delta t}\right| = \frac{BA}{2\pi/\omega} = \frac{B\omega(\pi R^2)}{2\pi} = \boxed{\frac{B\omega R^2}{2}}$$

    (b) Strategy The speed of the tip of the rod is $v = \omega R$.

    Solution Write the magnitude of the emf.

    $$|\mathcal{E}| = \frac{B\omega R^2}{2} = \frac{B(\omega R)R}{2} = \frac{BvR}{2}$$

    The motional emf of a rod of length R moving at constant velocity is $\mathcal{E} = vBR$, so our result

    > is half of the value of the motional emf of a rod moving at constant speed v; which is reasonable, since different points on the rotating rod have different speeds ranging from 0 to v.

29. **(a) Strategy** When the motor is running smoothly, the current and net emf are constant.

    Solution Compute the current when the motor is running smoothly.

    $$I = \frac{\mathcal{E}_{ext} - \mathcal{E}_{back}}{R} = \frac{24.00\ \text{V} - 18.00\ \text{V}}{8.00\ \Omega} = \boxed{0.750\ \text{A}}$$

    (b) Strategy Compute the current when the motor quits spinning and the back emf is zero. Consider how this may affect the motor.

    Solution Compute the current when there is no back emf.

    $$I = \frac{\mathcal{E}_{ext}}{R} = \frac{24.00\ \text{V}}{8.00\ \Omega} = \boxed{3.00\ \text{A}}$$

    This is four times the normal current. This much current flowing though the motor may damage it.

    > Tim should shut the trimmer off because the wires in the motor were not meant to sustain this much current. The wires will burn up if this current flows through them for very long.

31. **Strategy** Use Eq. (20-9).

    Solution Compute the secondary voltage amplitude.

    $$\mathcal{E}_2 = \frac{N_2}{N_1}\mathcal{E}_1 = \frac{200}{4000}(2.2\times10^3\ \text{V}) = \boxed{110\ \text{V}}$$

33. **Strategy** Use Eq. (20-9).

    Solution Compute the turns ratio and number of turns in the primary.

    (a) $\dfrac{N_2}{N_1} = \dfrac{\mathscr{E}_2}{\mathscr{E}_1} = \dfrac{8.5 \text{ V}}{170 \text{ V}} = \boxed{\dfrac{1}{20}}$

    (b) $N_1 = \dfrac{\mathscr{E}_1}{\mathscr{E}_2} N_2 = \dfrac{170 \text{ V}}{8.5 \text{ V}}(50) = \boxed{1000}$

35. **Strategy** Use Eq. (20-9).

    Solution Compute the turns ratio.
    $\dfrac{N_2}{N_1} = \dfrac{\mathscr{E}_2}{\mathscr{E}_1} = \dfrac{10.0 \text{ V}}{5.00 \text{ V}} = \boxed{2.00}$

37. **Strategy** Use Eq. (20-10).

    Solution Compute the voltage amplitude in the secondary coil.
    $\dfrac{\mathscr{E}_1}{\mathscr{E}_2} = \dfrac{N_1}{N_2}$, so $\mathscr{E}_2 = \dfrac{N_2}{N_1}\mathscr{E}_1 = \dfrac{300}{1800}(170 \text{ V}) = \boxed{28 \text{ V}}$.
    Compute the current amplitude in the primary coil.
    $\dfrac{I_2}{I_1} = \dfrac{N_1}{N_2}$, so $I_1 = \dfrac{N_2}{N_1}I_2 = \dfrac{300}{1800}(3.2 \text{ A}) = \boxed{0.53 \text{ A}}$.

39. **Strategy** The induced emf in a solid conductor subjected to a changing magnetic flux causes eddy currents to flow simultaneously along many different paths. These eddy currents dissipate energy. According to Lenz's law, the direction of an induced current in a loop always opposes the *change* in magnetic flux that induces the current.

    Solution

    (a) In the wall of the pipe, the magnetic field due to the falling magnet is directed upward. Above the magnet, the field decreases as the magnet falls. According to Lenz's law, eddy currents are generated to oppose this decrease. By the RHR, these eddy currents flow $\boxed{\text{CCW}}$ when viewed from the top of the pipe. Below the magnet, the field increases, so the eddy currents flow $\boxed{\text{CW}}$ to oppose this increase.

    (b) Since the eddy currents move in a resistive medium, they dissipate energy—the magnet's kinetic energy. Thus, the magnet's speed is decreased. The speed cannot be reduced to zero, though, since then there would be no eddy currents generated to slow the magnet (no changing flux). Therefore, the magnet must reach some terminal speed, as for a marble falling through honey. Below is a qualitative sketch of the speed of the magnet as a function of time.

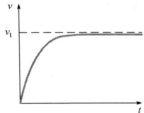

41. Strategy Use Eq. (20-12).

Solution Compute the mutual inductance.

$$M = \frac{N_2 \Phi_{21}}{I_1} = \frac{N_1 \Phi_{12}}{I_2} = \frac{(1)(2.35 \times 10^{-8}\ \text{T} \cdot \text{m}^2)}{0.75\ \text{A}} = \boxed{3.1 \times 10^{-8}\ \text{H}}$$

45. Strategy Use Eq. (20-15b).

Solution Compute the self-inductance.

$$L = \frac{\mu_0 N^2 \pi r^2}{\ell} = \frac{(4\pi \times 10^{-7}\ \text{H/m})(300.0)^2 \pi (0.012\ \text{m})^2}{0.060\ \text{m}} = \boxed{8.5 \times 10^{-4}\ \text{H}}$$

47. Strategy The emf through one winding is $\mathcal{E}/N$ where $\mathcal{E}$ is given by Eq. (20-16). Use Eq. (20-15b).

Solution

(a) Find the induced emf in one of the windings.

$$\frac{\mathcal{E}}{N} = -\frac{L}{N}\frac{\Delta I}{\Delta t} = -\frac{\mu_0 N^2 \pi \left(\frac{d}{2}\right)^2}{\ell N}\frac{\Delta I}{\Delta t} = -\frac{1}{4}\mu_0 \pi n d^2 \frac{\Delta I}{\Delta t}$$

$$= -\frac{1}{4}(4\pi \times 10^{-7}\ \text{H/m})\pi(160\ \text{cm}^{-1})(10^2\ \text{cm/m})(0.0075\ \text{m})^2(-35.0\ \text{A/s}) = \boxed{3.1 \times 10^{-5}\ \text{V}}$$

(b) Find the induced emf in the entire solenoid.

$$\mathcal{E} = N\left(\frac{\mathcal{E}}{N}\right) = n\ell\left(\frac{\mathcal{E}}{N}\right) = (160\ \text{cm}^{-1})(2.8\ \text{cm})(3.1 \times 10^{-5}\ \text{V}) = \boxed{14\ \text{mV}}$$

49. Strategy Use the magnetic field inside a long solenoid, the magnetic flux through a loop of area A, and the definition of self-inductance.

Solution

(a) $B = \mu_0 n I$ for a long solenoid.

(b) The magnetic flux through one turn is given by $\Phi = BA = B\pi r^2$.

(c) Since there are $N = n\ell$ turns, the total flux linkage is $N\Phi = NB\pi r^2$.

(d) The definition of self inductance is $N\Phi = LI$.

$LI = NB\pi r^2$, so $L = \dfrac{NB\pi r^2}{I}$.

Substitute for B.

$$L = \frac{N(\mu_0 n I)\pi r^2}{I} = (n\ell)\mu_0 n \pi r^2 = \mu_0 n^2 \pi r^2 \ell$$

This is Eq. (20-15a).

53. **Strategy** Since the inductors are in parallel, the emfs induced in each must be equal. An equivalent inductor replacing L_1 and L_2 would have the same induced emf. The current through the equivalent emf must be the sum of the currents through the individual inductors. The relation between induced emf and a changing current is given by Faraday's law.

Solution Calculate the equivalent inductance of two ideal inductors in parallel.

$\dfrac{\Delta I_{eq}}{\Delta t} = \dfrac{\Delta I_1}{\Delta t} + \dfrac{\Delta I_2}{\Delta t}$, $\mathscr{E}_{eq} = \mathscr{E}_1 = \mathscr{E}_2$, and $\mathscr{E} = -L\dfrac{\Delta I}{\Delta t}$. Find L_{eq} in terms of L_1 and L_2.

$L_{eq}\dfrac{\Delta I_{eq}}{\Delta t} = L_1\dfrac{\Delta I_1}{\Delta t} = L_2\dfrac{\Delta I_2}{\Delta t}$, so $\dfrac{\Delta I_2}{\Delta t} = \dfrac{L_1}{L_2}\dfrac{\Delta I_1}{\Delta t}$. $\dfrac{\Delta I_{eq}}{\Delta t} = \dfrac{\Delta I_1}{\Delta t} + \dfrac{\Delta I_2}{\Delta t} = \dfrac{\Delta I_1}{\Delta t}\left(1 + \dfrac{L_1}{L_2}\right)$, so

$L_{eq}\dfrac{\Delta I_{eq}}{\Delta t} = L_{eq}\dfrac{\Delta I_1}{\Delta t}\left(1 + \dfrac{L_1}{L_2}\right) = L_1\dfrac{\Delta I_1}{\Delta t}$.

Solve for L_{eq}.

$L_{eq}\left(1 + \dfrac{L_1}{L_2}\right) = L_{eq}\left(\dfrac{L_1 + L_2}{L_2}\right) = L_1$, so $\boxed{L_{eq} = \dfrac{L_1 L_2}{L_1 + L_2}}$.

57. **(a) Strategy and Solution** When the switch is opened, the 5.0-Ω resistor is no longer part of a complete circuit, so no current flows through it and the voltage across it is $\boxed{0}$.

(b) Strategy When the switch is opened, only the inductor and the 10.0-Ω resistor form a complete circuit. Immediately after the switch is opened, the inductor will maintain the current that was flowing through it before the switch was opened.

Solution In part (c) of Problem 56, the current was 1.2 A, so the voltage across the 10.0-Ω resistor is

$IR = (1.2 \text{ A})(10.0 \ \Omega) = \boxed{12 \text{ V}}$.

61. **(a) Strategy** Use the definition of resistance.

Solution Calculate the maximum current I_0.

$I_0 = \dfrac{\mathscr{E}_b}{R} = \dfrac{100.0 \text{ V}}{2.0 \ \Omega} = \boxed{50 \text{ A}}$

(b) Strategy According to Faraday's law, the magnitude of the induced emf around a loop is equal to the rate of change of the magnetic flux through the loop.

Solution

The attempt to suddenly stop the current would induce a huge emf in the windings of the electromagnet, possibly damaging it. It is likely that sparks would complete the circuit across the open switch. The shunt resistor is used to reduce the rate of change of the current, and allow the electromagnet to be safely shut off.

(c) Strategy Use Eq. (18-21a).

Solution Compute the maximum power dissipated in the shunt resistor.

$P_{max} = I_0^2 R = (50 \text{ A})^2 (20.0 \ \Omega) = \boxed{50 \text{ kW}}$

(d) Strategy Use Eq. (20-24).

Solution Solve for t when $I = 0.10$ A.

$$I = I_0 e^{-t/\tau}$$

$$\ln \frac{I}{I_0} = \ln e^{-t/\tau} = -\frac{t}{\tau}$$

$$t = -\tau \ln \frac{I}{I_0} = -\frac{L}{R_{\text{eq}}} \ln \frac{I}{I_0} = -\frac{8.0 \text{ H}}{22.0 \text{ }\Omega} \ln \frac{0.10 \text{ A}}{50 \text{ A}} = \boxed{2.3 \text{ s}}$$

(e) Strategy and Solution $\boxed{\text{Yes}}$, a larger shunt resistor would dissipate the energy stored in the electromagnet faster. $\boxed{\tau \propto \dfrac{1}{R_L + R_{\text{shunt}}}}$, so a larger shunt resistor decreases the time that it takes for the current in the circuit to decrease to zero. Thus, the time to dissipate the energy decreases.

65. Strategy Use Eqs. (20-24) and (20-17).

Solution Solve for t when $U = 0.10 U_0 = 0.10 \times \frac{1}{2} L I_0^2$.

$$U = \frac{1}{2} L[I(t)]^2 = \frac{1}{2} L(I_0 e^{-t/\tau})^2 = \frac{1}{2} L I_0^2 e^{-2tR_{\text{eq}}/L} = U_0 e^{-2tR_{\text{eq}}/L}, \text{ so } \ln \frac{U}{U_0} = \ln e^{-2tR_{\text{eq}}/L} = -\frac{2R_{\text{eq}}}{L} t \text{ and}$$

$$t = -\frac{L}{2R_{\text{eq}}} \ln \frac{U}{U_0} = -\frac{0.015 \text{ H}}{2(27 \times 10^3 \text{ }\Omega + 3.0 \times 10^3 \text{ }\Omega)} \ln 0.10 = \boxed{580 \text{ ns}}.$$

67. (a) Strategy When the current is no longer changing, the emf in the coil is zero.

Solution Compute the current in the coil.

$$I_0 = \frac{\mathscr{E}_b}{R} = \frac{6.0 \text{ V}}{33 \text{ }\Omega} = \boxed{180 \text{ mA}}$$

(b) Strategy Use Eq. (20-17).

Solution Calculate the energy stored in the coil.

$$U = \frac{1}{2} L I_0^2 = \frac{1}{2} L \left(\frac{\mathscr{E}_b}{R} \right)^2 = \frac{(0.15 \text{ H})(6.0 \text{ V})^2}{2(33 \text{ }\Omega)^2} = \boxed{2.5 \text{ mJ}}$$

(c) Strategy Use Eq. (18-21b).

Solution Calculate the rate at which energy is dissipated.

$$P = \frac{\mathscr{E}_b^2}{R} = \frac{(6.0 \text{ V})^2}{33 \text{ }\Omega} = \boxed{1.1 \text{ W}}$$

(d) Strategy and Solution Since the current is no longer changing, the induced emf is $\boxed{\text{zero}}$.

69. (a) Strategy $\vec{F} = I\vec{L} \times \vec{B}$ and $\vec{L}$ and $\vec{B}$ are always perpendicular for both sides. In side 2, current flows into the page. In side 4, current flows out of the page.

Solution The magnitude of the force is

$$F = ILB = \frac{\omega BA}{R} LB \sin \omega t = \frac{\omega B^2 AL}{R} \sin \omega t$$

where, according to the RHR, $\vec{F}_2$ is down and $\vec{F}_4$ is up, or

$$\boxed{\vec{F}_2 = \frac{\omega B^2 AL}{R} \sin \omega t \text{ down and } \vec{F}_4 = \frac{\omega B^2 AL}{R} \sin \omega t \text{ up}}.$$

(b) Strategy and Solution The magnetic forces on sides 1 and 3 are always parallel to the axis of rotation. Therefore, they do not cause a torque about the axis.

(c) Strategy Sum the torques using the forces found in part (a).

Solution $\Sigma \tau = (F_2 \sin \theta)r + (F_4 \sin \theta)r = 2Fr \sin \omega t$, since

$F = |\vec{F}_2| = |\vec{F}_4|$ and $\theta = \omega t$.

Thus, $\tau(t) = 2\left(\frac{\omega B^2 AL}{R} \sin \omega t \right) r \sin \omega t = \frac{2\omega r B^2 AL}{R} \sin^2 \omega t.$

Therefore, $\vec{\tau} = \boxed{\frac{2\omega r B^2 AL}{R} \sin^2 \omega t \text{ CCW}}.$

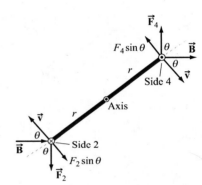

(d) Strategy and Solution

The torque is counterclockwise, and the angular velocity is clockwise, so the magnetic torque would tend to decrease the angular velocity.

73. Strategy Use Eqs. (20-12) and (20-14).

Solution

(a) By the definition of self-inductance, $N\Phi_{11} = LI_1$ where 1 refers to the toroid. (Let 2 refer to the single turn of wire.) Since the single turn has the same cross-sectional area and contains the same magnetic field, the flux through the turn is the same as the total flux through the toroid. So, $N\Phi_{11} = \Phi_{21} = MI_1 = LI_1$, or $M = \boxed{L}$.

(b) Since the flux through the loop doesn't change, so M still equals $\boxed{L}$.

77. Strategy The energy density in a magnetic field is $u_B = B^2/(2\mu_0)$.

Solution Compute the energy.

$$U = u_B V = \frac{B^2 V}{2\mu_0} = \frac{(0.45 \times 10^{-3} \text{ T})^2 (1.0 \text{ m}^3)}{2(4\pi \times 10^{-7} \text{ T} \cdot \text{m/A})} = \boxed{81 \text{ mJ}}$$

81. **(a) Strategy** Use Eqs. (19-17), (20-5), and (20-12).

    Solution Compute the mutual inductance.

    $$M = \frac{N_2 \Phi_{21}}{I_1} = \frac{N_2(B_1 A)}{I_1} = \frac{N_2(\pi r^2)}{I_1} \times \frac{\mu_0 N_1 I_1}{L} = \boxed{\frac{\mu_0 N_1 N_2 \pi r^2}{L}},$$

    where Φ_{21} is the flux through the coil of wire due to the solenoid.

    (b) Strategy Use Eq. (20-13).

    Solution Find the magnitude of the induced emf in the coil.

    $$|\mathscr{E}_{21}| = \left| -M \frac{\Delta I_1}{\Delta t} \right| = \boxed{\frac{\mu_0 N_1 N_2 \pi r^2}{L} \frac{\Delta I_1}{\Delta t}}$$

85. **Strategy** According to Faraday's law, the magnitude of the induced emf around a loop is equal to the rate of change of the magnetic flux through the loop. According to Lenz's law, the direction of an induced current in a loop always opposes the *change* in magnetic flux that induces the current.

    Solution As the bar magnet travels from 1 to 2, $\vec{\mathbf{B}}$ is increasing and to the left at the coil. As viewed from the left, a CW current is induced in the coil that generates a magnetic field to the right to oppose the increasing magnetic field due to the bar magnet. Thus, the current is negative and increasing in magnitude. As the bar magnet travels from 2 to 3, $\vec{\mathbf{B}}$ is decreasing and to the left at the coil. The induced current flows CCW to oppose the decreasing field. Thus, the current is positive and decreasing in magnitude.

    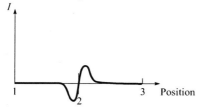

89. **(a) Strategy** We can model the airplane as a metal rod moving through a magnetic field. Then, the motional emf is $\mathscr{E} = vBL$, where B is the component of the field perpendicular to the motion of the plane and to the wing. Therefore, only the upwardly directed component of the magnetic field contributes to the motional emf.

    Solution Compute the motional emf.
    $$\mathscr{E} = vBL = (180 \text{ m/s})(0.38 \times 10^{-3} \text{ T})(46 \text{ m}) = \boxed{3.1 \text{ V}}.$$

    (b) Strategy Consider the magnetic forces on electrons in the plane's wings.

    Solution The force on an electron in the wing is $\vec{\mathbf{F}} = -e\vec{\mathbf{v}} \times \vec{\mathbf{B}}$. So, according to the RHR, the electrons are forced along the southernmost wing. Negative charge builds up on the southernmost wingtip and positive on the northernmost. Thus, the northernmost wingtip is positively charged.

Chapter 21

ALTERNATING CURRENT

Conceptual Questions

1. When an ac generator is turned on within a circuit containing an uncharged capacitor, current begins to flow and charge begins to accumulate on the capacitor's plates. As charge accumulates, the potential across the capacitor increases, thereby inducing an emf directed opposite to the emf of the generator. The flow of current in the circuit therefore decreases as the capacitor's potential increases. When the direction of the emf from the generator changes, the potential across the capacitor decreases and the magnitude of the current in the circuit increases. Therefore, the current in an ac circuit and the potential difference across a capacitor in that circuit are always out of phase.

5. Appliances such as light bulbs and electric heaters operate by converting the energy stored in electric current into heat energy. The operation of such devices depends only upon the energy dissipating properties of their resistive elements and thus operate equally well with ac or dc. Appliances requiring a transformer of some type will only operate with ac as they work on the principle of induction and therefore require a changing current.

9. The 500 W reported on the appliance is the average power consumption, given by $P_{av} = I_{rms}V_{rms} \cos\phi$. The power factor $\cos\phi = R/Z$ depends on the capacitance, inductance, and resistance of the circuit. Only if the power factor is one, or $R = Z$, would the average power consumption be 600 W.

13. Inserting a coil of wire with a soft iron core into the circuit in series with a light bulb would add inductance and increase the overall impedance of the circuit. The amplitude of the current in the circuit would be reduced and the bulb would be dimmed. Moving the soft-iron core in and out of the coil would vary the amount of dimming by changing the inductance of the coil.

17. Since the power dissipated in a light bulb is given by $P_{av} = V_{rms}^2/R$, the bulb would consume considerably less power with an rms voltage of 120 V (in fact 1/4 as much) and would burn less brightly.

Problems

1. **Strategy and Solution** The current reverses direction twice per cycle and there are 60 cycles per second, so the current reverses direction $\boxed{120 \text{ times per second}}$.

3. **Strategy** 1500 W is the average power dissipated by the heater and $P_{av} = I_{rms}V_{rms}$.

   Solution Calculate I, the peak current.

   $$I = \sqrt{2}I_{rms} = \sqrt{2}\left(\frac{P_{av}}{V_{rms}}\right) = \sqrt{2}\left(\frac{1500 \text{ W}}{120 \text{ V}}\right) = \boxed{18 \text{ A}}$$

5. **Strategy** Find the resistance of the hair dryer. Then, compute the power dissipated in Europe.

   Solution

   $P_{av} = \dfrac{V_{rms}^2}{R}$, so $R = \dfrac{V_{US}^2}{P_{US}}$. The power dissipated by the dryer in Europe is

   $P_R = \dfrac{V_E^2}{R} = \dfrac{V_E^2}{V_{US}^2}P_{US} = \left(\dfrac{240}{120}\right)^2 (1500\text{ W}) = \boxed{6000\text{ W}}$.

   | The heating element of the hair dryer will burn out because it is not designed for this amount of power. |

7. **(a) Strategy** 4200 W is the average power drawn by the heater.

   Solution Compute the rms current.
   $$I_{rms} = \dfrac{P_{av}}{V_{rms}} = \dfrac{4200\text{ W}}{120\text{ V}} = \boxed{35\text{ A}}$$

   (b) Strategy Since $P = V^2/R$ and $R_2 = R_1$, $P \propto V^2$. Form a proportion.

   Solution Calculate P_2.

   $\dfrac{P_2}{P_1} = \left(\dfrac{V_2}{V_1}\right)^2$, so $P_2 = \left(\dfrac{105}{120}\right)^2 (4200\text{ W}) = \boxed{3.2\text{ kW}}$.

9. **Strategy** Use the definition of rms.

   Solution Compute the amplitude.
   $\mathscr{E}_m = \sqrt{2}\mathscr{E}_{rms} = \sqrt{2}(4.0\text{ V}) = 5.7\text{ V}$

   The instantaneous sinusoidal emf oscillates between $\boxed{-5.7\text{ V and }5.7\text{ V}}$.

13. **Strategy and Solution** The peak current in the circuit is given by $I = \omega Q$ and $Q = CV$, so $I = \omega CV$. Therefore, if the capacitance is increased by a factor of 3.0 and the driving frequency is increased by a factor of 2.0, $\boxed{\text{the current is increased by a factor of 6.0}}$.

15. **Strategy** The reactance is $X_C = 1/(\omega C)$ where $\omega = 2\pi f$.

    Solution

    (a) Solve for f.
    $$X_C = \dfrac{1}{2\pi fC}, \text{ so } f = \dfrac{1}{2\pi X_C C} = \dfrac{1}{2\pi(6.63\times10^3\ \Omega)(0.400\times10^{-6}\text{ F})} = \boxed{60.0\text{ Hz}}.$$

    (b) Compute the reactance.
    $$X_C = \dfrac{1}{\omega C} = \dfrac{1}{2\pi(30.0\text{ Hz})(0.400\times10^{-6}\text{ F})} = \boxed{13.3\text{ k}\Omega}$$

17. Strategy $V_{rms} = I_{rms} X_C$ where $X_C = 1/(\omega C)$.

Solution Solve for the capacitance, C.

$$V_{rms} = I_{rms} X_C = \frac{I_{rms}}{\omega C}, \text{ so } C = \frac{I_{rms}}{2\pi f V_{rms}} = \frac{2.3 \times 10^{-3} \text{ A}}{2\pi (60.0 \text{ Hz})(115 \text{ V})} = \boxed{53 \text{ nF}}.$$

21. Strategy Use Eqs. (18-15), (21-6), and (21-7).

Solution

(a) Find I.

$$I = \frac{V}{X_{Ceq}} = \omega C_{eq} V = \frac{2\pi f V}{\frac{1}{C_1} + \frac{1}{C_2} + \frac{1}{C_3}} = \frac{2\pi (6300 \text{ Hz})(12.0 \text{ V})}{\frac{1}{2.0 \times 10^{-6} \text{ F}} + \frac{1}{3.0 \times 10^{-6} \text{ F}} + \frac{1}{6.0 \times 10^{-6} \text{ F}}} = 0.475 \text{ A}$$

Find V as a function of C.

$$V = I X_C = \frac{I}{\omega C} = \frac{0.475 \text{ A}}{2\pi (6300 \text{ Hz})C}$$

The table below gives the results for each capacitor.

C (µF)	V (V)
2.0	6.0
3.0	4.0
6.0	2.0

(b) From part (a), $I = \boxed{0.48 \text{ A}}$.

25. Strategy The inductive reactance is given by $X_L = \omega L$ and the inductance of a solenoid is given by $L = \mu_0 N^2 \pi r^2 / \ell$.

Solution Find the reactance of the solenoid.

$$X_L = \omega L = 2\pi f \left(\frac{\mu_0 N^2 \pi r^2}{\ell} \right) = \frac{2\pi^2 (15.0 \times 10^3 \text{ Hz})(4\pi \times 10^{-7} \text{ T} \cdot \text{m/A})(240)^2 (0.010 \text{ m})^2}{0.080 \text{ m}} = \boxed{27 \text{ }\Omega}$$

27. Strategy $V_{rms} = I_{rms} X_L$ where $X_L = \omega L = 2\pi f L$.

Solution Solve for f.

$$V_{rms} = I_{rms} X_L = I_{rms} (2\pi f L), \text{ so}$$
$$f = \frac{V_{rms}}{2\pi I_{rms} L} = \frac{151.0 \text{ V}}{2\pi (0.820 \text{ A})(4.00 \times 10^{-3} \text{ H})} = \boxed{7.33 \text{ kHz}}.$$

29. Strategy and Solution

(a) Since the current flows to the left (negative) and is increasing (positive), the sign of $\dfrac{\Delta i}{\Delta t}$ is $\boxed{\text{negative}}$.

(b) The current is increasing to the left, so the induced emf that opposes this increase is as shown $-\overset{-}{\text{ⲙⲙⲙ}}\overset{+}{}$.

(c) Eq. (21-8) is $v_L = L\dfrac{\Delta i}{\Delta t} = -L\left|\dfrac{\Delta i}{\Delta t}\right| < 0$ in this case. Since the current flows to the left and is increasing, the right side of the inductor is at a higher potential than the left side. So, v_L is negative, and Eq. (21-8) gives the correct sign for v_L.

(d) Since the current flows to the left (negative) and is decreasing (negative), the sign of $\dfrac{\Delta i}{\Delta t}$ is $\boxed{\text{positive}}$.

The current is decreasing to the left, so the induced emf that opposes this decrease is as shown $\overset{+}{-}\overset{}{\text{ⲙⲙⲙ}}\overset{-}{}$.

Eq. (21-8) is $v_L = L\dfrac{\Delta i}{\Delta t} = L\left|\dfrac{\Delta i}{\Delta t}\right| > 0$ in this case. Since the current flows to the left and is decreasing, the left side of the inductor is at a higher potential than the right side. So, v_L is positive, and Eq. (21-8) gives the correct sign for v_L.

33. Strategy $\mathscr{E}_{rms} = I_{rms}Z$ and $Z = \sqrt{R^2 + X_L^2}$. Only the resistance dissipates power, so use $P_{av} = I_{rms}^2 R$.

Solution Find I_{rms}.

$P_{av} = I_{rms}^2 R$, so $I_{rms} = \sqrt{\dfrac{P_{av}}{R}}$.

Find f.

$\mathscr{E}_{rms} = I_{rms}Z = I_{rms}\sqrt{R^2 + X_L^2} = I_{rms}\sqrt{R^2 + \omega^2 L^2}$

$\dfrac{\mathscr{E}_{rms}^2}{I_{rms}^2} = R^2 + \omega^2 L^2$

$\omega^2 L^2 = \dfrac{\mathscr{E}_{rms}^2}{I_{rms}^2} - R^2$

$\omega = \dfrac{1}{L}\sqrt{\left(\dfrac{\mathscr{E}_{rms}}{I_{rms}}\right)^2 - R^2}$

$f = \dfrac{1}{2\pi L}\sqrt{\left(\dfrac{\mathscr{E}_{rms}}{\sqrt{P_{av}/R}}\right)^2 - R^2} = \dfrac{1}{2\pi L}\sqrt{\dfrac{\mathscr{E}_{rms}^2 R}{P_{av}} - R^2} = \dfrac{1}{2\pi(0.0250\ \text{H})}\sqrt{\dfrac{(110\ \text{V})^2(25.0\ \Omega)}{50.0\ \text{W}} - (25.0\ \Omega)^2}$

$= \boxed{470\ \text{Hz}}$

25.0 mH
25.0 Ω
110 V

37. Strategy Use Eq. (21-14b) with $X_L = 0$, and Eq. (21-7).

Solution Find the impedance.

$$Z = \sqrt{R^2 + X_C^2} = \sqrt{R^2 + \frac{1}{\omega^2 C^2}} = \sqrt{(300.0\ \Omega)^2 + \frac{1}{4\pi^2(159\ \text{Hz})^2(2.5\times10^{-6}\ \text{F})^2}} = \boxed{500\ \Omega}$$

39. (a) Strategy The power factor is $\cos\phi = R/Z$ where $Z = \sqrt{R^2 + (X_L - X_C)^2}$.

Solution Find the power factor.

$$\cos\phi = \frac{R}{Z} = \frac{R}{\sqrt{R^2 + \left(\omega L - \frac{1}{\omega C}\right)^2}} = \frac{40.0\ \Omega}{\sqrt{(40.0\ \Omega)^2 + \left[(1.00\times10^4\ \text{rad/s})(0.0220\ \text{H}) - \frac{1}{(1.00\times10^4\ \text{rad/s})(0.400\times10^{-6}\ \text{F})}\right]^2}}$$

$$= \boxed{0.800}$$

(b) Strategy In Example 21.4, I was found to be 2.0 mA. Use Eq. (21-4).

Solution Power is not dissipated in the inductor or capacitor (assuming they are ideal), so $\boxed{P_{\text{av, C}} = P_{\text{av, L}} = 0}$. Compute the average power dissipated in the resistor.

$$P_{\text{av}} = I_{\text{rms}}^2 R = \left(\frac{I}{\sqrt{2}}\right)^2 R = \frac{1}{2}(0.0020\ \text{A})^2(40.0\ \Omega) = \boxed{P_{\text{av, R}} = 8.0\times10^{-5}\ \text{W}}$$

41. Strategy The average power is given by $P_{\text{av}} = I_{\text{rms}}\mathcal{E}_{\text{rms}}\cos\phi$ where $I_{\text{rms}} = \mathcal{E}_{\text{rms}}/Z$ and $\cos\phi = R/Z$.

Solution Find the average power dissipated.

$$P_{\text{av}} = I_{\text{rms}}\mathcal{E}_{\text{rms}}\cos\phi = \frac{\mathcal{E}_{\text{rms}}}{Z}\mathcal{E}_{\text{rms}}\frac{R}{Z} = \left(\frac{\mathcal{E}_{\text{rms}}}{Z}\right)^2 R = \frac{(\mathcal{E}/\sqrt{2})^2 R}{R^2 + \left(\omega L - \frac{1}{\omega C}\right)^2}$$

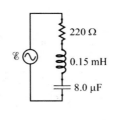

$$= \frac{(12\ \text{V})^2(220\ \Omega)}{2\left\{(220\ \Omega)^2 + \left[2\pi(2500\ \text{Hz})(0.15\times10^{-3}\ \text{H}) - \frac{1}{2\pi(2500\ \text{Hz})(8.0\times10^{-6}\ \text{F})}\right]^2\right\}} = \boxed{0.33\ \text{W}}$$

45. (a) Strategy Use Eqs. (21-14) with $X_L = 0$, Eqs. (21-6) and (21-7), and Ohm's law.

Solution Find I.

$$I = \frac{\mathcal{E}_m}{Z} = \frac{\mathcal{E}_m}{\sqrt{R^2 + X_C^2}} = \frac{\mathcal{E}_m}{\sqrt{R^2 + \frac{1}{\omega^2 C^2}}} = \frac{2.0\ \text{V}}{\sqrt{(5.00\times10^3\ \Omega)^2 + \frac{1}{4\pi^2(120\ \text{Hz})^2(0.48\times10^{-6}\ \text{F})^2}}}$$

$$= 3.5\times10^{-4}\ \text{A}$$

Find the voltage amplitudes.

$$V_C = IX_C = \frac{I}{\omega C} = \frac{3.5\times10^{-4}\ \text{A}}{2\pi(120\ \text{Hz})(0.48\times10^{-6}\ \text{F})} = \boxed{0.97\ \text{V}}$$

$$V_R = IR = (3.5\times10^{-4}\ \text{A})(5.00\times10^3\ \Omega) = \boxed{1.8\ \text{V}}$$

(b) Strategy and Solution No, the voltage amplitudes do not add to give the amplitude of the source voltage; the voltages across the resistor and capacitor are 90° out of phase. Similar to the impedance, the source voltage is the square root of the sum of squares, $\boxed{\mathscr{E}_m = \sqrt{V_C^2 + V_R^2}}$, as can be seen in a phasor diagram.

(c) Strategy Use the results of parts (a) and (b).

Solution Sketch the phasor diagram.

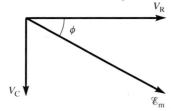

47. (a) Strategy Set $V_R = IR$ equal to $V_C = IX_C = I/(\omega C) = I/(2\pi f C)$ and solve for the frequency.

Solution Find the frequency.

$$V_R = IR = V_C = IX_C = \frac{I}{\omega C} = \frac{I}{2\pi f C}, \text{ so } f = \frac{1}{2\pi RC} = \frac{1}{2\pi(3300\ \Omega)(2.0\times10^{-6}\ \text{F})} = \boxed{24\ \text{Hz}}.$$

(b) Strategy In a phasor diagram, the source is the hypotenuse of a right triangle and V_R and V_C are the legs, so $\mathscr{E}_m = \sqrt{V_R^2 + V_C^2}$ according to the Pythagorean theorem.

Solution Find $V_C/\mathscr{E}_m = V_R/\mathscr{E}_m$.

$$\mathscr{E}_m{}^2 = V_R^2 + V_C^2 = V_R^2 + V_R^2 = 2V_R^2, \text{ so } \frac{V_R^2}{\mathscr{E}_m{}^2} = \frac{1}{2} \text{ and } \frac{V_R}{\mathscr{E}_m} = \frac{1}{\sqrt{2}} = \frac{V_C}{\mathscr{E}_m}.$$

Therefore, the rms voltages across the components are not half of the rms voltage of the source: $\boxed{\dfrac{V_R}{\mathscr{E}_m} = \dfrac{V_C}{\mathscr{E}_m} = \dfrac{1}{\sqrt{2}}}$.

(c) Strategy The voltage across a capacitor lags the current through it, so the source voltage lags the current. Use the power factor to find ϕ.

Solution Solve for the phase angle.

$\cos\phi = \dfrac{R}{Z}$, so

$$\phi = \cos^{-1}\frac{R}{Z} = \cos^{-1}\frac{R}{\sqrt{R^2 + X_C^2}} = \cos^{-1}\frac{IR}{\sqrt{(IR)^2 + (IX_C)^2}} = \cos^{-1}\frac{V_R}{\sqrt{V_R^2 + V_C^2}} = \cos^{-1}\frac{V_R}{\mathscr{E}_m} = \cos^{-1}\frac{1}{\sqrt{2}} = \frac{\pi}{4}.$$

Therefore, $\boxed{I \text{ leads } \mathscr{E} \text{ by } \dfrac{\pi}{4} \text{ rad} = 45°}$.

(d) Strategy Use the power factor to find Z.

Solution Find the impedance.

$$\cos\phi = \frac{R}{Z}, \text{ so } Z = \frac{R}{\cos\phi} = \frac{3.3 \text{ k}\Omega}{\cos\frac{\pi}{4}} = \boxed{4.7 \text{ k}\Omega}.$$

49. **Strategy** $\tau = RC$ and $Z = \sqrt{R^2 + 1/(\omega C)^2}$.

Solution Substitute $C = \tau/R$ into Z and solve for R.

$$Z = \sqrt{R^2 + \frac{R^2}{\tau^2\omega^2}} = R\sqrt{1 + \frac{1}{\tau^2\omega^2}}, \text{ so}$$

$$R = \frac{Z}{\sqrt{1 + \frac{1}{4\pi^2\tau^2 f^2}}} = \frac{350 \ \Omega}{\sqrt{1 + \frac{1}{4\pi^2(0.25\times10^{-3} \text{ s})^2(1250 \text{ Hz})^2}}} = \boxed{310 \ \Omega}$$

Calculate C.

$$C = \frac{\tau}{R} = \frac{0.25\times10^{-3} \text{ s}}{312 \ \Omega} = \boxed{8.0\times10^{-7} \text{ F}}$$

53. **Strategy and Solution** At resonance, the reactances are equal, so the impedance is equal to the resistance. Thus,
$$Z = R = \boxed{500.0 \ \Omega}.$$

57. **(a) Strategy** Use Eq. (21-18).

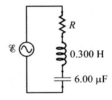

Solution Compute the resonant frequency.

$$\omega_0 = \sqrt{\frac{1}{LC}} = \sqrt{\frac{1}{(0.300 \text{ H})(6.00\times10^{-6} \text{ F})}} = \boxed{745 \text{ rad/s}}$$

(b) Strategy At resonance, $X_L = X_C$. Use Eq. (21-14a).

Solution Compute the resistance.

$$R = Z = \frac{\mathscr{E}_m}{I} = \frac{440 \text{ V}}{0.560 \text{ A}} = \boxed{790 \ \Omega}$$

(c) Strategy At resonance, the voltages across the capacitor and inductor are 180° out of phase and equal in magnitude, so they cancel.

Solution The peak voltage across the resistor is $V_R = \mathscr{E}_m = \boxed{440 \text{ V}}$.

Across the inductor, the peak voltage is

$$V_L = IX_L = I\omega_0 L = \frac{IL}{\sqrt{LC}} = I\sqrt{\frac{L}{C}} = (0.560 \text{ A})\sqrt{\frac{0.300 \text{ H}}{6.00\times10^{-6} \text{ F}}} = \boxed{125 \text{ V}}.$$

Since $X_L = X_C$, $V_C = V_L = \boxed{125 \text{ V}}$.

61. **Strategy** Use Eqs. (21-7), (21-10), and (21-14b).

Solution

(a) Find the impedance of the tweeter branch of the circuit.

$$Z = \sqrt{R^2 + X_C^2} = \sqrt{R^2 + \frac{1}{\omega^2 C^2}} = \sqrt{(8.0\ \Omega)^2 + \frac{1}{4\pi^2 (252\ \text{Hz})^2 (560 \times 10^{-6}\ \text{F})^2}} = \boxed{8.1\ \Omega}$$

(b) If the current amplitudes in the two branches are the same, the impedance must be the same in each branch, so $Z = \boxed{8.1\ \Omega}$.

(c) In the woofer branch, $Z = \sqrt{R^2 + X_L^2} = 8.08\ \Omega$. Solve for L.

$$Z^2 = R^2 + \omega^2 L^2, \text{ so } L = \frac{\sqrt{Z^2 - R^2}}{\omega} = \frac{\sqrt{(8.08\ \Omega)^2 - (8.0\ \Omega)^2}}{2\pi (252\ \text{Hz})} = \boxed{7 \times 10^{-4}\ \text{H}}.$$

(d) The crossover frequency occurs when the impedances of the branches are equal, which means that $X_C = X_L$.

$$X_C = \frac{1}{\omega C} = X_L = \omega L, \text{ so } \omega = \frac{1}{\sqrt{LC}}.$$

Since $\omega = 2\pi f$, the crossover frequency is $\boxed{f_{\text{co}} = \dfrac{1}{2\pi \sqrt{LC}}}$.

65. (a) **Strategy** Use Ohm's law.

Solution Compute the current amplitude.

$$I = \frac{V_R}{R} = \frac{10.0\ \text{V}}{100\ \Omega} = \boxed{0.10\ \text{A}}$$

(b) **Strategy** $V_C = IX_C$ and $X_C = 1/(\omega C)$.

Solution Compute the capacitance.

$$V_C = IX_C = I\frac{1}{\omega C}, \text{ so } C = \frac{I}{\omega V_C} = \frac{0.10\ \text{A}}{2\pi (60\ \text{Hz})(7.0\ \text{V})} = \boxed{38\ \mu\text{F}}.$$

(c) **Strategy** $V_L = IX_L$ and $X_L = \omega L$.

Solution Compute the inductance.

$$V_L = IX_L = I\omega L, \text{ so } L = \frac{V_L}{\omega I} = \frac{15.0\ \text{V}}{2\pi (60\ \text{Hz})(0.10\ \text{A})} = \boxed{0.40\ \text{H}}.$$

69. Strategy Use Eqs. (20-10) and (21-4).

Solution

(a) Calculate the turns ratio.

$$\frac{N_2}{N_1} = \frac{\mathcal{E}_2}{\mathcal{E}_1} = \frac{240\times10^3 \text{ V}}{420 \text{ V}} = \boxed{570}$$

(b) Calculate the rms current using the turns ratio.

$$I_1 = \frac{N_2}{N_1} I_2 = 570(60.0\times10^{-3} \text{ A}) = \boxed{34 \text{ A}}$$

(c) Calculate the average power.

$$P_{av} = I_{rms}V_{rms} = (0.0600 \text{ A})(240,000 \text{ V}) = \boxed{14 \text{ kW}}$$

73. Strategy Use Eq. (21-7).

Solution Find the capacitance.

$$X_C = \frac{1}{\omega C}, \text{ so } C = \frac{1}{\omega X_C} = \frac{1}{2\pi(520 \text{ Hz})(6.20 \text{ }\Omega)} = \boxed{49 \text{ }\mu\text{F}}.$$

77. Strategy Use Eq. (21-18).

Solution Find the maximum capacitance required.

$$\omega_0 = 2\pi f_0 = \frac{1}{\sqrt{LC}}, \text{ so } C = \frac{1}{4\pi^2 f_0^2 L} = \frac{1}{4\pi^2(0.52\times10^6 \text{ Hz})^2(2.4\times10^{-4} \text{ H})} = \boxed{390 \text{ pF}}.$$

81. Strategy Use Eqs. (21-4) and (21-17).

Solution

(a) The average power is $P_{av} = I_{rms}\mathcal{E}_{rms}$. Find I_{rms}.

$$I_{rms} = \frac{P_{av}}{\mathcal{E}_{rms}} = \frac{12\times10^6 \text{ W}}{250\times10^3 \text{ V}} = \boxed{48 \text{ A}}$$

(b) The average power is $P_{av} = I_{rms}\mathcal{E}_{rms}\cos\phi$. Find I_{rms}.

$$I_{rms} = \frac{P_{av}}{\mathcal{E}_{rms}\cos\phi} = \frac{12\times10^6 \text{ W}}{(250\times10^3 \text{ V})(0.86)} = \boxed{56 \text{ A}}$$

(c) Since the current is greater, $\boxed{\text{the power lost in transmission is greater}}$ due to I^2R losses in the transmission line. The power company would want to charge more to make up for this loss.

85. (a) Strategy $\mathcal{E}(t) = (286 \text{ V})\sin[(390 \text{ rad/s})t]$, so $\omega = 390$ rad/s. Model the inductor as an *RL* series circuit.

Solution Find the impedance.

$$Z = \sqrt{R^2 + X_L^2} = \sqrt{R^2 + \omega^2 L^2} = \sqrt{(30.0 \text{ }\Omega)^2 + (390 \text{ rad/s})^2(0.0400 \text{ H})^2} = \boxed{33.8 \text{ }\Omega}$$

(b) Strategy Use the definition of rms.

Solution The peak voltage across the inductor including its internal resistance is equal to the peak source voltage, so $V_L = \mathcal{E}_m = \boxed{286 \text{ V}}$.

Calculate V_{rms}.

$$V_{rms} = \frac{V_L}{\sqrt{2}} = \frac{286 \text{ V}}{\sqrt{2}} = \boxed{202 \text{ V}}$$

(c) Strategy Use Eq. (21-14a).

Solution Calculate the peak current, *I*.

$$I = \frac{\mathcal{E}_m}{Z} = \frac{286 \text{ V}}{33.8 \text{ }\Omega} = \boxed{8.46 \text{ A}}$$

(d) Strategy Use Eqs. (21-16) and (21-17).

Solution Find the average power dissipated.

$$P_{av} = I_{rms}\mathcal{E}_{rms}\cos\phi \text{ and } \cos\phi = \frac{R}{Z}, \text{ so } P_{av} = \frac{8.46 \text{ A}}{\sqrt{2}}(202 \text{ V})\frac{30.0 \text{ }\Omega}{33.8 \text{ }\Omega} = \boxed{1.07 \text{ kW}}.$$

(e) Strategy $i(t) = I_{peak}\sin(\omega t - \phi)$ since the current through an inductor lags the voltage across it. Use Eq. (21-16).

Solution Find the phase angle.

$$\cos\phi = \frac{R}{Z}, \text{ so } \phi = \cos^{-1}\frac{R}{Z} = \cos^{-1}\frac{30.0 \text{ }\Omega}{33.814 \text{ }\Omega} = 0.480 \text{ rad}.$$

Therefore, the expression for the current is $\boxed{i(t) = (8.46 \text{ A})\sin\left[(390 \text{ rad/s})t - 0.480 \text{ rad}\right]}$.

REVIEW AND SYNTHESIS: CHAPTERS 19–21

Review Exercises

1. **Strategy** The maximum torque occurs when the plane of the loop is parallel to the magnetic field of the solenoid. Use Eqs. (19-13a) and (19-17).

   Solution Find the maximum possible magnetic torque on the loop.
   $$\tau = N_1 I_1 A_1 B_s = N_1 I_1 A_1 \mu_0 n_s I_s = (100)(2.20 \text{ A})\pi(0.0800 \text{ m})^2(4\pi\times10^{-7} \text{ T}\cdot\text{m/A})(8500 \text{ m}^{-1})(25.0 \text{ A})$$
   $$= \boxed{1.2 \text{ N}\cdot\text{m}}$$
   Since the magnetic field inside the solenoid is along the axis of the solenoid, the magnetic torque on the loop is at its maximum value $\boxed{\text{when the plane of the loop is parallel to the axis of the solenoid}}$.

5. **Strategy** Use Eq. (19-14) and $\vec{F}_B = q\vec{v}\times\vec{B}$. Let the positive y-direction be up.

   Solution According to the RHR, the magnetic field generated by the power line at point P is in the positive y-direction. Looking at the side view, we see that the angle between the velocity of the muon and the magnetic field is $180° - 25° = 155°$. The charge of the muon is negative, so according to the RHR and $\vec{F}_B = q\vec{v}\times\vec{B}$, the direction of the force on the muon is out of the plane of the paper in the side view (or to the right in the end-on view). Compute the magnitude of the force.
   $$F_B = evB\sin\theta = ev\frac{\mu_0 I}{2\pi r}\sin\theta = \frac{(1.602\times10^{-19} \text{ C})(7.0\times10^7 \text{ m/s})(4\pi\times10^{-7} \text{ T}\cdot\text{m/A})(16.0 \text{ A})\sin 155°}{2\pi(0.850 \text{ m})}$$
   $$= 1.8\times10^{-17} \text{ N}$$
   Thus, $\vec{F}_B = \boxed{1.8\times10^{-17} \text{ N out of the plane of the paper in the side view (or to the right in the end on view)}}$.

9. **Strategy** Use Lenz's law and $\vec{F} = I\vec{L}\times\vec{B}$.

   Solution According to the RHR, the magnetic field points into the page at the loop.

   (a) $\boxed{\text{A counterclockwise current is induced because the flux through the loop is increasing as it nears the wire}}$. This counterclockwise current generates a field directed out of the page. According to the RHR, the magnetic force on the part of the loop closest to the wire is away from the wire and that on the part of the loop farthest from the wire is toward the wire. Since the magnetic field is greater nearer the wire, $\boxed{\text{the net force on the loop is away from the wire}}$.

   (b) $\boxed{\text{No current is induced because there is no change in flux through the loop}}$. Since there is no current in the loop, $\boxed{\text{there is no magnetic force acting on the loop}}$.

   (c) $\boxed{\text{A clockwise current is induced because the flux through the loop is decreasing as it moves away from the wire}}$. This clockwise current generates a field directed into the page. According to the RHR, the magnetic force on the part of the loop closest to the wire is toward the wire and that on the part of the loop farthest from the wire is away from the wire. Since the magnetic field is greater nearer the wire, $\boxed{\text{the net force on the loop is toward the wire}}$.

13. (a) **Strategy** At resonance, $X_C = X_L$, so $I \propto R^{-1}$. Thus, when the resistance is doubled, the current is cut in half. Use Eq. (21-4).

Solution If the initial power dissipated is $P_i = I_i^2 R_i$, then the final power dissipated is

$$P_f = I_f^2 R_f = \left(\frac{I_i}{2}\right)^2 (2R_i) = \frac{1}{2}I_i^2 R_i = \frac{1}{2}P_i.$$ Therefore, $\boxed{\text{the power is cut in half}}$ when the resistance of an *RLC* circuit is doubled.

(b) **Strategy** When the circuit is not at resonance, the situation is more complicated because the current is inversely proportional to the impedance Z, and the impedance is not simply equal to the resistance. Use Eq. (21-14b).

Solution Find the initial and final impedances in terms of the initial resistance.

$$Z_i = \sqrt{R^2 + (X_L - X_C)^2} = \sqrt{R^2 + (2R - R)^2} = \sqrt{2}R$$

$$Z_f = \sqrt{(2R)^2 + (X_L - X_C)^2} = \sqrt{4R^2 + (2R - R)^2} = \sqrt{5}R$$

Form a proportion to find the final current in terms of the initial current.

$$\frac{I_f}{I_i} = \frac{Z_i}{Z_f} = \frac{\sqrt{2}R}{\sqrt{5}R} = \sqrt{\frac{2}{5}}, \text{ so } I_f = \sqrt{\frac{2}{5}}I_i.$$

Form a proportion to find the final power dissipated in terms of the initial power dissipated.

$$\frac{P_f}{P_i} = \frac{I_f^2 R_f}{I_i^2 R_i} = \frac{\left(\sqrt{\frac{2}{5}}I_i\right)^2 (2R)}{I_i^2 R} = \frac{4}{5}, \text{ so } P_f = \frac{4}{5}P_i.$$

Thus, $\boxed{\text{the power is } 4/5 \text{ of its original value}}$.

17. **Strategy** Since the measurement is made relatively close to the electron beam, the beam can be approximated as a long, straight wire. Use Eq. (19-14) and the definition of current.

Solution Let N be the number of electrons, then the amount of charge passing the point in 1.30 microseconds is $\Delta q = Ne$. Find the magnetic field strength Kieran measures.

$$B = \frac{\mu_0 I}{2\pi r} = \frac{\mu_0}{2\pi r}\left(\frac{\Delta q}{\Delta t}\right) = \frac{\mu_0}{2\pi r}\left(\frac{Ne}{\Delta t}\right) = \frac{(4\pi \times 10^{-7} \text{ T}\cdot\text{m/A})(1.40 \times 10^{11})(1.602 \times 10^{-19} \text{ C})}{2\pi(0.0200 \text{ m})(1.30 \times 10^{-6} \text{ s})} = \boxed{1.73 \times 10^{-7} \text{ T}}$$

21. **(a) Strategy** The kinetic energy of an ion is equal to the potential energy gained.

Solution Find v as a function of m.

$$\frac{1}{2}mv^2 = e\Delta V, \text{ so } v = \sqrt{\frac{2e\Delta V}{m}} \propto \frac{1}{\sqrt{m}}.$$

Form a proportion to find the speed of the ions.

$$\frac{v_{235}}{v_{238}} = \sqrt{\frac{m_{238}}{m_{235}}} = \sqrt{\frac{m}{0.98737m}} = \frac{1}{\sqrt{0.98737}} = \frac{v_{235}}{v}, \text{ so } v_{235} = \frac{v}{\sqrt{0.98737}} = \boxed{1.00638v}.$$

(b) Strategy For an ion with speed v, the electric and magnetic forces are equal in magnitude and opposite in direction. For an ion with speed greater than v, the magnetic force dominates. In this case, the ion has a positive charge, so the magnetic force deflects it downward.

Solution The sketch is shown.

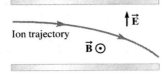

(c) Strategy The U-235 ions move in circular paths with diameters smaller than D, since the magnetic force on them is relatively stronger.

Solution Apply Newton's second law to the circular path.

$$\Sigma F_r = evB = m\frac{v^2}{r} = m\frac{2v^2}{D}, \text{ so } \frac{mv}{D} = \frac{eB}{2} = \frac{m_{235}v_{235}}{D_{235}}, \text{ since } e \text{ and } B \text{ are constants.}$$

$$D_{235} = \frac{m_{235}v_{235}}{mv}D = \frac{0.98737m\dfrac{v}{\sqrt{0.98737}}}{mv}D = \sqrt{0.98737}D = \boxed{0.99366D}$$

25. **Strategy** Use Eqs. (20-15b), (21-17) with $\cos\phi = R/Z$, and (14-4).

Solution

(a) Find the inductance of the coil.

$$L = \frac{\mu_0 N^2 \pi r^2}{\ell} = \frac{(4\pi \times 10^{-7} \text{ T} \cdot \text{m/A})(18)^2 \pi (0.025 \text{ m})^2}{0.010 \text{ m}} = 8.0 \times 10^{-5} \text{ H}$$

Find the average power dissipated.

$$P_{av} = I_{rms}\mathscr{E}_{rms}\frac{R}{Z} = \frac{\mathscr{E}_{rms}}{Z}\mathscr{E}_{rms}\frac{R}{Z} = \frac{\mathscr{E}^2_{rms}R}{Z^2} = \frac{\mathscr{E}^2_{rms}R}{R^2 + (2\pi f L)^2}$$

$$= \frac{[(340 \text{ V})/\sqrt{2}]^2(1.0 \text{ }\Omega)}{(1.0 \text{ }\Omega)^2 + 4\pi^2(50 \times 10^3 \text{ Hz})^2(8.0 \times 10^{-5} \text{ H})^2} = \boxed{91 \text{ W}}$$

(b) Find the average power required to heat the water.

$$P_{av} = \frac{Q}{\Delta t} = \frac{mc\Delta T}{\Delta t} = \frac{(1.0 \text{ L})(1.0 \text{ kg/L})[4186 \text{ J/(kg} \cdot \text{°C)}](80\text{°C})}{(5.0 \text{ min})(60 \text{ s/min})} = \boxed{1.1 \text{ kW}}$$

MCAT Review

1. **Strategy and Solution** The current flows counterclockwise through the apparatus. By the RHR, the magnetic field generated by the current is directed upward. The correct answer is $\boxed{\text{A}}$.

2. **Strategy** Refer to the data in the table.

    Solution Two rows of data are given for a mass of 0.01 kg. The current is increased by 150% from 10.0 A to 15.0 A. Due to the increase in the current, the exit speed of the projectile increases by 150% from 2.0 km/s to 3.0 km/s. Thus, the exit speed is directly proportional to the current. So, if the current were decreased by a factor of two, the exit speed would be decreased by a factor of two, as well. The correct answer is $\boxed{\text{D}}$.

3. **Strategy and Solution** Since, for a given current, the force is constant along the entire length of the railgun, lengthening the rails would increase the exit speed because of the longer distance over which the force is present. The correct answer is $\boxed{\text{D}}$.

4. **Strategy and Solution** Since the resistance of the rails is directly proportional to their resistivity, lowering the resistivity of the rails would decrease the power required to maintain the current that flows through them. The correct answer is $\boxed{\text{B}}$.

5. **Strategy** The average power is equal to the change in kinetic energy divided by the time interval.

    Solution Find the average power supplied by the railgun.
    $$P_{av} = \frac{\Delta K}{\Delta t} = \frac{\frac{1}{2}mv^2 - 0}{\Delta t} = \frac{mv^2}{2\Delta t} = \frac{(0.10 \text{ kg})(10.0 \text{ m/s})^2}{2(2.0 \text{ s})} = 2.5 \text{ W}$$
    The correct answer is $\boxed{\text{B}}$.

6. **Strategy and Solution** $F = ma \propto I^2$, so $a \propto I^2/m$. The speed of the projectile is directly proportional to its acceleration, so $v \propto I^2/m$. Form a proportion using data from the table to find the approximate speed of the 0.08-kg projectile.
    $$\frac{v_2}{v_1} = \frac{I_2^2/m_2}{I_1^2/m_1} = \left(\frac{I_2}{I_1}\right)^2 \frac{m_1}{m_2}, \text{ so } v_2 = \left(\frac{I_2}{I_1}\right)^2 \frac{m_1}{m_2} v_1. \text{ For (1), we use the data given in the third row of the table.}$$
    $$v_2 = \left(\frac{I_2}{I_1}\right)^2 \frac{m_1}{m_2} v_1 = \left(\frac{20.0}{10.0}\right)^2 \frac{0.02}{0.08}(1.4 \text{ km/s}) = 1.4 \text{ km/s}$$
    The correct answer is $\boxed{\text{C}}$.

7. **Strategy and Solution** The power from the power plant is given by $P = IV$, so, for a given amount of power, increasing the voltage decreases the current required. Since the power lost as heat is given by $P = I^2R$, reducing the current reduces the power lost as heat. The correct answer is $\boxed{\text{A}}$.

8. **Strategy and Solution** The magnetic field lines due to the section of current-carrying wire are circles (which is the only possibility, given the symmetry of the situation). The direction of the field is determined by the RHR. Pointing the thumb of the right hand in the direction of the current in the wire, then curling the fingers inward toward the palm, the direction of the field is indicated by the direction of the curl of the fingers—in this case counterclockwise, as seen from above. The correct answer is $\boxed{\text{D}}$.

35. **Strategy** Use Eqs. (22-10), (22-12), and (22-13).

    Solution Find the intensity of the beam.
    $$I = \frac{\langle P \rangle}{A} = \frac{10.0 \times 10^{-3} \text{ W}}{\frac{1}{4}\pi(0.85 \times 10^{-2} \text{ m})^2} = 180 \text{ W/m}^2$$
    Find the rms value of the electric field.
    $$I = \langle u \rangle c = (\epsilon_0 E_{rms}{}^2)c, \text{ so } E_{rms} = \sqrt{\frac{I}{\epsilon_0 c}} = \sqrt{\frac{180 \text{ W/m}^2}{[8.854 \times 10^{-12} \text{ C}^2/(\text{N} \cdot \text{m}^2)](3.00 \times 10^8 \text{ m/s})}} = \boxed{260 \text{ V/m}}.$$

37. **Strategy** The energy collected by a solar panel is $U = IA\Delta t$, where I is the intensity of the radiation at the panel, A is the area of the panel, and Δt is the time the panel is exposed to the radiation. The intensity of the radiation at the panel is $I = \langle P \rangle / (4\pi r^2)$, where $\langle P \rangle$ is the average power of the Sun and r is the distance from the Sun to the solar panel. Let $U_1 = U_2 = U = 1.4$ kJ; then the time interval Δt_1 during which the first plate absorbs 1.4 kJ of energy is 1.0 s.

    Solution Set the energy collected by the panels equal to find the time required for the second panel to absorb the same amount of energy as the first.
    $$U_2 = I_2 A \Delta t_2 = \frac{\langle P \rangle}{4\pi r_2{}^2} A \Delta t_2 = U_1 = I_1 A \Delta t_1 = \frac{\langle P \rangle}{4\pi r_1{}^2} A \Delta t_1, \text{ so } \Delta t_2 = \left(\frac{r_2}{r_1}\right)^2 \Delta t_1 = \left(\frac{1.55}{1.00}\right)^2 (1.0 \text{ s}) = \boxed{2.4 \text{ s}}.$$

39. **Strategy** Intensity is related to the average power radiated by $I = \langle P \rangle / A$ where $A = 4\pi r^2$ and $r = 14 \times 10^6$ ly.

    Solution Find the rate at which the star radiates EM energy.
    $$\langle P \rangle = 4\pi r^2 I = 4\pi(14 \times 10^6 \text{ ly})^2 (9.461 \times 10^{15} \text{ m/ly})^2 (4 \times 10^{-21} \text{ W/m}^2) = \boxed{9 \times 10^{26} \text{ W}}$$

41. **Strategy and Solution** Suppose you have an EM wave traveling in vacuum. The relation $E = cB$ holds at all points and all times. The electric energy density can therefore be written as
    $$\frac{1}{2}\epsilon_0 E^2 = \frac{1}{2}\epsilon_0 (cB)^2 = \frac{1}{2}\epsilon_0 c^2 B^2 = \frac{1}{2}\epsilon_0 \frac{1}{\epsilon_0 \mu_0} B^2 = \frac{1}{2\mu_0} B^2.$$

45. **Strategy** Since the light is initially unpolarized, the intensity of the light after passing through the first polarizer is half the initial intensity. Use Eq. (22-16b).

    Solution Let the initial intensity be $I_1 = I_0/2$. The intensity after passing through the second polarizer is given by $I_2 = I_1 \cos^2 \theta$. Combining these equations, and using $\theta = 45°$, we have
    $$I_2 = \frac{1}{2}I_0 \cos^2 45° = 0.25 I_0.$$
    Therefore, $\boxed{0.25}$ of the incident intensity is transmitted.

49. Strategy Use Eq. (22-16b).

Solution

(a) If we try to use one sheet, the resulting intensity is $\boxed{I_1 = I_0 \cos^2 90.0° = 0}$, so at least two sheets must be used.

(b) The transmitted intensity of the first sheet is $I_1 = I_0 \cos^2 45.0° = 0.500 I_0$.

The transmitted intensity of the second sheet is $I_2 = I_1 \cos^2 45.0° = 0.500 I_0 (0.500) = \boxed{0.250 I_0}$.

(c) The transmitted intensity of the four sheets combined is $I_4 = I_0 (\cos^2 22.5°)^4 = I_0 \cos^8 22.5° = \boxed{0.531 I_0}$.

53. Strategy and Solution The incident light is horizontal, as would occur shortly after sunrise. It is unpolarized, so the molecules in the atmosphere become oscillating dipoles, which oscillate in random directions perpendicular to the incident wave. As an oscillating dipole, the molecules radiate EM waves. Horizontal oscillations of the dipoles do not radiate in the horizontal direction. So, $\boxed{\text{yes}}$, the light is polarized in the $\boxed{\text{up-down}}$ direction (which is perpendicular to both the direction of the incident light and the direction of the scattered light).

55. Strategy The calculation of the relative velocity is the same as in Example 22.9, except that v_{rel} is negative (source and observer receding).

Solution The speeder is going faster than the police car. Since the police car is going 38.0 m/s, the speeder is going 38.0 m/s + 7.0 m/s = $\boxed{45.0 \text{ m/s}}$.

57. Strategy Use the Doppler shift formula, Eq. (22-17), and $\lambda f = c$. Note that $v_{\text{rel}} < 0$, since the star is moving away.

Solution Find the wavelength of the light emitted by the star.

$$f_o = \frac{c}{\lambda_o} = f_s \sqrt{\frac{1 + v_{\text{rel}}/c}{1 - v_{\text{rel}}/c}} = \frac{c}{\lambda_s}\sqrt{\frac{1 + v_{\text{rel}}/c}{1 - v_{\text{rel}}/c}}, \text{ so } \lambda_o = \lambda_s \sqrt{\frac{1 - v_{\text{rel}}/c}{1 + v_{\text{rel}}/c}} = (480 \text{ nm})\sqrt{\frac{1 + 2.4/3.0}{1 - 2.4/3.0}} = \boxed{1440 \text{ nm}}.$$

59. Strategy Use the Doppler shift formula, Eq. (22-17), and the binomial approximations, since the relative speed is small compared to c. The cars are receding, so the relative speed is negative.

Solution

(a) Since the speeder is moving away from the police car, the frequency of the microwaves observed by the police is less than that of the emitted microwaves. The microwaves must travel farther for a fixed speed. Therefore, $\boxed{f_1}$ is larger.

(b) Find the frequency of the pulse as received by the speeder.

$$f_{\text{at speeder}} = f_{\text{as}} = f_1 \sqrt{\frac{1 + v_{\text{rel}}/c}{1 - v_{\text{rel}}/c}}$$

Find the frequency of the pulse as received by the police.

$$f_2 = f_{\text{as}}\sqrt{\frac{1 + v_{\text{rel}}/c}{1 - v_{\text{rel}}/c}} = f_1\left(\frac{1 + v_{\text{rel}}/c}{1 - v_{\text{rel}}/c}\right) \approx f_1\left(1 + \frac{v_{\text{rel}}}{c}\right)\left(1 + \frac{v_{\text{rel}}}{c}\right) \approx f_1\left[1 + \left(\frac{v_{\text{rel}}}{c}\right)^2\right] \approx f_1\left(1 + \frac{2v_{\text{rel}}}{c}\right), \text{ so}$$

$$f_2 - f_1 \approx f_1\left(1 + \frac{2v_{\text{rel}}}{c}\right) - f_1 = f_1\left(1 + \frac{2v_{\text{rel}}}{c} - 1\right) = \frac{2v_{\text{rel}} f_1}{c} = \frac{2(-43.0 \text{ m/s})(36.0 \times 10^9 \text{ Hz})}{2.998 \times 10^8 \text{ m/s}} = \boxed{-10.3 \text{ kHz}}.$$

61. **Strategy** A Doppler shift of this magnitude almost certainly requires a relativistic relative velocity. Use Eq. (22-17).

    Solution $v_{rel} > 0$ since the source of the light is stationary (observer approaching the source). Let $v_{rel} = v$ for simplicity. Find v.

    $$f_0 = f_s \sqrt{\frac{1 + \frac{v}{c}}{1 - \frac{v}{c}}}$$

    $$\left(\frac{f_0}{f_s}\right)^2 = \frac{1 + \frac{v}{c}}{1 - \frac{v}{c}}$$

    $$\left(\frac{f_0}{f_s}\right)^2 - \left(\frac{f_0}{f_s}\right)^2 \frac{v}{c} = 1 + \frac{v}{c}$$

    $$\left(\frac{f_0}{f_s}\right)^2 - 1 = \frac{v}{c}\left[1 + \left(\frac{f_0}{f_s}\right)^2\right]$$

    $$v = c\frac{(f_0/f_s)^2 - 1}{(f_0/f_s)^2 + 1} = c\frac{(\lambda_s/\lambda_o)^2 - 1}{(\lambda_s/\lambda_o)^2 + 1} = (3.00 \times 10^8 \text{ m/s})\frac{(630/530)^2 - 1}{(630/530)^2 + 1} = \boxed{5 \times 10^7 \text{ m/s}}$$

65. **Strategy** The frequency, wavelength, and speed of EM radiation are related by $\lambda f = c$.

    Solution Compute the wavelength.

    $$\lambda = \frac{c}{f} = \frac{3.00 \times 10^8 \text{ m/s}}{2.0 \times 10^9 \text{ Hz}} = 0.15 \text{ m} = 15 \text{ cm}$$

    The maximum length is half the wavelength.

    $$\frac{1}{2} \times 15 \text{ cm} = \boxed{7.5 \text{ cm}}$$

69. **Strategy** The frequency, wavelength, and speed of EM radiation are related by $\lambda f = c$.

    Solution The corresponding frequencies are

    $$f_1 = \frac{c}{\lambda_1} = \frac{3.00 \times 10^8 \text{ m/s}}{190 \text{ m}} = 1.58 \times 10^6 \text{ Hz and } f_2 = \frac{c}{\lambda_2} = \frac{3.00 \times 10^8 \text{ m/s}}{550 \text{ m}} = 5.45 \times 10^5 \text{ Hz.}$$

    The number of 10 kHz bands that will fit in this range is $\dfrac{1.58 \times 10^6 \text{ Hz} - 5.45 \times 10^5 \text{ Hz}}{10 \times 10^3 \text{ Hz}} = \boxed{100}$.

73. **(a) Strategy** The heat that enters the water is given by $\Delta Q = mc_w \Delta T$.

Solution Find the rate of energy absorption, $\Delta Q/\Delta t$.

$$\frac{\Delta Q}{\Delta t} = \frac{mc_w \Delta T}{\Delta t} = \frac{(0.35 \text{ kg})[4186 \text{ J/(kg} \cdot \text{K})](75.0 \text{ K})}{2.00(60.0 \text{ s})} = \boxed{0.92 \text{ kW}}$$

(b) Strategy Use Eq. (22-12).

Solution Compute the average intensity of the microwaves in the waveguide.

$$I = \frac{\langle P \rangle}{A} = \frac{mc_w \Delta T}{A\Delta t} = \frac{(0.35 \text{ kg})[4186 \text{ J/(kg} \cdot \text{K})](75.0 \text{ K})}{(88.0 \times 10^{-4} \text{ m}^2)2.00(60.0 \text{ s})} = \boxed{1.0 \times 10^5 \text{ W/m}^2}$$

(c) Strategy Use Eqs. (22-10) and (22-13), and the fact that $E_{rms} = cB_{rms}$.

Solution Find the rms electric field inside the waveguide.

$$I = \langle u \rangle c = \epsilon_0 E_{rms}^2 c = \frac{mc_w \Delta T}{A\Delta t}, \text{ so}$$

$$E_{rms} = \sqrt{\frac{mc_w \Delta T}{\epsilon_0 cA\Delta t}} = \sqrt{\frac{(0.35 \text{ kg})[4186 \text{ J/(kg} \cdot \text{K})](75.0 \text{ K})}{[8.854 \times 10^{-12} \text{ C}^2/(\text{N} \cdot \text{m}^2)](3.00 \times 10^8 \text{ m/s})(88.0 \times 10^{-4} \text{ m}^2)2.00(60.0 \text{ s})}}$$

$$= \boxed{6.3 \text{ kV/m}}.$$

Find the rms magnetic field inside the waveguide.

$$B_{rms} = \frac{E_{rms}}{c} = \sqrt{\frac{(0.35 \text{ kg})[4186 \text{ J/(kg} \cdot \text{K})](75.0 \text{ K})}{[8.854 \times 10^{-12} \text{ C}^2/(\text{N} \cdot \text{m}^2)](3.00 \times 10^8 \text{ m/s})^3(88.0 \times 10^{-4} \text{ m}^2)2.00(60.0 \text{ s})}} = \boxed{2.1 \times 10^{-5} \text{ T}}$$

77. **Strategy** Let the distance between the mountains be d. Then, $2d = c\Delta t$, where Δt is the time it takes light to make the round trip minus the reaction time.

Solution Find d.

$$2d = c\Delta t, \text{ so } d = \frac{1}{2}c\Delta t = \frac{1}{2}(3.00 \times 10^8 \text{ m/s})(0.35 \text{ s} - 0.25 \text{ s}) = \boxed{15,000 \text{ km}}.$$

For comparison, the circumference of Earth is about 40,000 km.

Therefore, $\boxed{\text{this scenario is certainly not feasible since it is almost halfway around the entire Earth}}$.

Chapter 23

REFLECTION AND REFRACTION OF LIGHT

Conceptual Questions

1. In specular reflection, radiation incident on a surface at a given angle always reflects at the same angle. Specular reflection occurs for materials with surface features that are small compared to the wavelength of the radiation. For visible light, such materials include polished mirrors, lenses, and metals. In diffuse reflection, radiation incident on a surface at a given angle is not necessarily reflected at the same angle. Diffuse reflection occurs for materials with surface features that are large compared to the wavelength of the reflected radiation. Examples of such materials for visible light include rock, wood, and most other items encountered in the natural world.

5. The rough surface of "nonglare glass" results in diffuse reflection instead of specular reflection that is indicative of smooth glass. As a result, light rays from a bright source are reflected in various directions so that no focused image of the source is formed.

9. **(a)** The image is upright.

   (b) The image is inverted.

   (c) The real image is formed at a distance greater than $2f$, and therefore, cannot be seen.

13. A virtual image would be formed on the left side of the lens, while a real image would be formed on the right side.

17. An image formed by a projector needs to be located on the viewing screen, where the light is diffusely reflected allowing the image to be seen from any angle. Since the light from the projector actually comes from the location of the image, it is a real image. For a camera, the image must be focused on the film so that the exposed film forms a copy of the actual image. Thus, the camera must form a real image as well. The lens in the eye works like a camera, forming a real image on the retina.

21. The amount by which a lens causes light rays to converge depends upon the refraction of the rays as they enter the lens. The greater the refraction, the more the rays will converge, and the smaller the focal length will be. Glycerine ($n \approx 1.47$) has a larger index of refraction than air ($n \approx 1.00$), so the refraction of the rays entering the dense flint glass ($n \approx 1.66$) will be reduced. Therefore, the focal length will increase.

Problems

1. **Strategy** Every point on a wavefront is considered a source of spherical wavelets. A surface tangent to the wavelets at a later time is the wavefront at that time.

   Solution The wavefronts due to an isotropic point source are spherical.

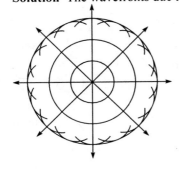

3. Strategy Every point on a wavefront is considered a source of spherical wavelets. A surface tangent to the wavelets at a later time is the wavefront at that time.

Solution The planar wavefront incident on the reflecting wall at normal incidence is transmitted through the opening and reflected from the reflecting wall beyond the opening.

On the incident side are two planar waves.
On the transmitted side is one hemispherical wave.

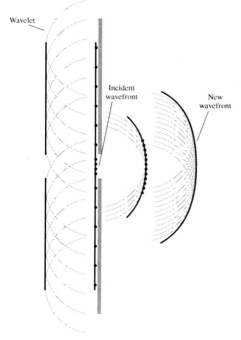

5. Strategy With respect to the normal at the point of incidence, the angle of incidence equals the angle of reflection and the reflected ray lies in the same plane, the plane of incidence, as the incident ray and the normal. Every point on a wavefront is considered a source of spherical wavelets. A surface tangent to the wavelets at a later time is the wavefront at that time.

Solution The ray diagram and wavefronts for a spherical wave (from a point source) reflecting from a planar surface.

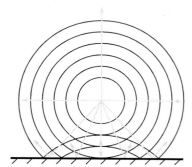

7. Strategy Redraw the figure, including the angle of incidence and the angle of reflection. Use the laws of reflection.

Solution From the figure, we see that the angle of incidence is $\theta_i = 90° - 50° = 40°$. Using the laws of reflection, we find that the angle of reflection is $\theta_r = \theta_i = 40°$. The angles θ_i, θ_r, and δ must add to 180°. Solve for δ.

$\theta_i + \theta_r + \delta = 40° + 40° + \delta = 180°$, so $\delta = \boxed{100°}$.

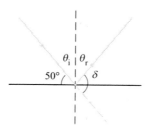

9. **Strategy** Draw a figure after Figure 23.7, labeling angles and lengths as necessary.

Solution

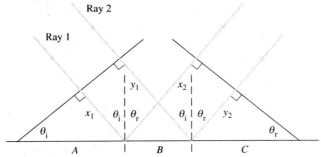

Since the time of travel from one wavefront to the other is the same for both rays, the distance traveled must also be the same.

Distance traveled by ray 1 = Distance traveled by ray 2

$$x_1 + x_2 = y_1 + y_2 \quad (1)$$

These distances can be expressed in terms of θ_i, θ_r, A, B, and C.

$x_1 = A\sin\theta_i$, $x_2 = (B+C)\sin\theta_r$, $y_1 = (A+B)\sin\theta_i$, and $y_2 = C\sin\theta_r$.

Substitute into (1) and simplify.

$$A\sin\theta_i + (B+C)\sin\theta_r = (A+B)\sin\theta_i + C\sin\theta_r$$
$$A\sin\theta_i + B\sin\theta_r + C\sin\theta_r = A\sin\theta_i + B\sin\theta_i + C\sin\theta_r$$
$$B\sin\theta_r = B\sin\theta_i$$
$$\sin\theta_r = \sin\theta_i$$
$$\theta_r = \theta_i$$

11. **Strategy** Use Snell's law, Eq. (23-4).

Solution Find the angle the Sun's rays in air make with the vertical.

$$n_i \sin\theta_i = n_t \sin\theta_t, \text{ so } \theta_i = \sin^{-1}\left(\frac{n_t}{n_i}\sin\theta_t\right) = \sin^{-1}\left(\frac{1.333}{1.000}\sin 42.0°\right) = \boxed{63.1°}.$$

13. **Strategy** Due to refraction of the light coming from the dolphin, the speed of the dolphin appears to be less than the actual speed by a factor of $1/n$.

Solution The dolphin appears to be moving at a speed of

$$\frac{15 \text{ m/s}}{1.33} = \boxed{11 \text{ m/s}}.$$

15. Strategy Draw a diagram. Use Snell's law, Eq. (23-4).

Solution Find θ_1.

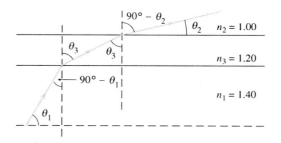

Relate θ_1 and θ_3.

$n_1 \sin(90° - \theta_1) = n_3 \sin\theta_3$

Relate θ_2 and θ_3.

$n_2 \sin(90° - \theta_2) = n_3 \sin\theta_3$

Eliminate $n_3 \sin\theta_3$ and solve for θ_1.

$n_1 \sin(90° - \theta_1) = n_2 \sin(90° - \theta_2)$

$$90° - \theta_1 = \sin^{-1}\left[\frac{n_2}{n_1}\sin(90° - \theta_2)\right]$$

$$\theta_1 = 90° - \sin^{-1}\left[\frac{1.00}{1.40}\sin(90° - 5.00°)\right] = \boxed{44.6°}$$

17. Strategy Draw a diagram. Use Snell's law, Eq. (23-4).

Solution Initially, the coin is just hidden from view, so the angle of the observer's eye with respect to the vertical is $\theta_1 = \tan^{-1}\frac{6.5}{8.9}$. After the water is poured into the mug, the observer can just see the near end of the coin. Find θ_2.

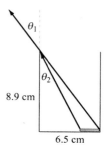

$$n_1 \sin\theta_1 = n_1 \sin\left(\tan^{-1}\frac{6.5}{8.9}\right) = n_2 \sin\theta_2, \text{ so } \theta_2 = \sin^{-1}\left[\frac{n_1}{n_2}\sin\left(\tan^{-1}\frac{6.5}{8.9}\right)\right].$$

From the diagram, we see that $\tan\theta_2 = \dfrac{6.5\ \text{cm} - d}{8.9\ \text{cm}}$, where d is the diameter of the coin.

Find d.

$$\frac{6.5\ \text{cm} - d}{8.9\ \text{cm}} = \tan\left\{\sin^{-1}\left[\frac{n_1}{n_2}\sin\left(\tan^{-1}\frac{6.5}{8.9}\right)\right]\right\}, \text{ so}$$

$$d = 6.5\ \text{cm} - (8.9\ \text{cm})\tan\left\{\sin^{-1}\left[\frac{1.000}{1.333}\sin\left(\tan^{-1}\frac{6.5}{8.9}\right)\right]\right\} = \boxed{2.1\ \text{cm}}.$$

19. **Strategy** Draw a diagram. Use geometry and Snell's law, Eq. (23-4).

Solution Find θ_1 in terms of β.

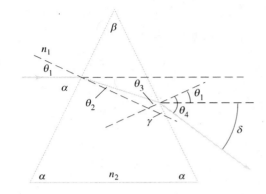

$$\beta + 2\alpha = 180° \rightarrow \alpha = \frac{180° - \beta}{2} = 90° - \frac{\beta}{2}$$

$$\theta_1 = 90° - \alpha = 90° - \left(90° - \frac{\beta}{2}\right) = \frac{\beta}{2}$$

Find θ_2.

$$n_1 \sin\theta_1 = n_2 \sin\theta_2, \text{ so } \sin\theta_2 = \frac{n_1}{n_2}\sin\theta_1 = \frac{n_1}{n_2}\sin\frac{\beta}{2}.$$

Since β is small, use the approximation $\sin\theta \approx \theta$.

$$\sin\theta_2 \approx \frac{n_1}{n_2}\frac{\beta}{2}$$

Since $n_1 \sim n_2$ and $\frac{\beta}{2}$ is small, $\sin\theta_2$ is small, so apply the approximation for θ_2.

$$\theta_2 \approx \frac{n_1}{n_2}\frac{\beta}{2}$$

Using $n_1 = 1$ and $n_2 = n$ gives $\theta_2 = \frac{\beta}{2n}$.

Find θ_3 in terms of n and β.
From the figure, $90° + \beta + \gamma = 180°$, which implies $\gamma = 90° - \beta$.

$$\gamma + (90° + \theta_3) + \theta_2 = 90° - \beta + 90° + \theta_3 + \theta_2 = 180°, \text{ so } \theta_3 = \beta - \theta_2 = \beta - \frac{\beta}{2n} = \beta\left(1 - \frac{1}{2n}\right).$$

Use Snell's law to find θ_4, with $n_1 = 1$ and $n_2 = n$.

$$\sin\theta_4 = n\sin\theta_3 = n\sin\left[\beta\left(1 - \frac{1}{2n}\right)\right]$$

Now, $\beta\left(1 - \frac{1}{2n}\right) < \beta$, so it is small, and θ_4 is small as well. Thus, $\theta_4 \approx n\beta\left(1 - \frac{1}{2n}\right) = \beta\left(n - \frac{1}{2}\right).$

Now find δ.

$$\delta + \theta_1 = \theta_4, \text{ so } \delta = \theta_4 - \theta_1 = \beta\left(n - \frac{1}{2}\right) - \frac{\beta}{2} = \beta n - \frac{\beta}{2} - \frac{\beta}{2} = \boxed{\beta(n-1)}.$$

21. **Strategy** Draw a diagram. Use geometry and Snell's law, Eq. (23-4).

Solution For the longest visible wavelengths, $n_2 = 1.517$.
$n_1 \sin \theta_1 = n_2 \sin \theta_2$, so

$$\theta_2 = \sin^{-1}\left(\frac{n_1}{n_2}\sin \theta_1\right) = \sin^{-1}\left(\frac{1.000}{1.517}\sin 55.0°\right) = 32.7°.$$

Find θ_3.

$60.0° + 90.0° + \alpha = 180.0°$, so $\alpha = 30.0°$.

$\theta_2 + (90.0° + \theta_3) + \alpha = 180.0°$

$\quad \theta_2 + \theta_3 + 30.0° = 90.0°$

$\quad\quad \theta_3 = 60.0° - \theta_2 = 60.0° - 32.7° = 27.3°$

Find θ_4.

$n_2 \sin \theta_3 = n_1 \sin \theta_4$, so

$$\theta_4 = \sin^{-1}\left(\frac{n_2}{n_1}\sin \theta_3\right) = \sin^{-1}\left(\frac{1.517}{1.000}\sin 27.3°\right) = 44.1°.$$

For the shortest visible wavelengths, set $n_2 = 1.538$ and follow the same process. Find θ_2.

$$\theta_2 = \sin^{-1}\left(\frac{1.000}{1.538}\sin 55.0°\right) = 32.18°$$

Find θ_3.

$\theta_3 = 60.0° - 32.18° = 27.82°$

Find θ_4.

$$\theta_4 = \sin^{-1}\left(\frac{1.538}{1.000}\sin 27.82°\right) = 45.9°$$

The range of refraction angles is $\boxed{44.1° \le \theta \le 45.9°}$.

23. **Strategy** From Table 23.1, the index of refraction for diamond is 2.419, for air it is 1.000, and for water it is 1.333. Total internal reflection occurs when the angle of incidence is greater than or equal to the critical angle. Use Eq. (23-5a).

Solution

(a) Calculate the critical angle for diamond surrounded by air.

$$\theta_c = \sin^{-1}\frac{n_t}{n_i} = \sin^{-1}\frac{1.000}{2.419} = \boxed{24.42°}$$

(b) Calculate the critical angle for diamond under water.

$$\theta_c = \sin^{-1}\frac{n_t}{n_i} = \sin^{-1}\frac{1.333}{2.419} = \boxed{33.44°}$$

(c) | Under water, the larger critical angle means that fewer light rays are totally reflected at the bottom surfaces of the diamond. Thus, less light is reflected back toward the viewer. |

25. Strategy Total internal reflection occurs when the angle of incidence is greater than or equal to the critical angle. Use Eqs. (23-5).

Solution

(a) Find the index of refraction of the glass.

$$\theta_c = \sin^{-1}\frac{n_t}{n_i},\ \text{so}\ \sin\theta_c = \frac{n_t}{n_i}\ \text{and}\ n_i = \frac{1.000}{\sin 40.00°} = \boxed{1.556}.$$

(b) $\boxed{\text{No}}$; rays from the defect could reach all points above the glass since $\boxed{\text{for}\ 0 \le \theta_i \le \theta_c,\ 0 \le \theta_t \le 90°}$.

29. Strategy Total internal reflection occurs when the angle of incidence is greater than or equal to the critical angle. Use Eqs. (23-5) and Snell's law, Eq. (23-4).

Solution When the light is incident on the Plexiglas tank, some is transmitted at angle θ_1, so $n\sin\theta_i = n_1\sin\theta_1$ where $n = 1.00$ for air and $n_1 = 1.51$ for Plexiglas. At the Plexiglas carbon tetrachloride interface, θ_1 is the incident angle and θ_2 is the transmitted angle, so $n_2\sin\theta_2 = n_1\sin\theta_1$ where $n_2 = 1.461$ for carbon tetrachloride. The ray passes through the carbon tetrachloride and is incident on the bottom tank-liquid interface at angle θ_2.

Here the light must experience total internal reflection, so $\theta_2 = \theta_c = \sin^{-1}\frac{n_1}{n_2}$, or $\frac{n_1}{n_2} = \sin\theta_2$. Find θ_i.

$$n\sin\theta_i = n_1\sin\theta_1 = n_2\sin\theta_2 = n_2\frac{n_1}{n_2} = n_1,\ \text{so}\ \theta_i = \sin^{-1}\frac{n_1}{n} = \sin^{-1}\frac{1.51}{1.00} = \sin^{-1}1.51,\ \text{or}\ \sin\theta_i = 1.51,$$

which is impossible since $\sin\theta \le 1$ for all θ. Thus, there is $\boxed{\text{no}}$ angle θ for which light is transmitted into the carbon tetrachloride but not into the Plexiglas at the bottom of the tank.

33. (a) Strategy The reflected light is totally polarized when the angle of incidence equals Brewster's angle. Use Eq. (23-6).

Solution Compute Brewster's angle.

$$\theta_B = \tan^{-1}\frac{n_t}{n_i} = \tan^{-1}\frac{1.309}{1.000} = 52.62°$$

The angle with respect to the horizontal is the complement of this angle.

$$\theta = 90° - 52.62° = \boxed{37.38°}$$

(b) Strategy and Solution For Brewster's angle, the reflected light is polarized $\boxed{\text{perpendicular to the plane of incidence.}}$

(c) Strategy When the angle of incidence is Brewster's angle, the incident and transmitted rays are complementary.

Solution Find the angle of transmission.

$$\theta_t = 90° - \theta_i = 90° - 52.62° = 37.38°$$

The angle below the horizontal is the complement of this angle.

$$90° - 37.38° = \boxed{52.62°}$$

35. **Strategy** The equation derived in Example 23.4 can be used for this problem (with $n_w = n_d$) since $n_d > n_a$.

    Solution Find the depth of the defect.
    $$\frac{\text{apparent depth}}{\text{actual depth}} = \frac{n_a}{n_d} = \frac{1.000}{2.419}, \text{ so actual depth} = \text{apparent depth} \cdot 2.419 = 2.0 \text{ mm} \cdot 2.419 = \boxed{4.8 \text{ mm}}.$$

37. **Strategy** Use Snell's law with $n_a = 1.000$, $n_w = 1.333$, and $\theta_a = 90°$.

    Solution Find θ_w.
    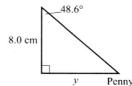
    $$n_w \sin \theta_w = n_a \sin \theta_a, \text{ so}$$
    $$\theta_w = \sin^{-1}\left(\frac{n_a}{n_w} \sin 90°\right) = \sin^{-1}\left[\frac{n_a}{n_w}(1)\right] = \sin^{-1}\left(\frac{1.000}{1.333}\right) = 48.6°.$$
    A right triangle is formed by the actual location of the penny, the apparent location of
    the penny, and the bottom of the bowl directly below the apparent location. Let y be the distance from the latter
    location to the actual location of the penny, then $y = (8.0 \text{ cm}) \tan 48.6° = 9.1 \text{ cm}$. Since the penny appears to be
    3.0 cm from the edge of the bowl, the horizontal distance between the penny and the edge of the bowl is 9.1 cm +
    3.0 cm = $\boxed{12.1 \text{ cm}}$.

39. **Strategy** For a plane mirror, a point source and its image are at the same distance from the mirror (on opposite
    sides) and both lie on the same normal line. Draw a ray diagram. Use geometry and the laws of reflection.

    Solution Suppose the mirror is hung at the proper height (see the figure).
    The top of Daniel's head is at point A, his eyes are at point B, and his
    shoes are at point D. Lines AD and CE are perpendicular, and $\theta_i = \theta_r$,
    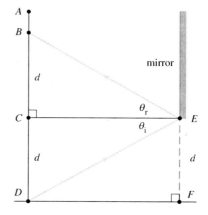
    so triangles BCE and DCE are congruent and $BC = CD = \frac{1}{2}BD$.

    Similarly, triangles CDE and FED are congruent, so
    $$EF = CD = \frac{1}{2}BD = \frac{1}{2} \cdot 1.82 \text{ m} = \boxed{0.91 \text{ m}}.$$

41. **Strategy** For a plane mirror, a point source and its image are at the same distance from the mirror (on opposite
    sides) and both lie on the same normal line.

    Solution Since Gustav is holding the match, the distance to the image is twice the distance from Gustav to the
    mirror, so the distance from Gustav to the mirror is half this distance.
    $$\frac{1}{2} \cdot 4 \text{ m} = \boxed{2 \text{ m}}$$

43. **Strategy** For a plane mirror, a point source and its image are at the same distance from the mirror (on opposite
    sides) and both lie on the same normal line.

    Solution Maurizio sees three images by looking straight into each mirror. He sees three other images by looking
    where each pair of mirrors meet (left wall and right wall, left wall and ceiling, right wall and ceiling). He sees one
    more image by looking at the corner where all three mirrors meet. $\boxed{\text{He sees 7 images total.}}$

73. **(a) Strategy** The image is virtual, so the image distance is negative. Use the magnification equation.

Solution Find the object distance.

$$m = \frac{h'}{h} = -\frac{q}{p}, \text{ so } p = -\frac{h}{h'}q = -\frac{8.0 \text{ cm}}{3.5 \text{ cm}}(-4.0 \text{ cm}) = \boxed{9.1 \text{ cm}}.$$

(b) Strategy and Solution The image is upright, virtual, smaller than the object, and closer to the mirror than the object. The mirror is $\boxed{\text{convex}}$.

(c) Strategy The radius of curvature is twice the absolute value of the focal length. Use the mirror equation.

Solution Find the focal length of the mirror.

$$\frac{1}{p} + \frac{1}{q} = \frac{1}{f}, \text{ so } f = \left(\frac{1}{p} + \frac{1}{q}\right)^{-1} = \left(-\frac{h'}{hq} + \frac{1}{q}\right)^{-1} = q\left(1 - \frac{h'}{h}\right)^{-1} = (-4.0 \text{ cm})\left(1 - \frac{3.5}{8.0}\right)^{-1} = \boxed{-7.1 \text{ cm}}.$$

Compute the radius of curvature.

$$R = 2|f| = 2(7.1 \text{ cm}) = \boxed{14 \text{ cm}}$$

77. **Strategy** Use the mirror and magnification equations.

Solution

(a) The image appears to be behind the mirror, where no light passes through it, so it is a $\boxed{\text{virtual}}$ image.

(b) The image is virtual, so $q = -|q| < 0$. The image is upright and diminished, so $0 < m < 1$. This implies $|q| < p$

since $m = -\frac{q}{p} = \frac{|q|}{p} < 1$. Using the mirror equation, we have

$$\frac{1}{q} + \frac{1}{p} = -\frac{1}{|q|} + \frac{1}{p} = \frac{1}{f}, \text{ so } \frac{p}{|q|} - 1 = -\frac{p}{f}.$$

Since $\frac{|q|}{p} < 1$, $\frac{p}{|q|} > 1$. So, $\frac{p}{|q|} - 1 > 0$, and $-\frac{p}{f} > 0$, or $f < 0$ since $p > 0$. Since the focal length is negative, the mirror is $\boxed{\text{convex}}$.

(c) In a convex mirror, the image is upright and smaller than the object, so $m < 1$ and $pm < p$. But $pm = -q$, so $-q < p$. Since q is negative, $|q| < p$ and the image is closer to the mirror than the object.

(d) $\boxed{\text{The image seems to be farther away than the object because it appears smaller than the object.}}$

79. **Strategy and Solution** For a plane mirror, the image distance equals the object distance. Since speed is distance divided by time, the image speed equals the object speed: $\boxed{0.8 \text{ m/s}}$.

81. Strategy Use the laws of reflection.

Solution In (1), a ray strikes a plane mirror. The angles of incidence and reflection are both θ. In (2), the mirror has been rotated relative to the normal by an angle α. The angle of incidence must equal the angle of reflection, so $\theta_2 = \theta + \alpha$. Then $\theta_3 = \alpha + \theta_2 = \alpha + \theta + \alpha = \theta + 2\alpha$.
Since the angle of the original reflected ray was θ, the reflected ray has rotated through an angle of 2α.

(1)

(2)

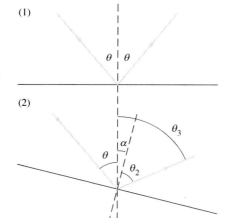

85. Strategy Use the thin lens and magnification equations.

Solution

(a) Since the image is upright, $m = -q/p > 0$, so $q < 0$ and the image is $\boxed{\text{virtual}}$.

(b) Find the image location.
$$m = \frac{h'}{h} = \frac{-q}{p}, \text{ so } q = \frac{-ph'}{h} = -\frac{4.5 \text{ cm} \cdot 2.4 \text{ cm}}{0.060 \text{ cm}} = -180 \text{ cm.}$$
The image is $\boxed{180 \text{ cm behind the lens.}}$

(c) Find the focal length.
$$\frac{1}{f} = \frac{1}{p} + \frac{1}{q}, \text{ so } f = \frac{1}{\frac{1}{p} + \frac{1}{q}} = \frac{1}{\frac{1}{4.5 \text{ cm}} - \frac{1}{180 \text{ cm}}} = \boxed{4.6 \text{ cm}}.$$
The focal length is positive, so the lens is $\boxed{\text{converging}}$.

89. Strategy Use the laws of reflection.

Solution For the first reflection, $\theta_i = \theta_r$, so $\alpha = \theta_i = \boxed{34°}$. For the second reflection, we have a right triangle with angle α and the second angle of incidence, θ_{i2}. So, $\theta_{i2} + \alpha + 90° = 180°$, or $\theta_{i2} = 90° - \alpha = 56°$. Thus, $\beta = \theta_{i2} = \boxed{56°}$.

93. Strategy The critical angle must be less than $50.0°$. Use Eq. (23-5a).

Solution Solve for n_t, the index of refraction of the liquid.
$$\theta_c = \sin^{-1}\frac{n_t}{n_i} < 50.0°, \text{ so } \sin 50.0° > \frac{n_t}{n_i} \text{ or } n_t < n_i \sin 50.0° = 1.7 \sin 50.0° = 1.3. \text{ Therefore, } \boxed{n_{\text{liquid}} < 1.3}.$$

97. **Strategy** Use Snell's law, Eq. (23-4). Draw a diagram.

   Solution According to the figure, $\theta_1 = 90° - 60.0° = 30.0°$.
   Applying Snell's law, $n_1 \sin \theta_1 = n_2 \sin \theta_2$ and
   $n_2 \sin \theta_2 = n_3 \sin \theta_3$, so $n_1 \sin \theta_1 = n_3 \sin \theta_3$.
   Find θ_3, the angle of refraction in the glass.

   $$\theta_3 = \sin^{-1}\left(\frac{n_1}{n_3} \sin \theta_1\right) = \sin^{-1}\left(\frac{1.000}{1.517} \sin 30.0°\right) = \boxed{19.2°}$$

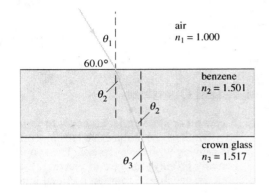

101. **Strategy** Redraw the diagram, labeling the vertices of similar triangles.

   Solution In the figure, triangle ABF and triangle ACG are
   similar, so
   $$\frac{h}{f} = \frac{-d}{D}, \text{ so } f = -\frac{h}{d}D.$$
   For paraxial rays, the slope of the d vs. h graph is constant.
   The middle three data points reflect this case. Find the slope.
   $$m = \frac{\Delta d}{\Delta h} = \frac{-1.0 - 1.0}{0.5 - (-0.5)} = \frac{-2.0}{1.0} = -2.0$$

   For a constant slope, $\dfrac{\Delta d}{\Delta h} = \dfrac{d}{h}$, so

   $$f = -\frac{1}{-2.0}D = 0.50(1.0 \text{ m}) = \boxed{50 \text{ cm}}.$$

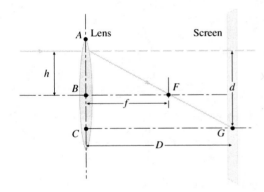

Chapter 24

OPTICAL INSTRUMENTS

Conceptual Questions

1. A camera or slide projector needs to form a real image of a real object. This can only be done with a converging lens. Similarly, the objective lens in a microscope or telescope must be converging so that it can form a real image for viewing with the eyepiece. The eyepiece used to view the image forms a virtual image at a large distance (usually infinity) for ease of viewing. Either a converging or diverging lens can be used as the eyepiece, since both are capable of forming a virtual image.

5. One of the greatest factors limiting the resolution of Earth-based telescopes is the variation in air temperature and density in the atmosphere, which limits the amount of detail that can be seen in astronomical objects. Putting telescopes on mountaintops helps reduce this effect by decreasing the amount of atmosphere that the light rays must pass through.

9. Chromatic aberration is caused by dispersion—the varying value of the index of refraction as a function of frequency. Chromatic aberration in lenses may be reduced by the use of a compound lens made of several lenses with different dispersion relations so that the aberrations caused by one lens offset those caused by another.

13. Our eyes are designed to focus properly on objects when rays emerging from those objects are incident on our eyes from air. When in water, the rays are incident from a medium which has a higher index of refraction than air. A nearsighted person may see more clearly underwater because light is refracted less—the effect is identical to wearing diverging lenses in air.

17. The defect in this eyeball is called myopia (nearsightedness).

Problems

1. **Strategy** Use the lens and total transverse magnification equations.

   Solution

   (a) Find the image due to the first lens.
   $$\frac{1}{p_1} + \frac{1}{q_1} = \frac{1}{f_1}, \text{ so } q_1 = \frac{1}{\frac{1}{f_1} - \frac{1}{p_1}} = \frac{1}{\frac{1}{5.0 \text{ cm}} - \frac{1}{12.0 \text{ cm}}} = 8.6 \text{ cm}.$$
   Find the object distance for the second lens.
   $$p_2 = s - q_1 = 2.0 \text{ cm} - 8.6 \text{ cm} = -6.6 \text{ cm}$$
   Find the location of the final image.
   $$\frac{1}{p_2} + \frac{1}{q_2} = \frac{1}{f_2}, \text{ so } q_2 = \frac{1}{\frac{1}{f_2} - \frac{1}{p_2}} = \frac{1}{\frac{1}{4.0 \text{ cm}} - \frac{1}{-6.6 \text{ cm}}} = 2.5 \text{ cm}.$$
   The final image is $\boxed{2.5 \text{ cm past the 4.0-cm lens}}$. The image is $\boxed{\text{real}}$ since q_2 is positive.

   (b) Compute the overall magnification.
   $$m = m_1 m_2 = -\frac{q_1}{p_1}\left(-\frac{q_2}{p_2}\right) = -\frac{8.6 \text{ cm}}{12.0 \text{ cm}} \times -\frac{2.5 \text{ cm}}{-6.6 \text{ cm}} = \boxed{-0.27}$$

3. **(a) Strategy** Use the lens equations.

    Solution Find the image location of the first lens.

    $$\frac{1}{p_1} + \frac{1}{q_1} = \frac{1}{f_1}, \text{ so } q_1 = \left(\frac{1}{f_1} - \frac{1}{p_1}\right)^{-1}. \text{ So, the object distance for the second lens is}$$

    $$p_2 = s - q_1 = s - \left(\frac{1}{f_1} - \frac{1}{p_1}\right)^{-1} = 0.880 \text{ m} - \left(\frac{1}{0.250 \text{ m}} - \frac{1}{1.100 \text{ m}}\right)^{-1} = 0.556 \text{ m.}$$

    Find the focal length of the second lens.

    $$\frac{1}{p_2} + \frac{1}{q_2} = \frac{1}{f_2}, \text{ so } f_2 = \left(\frac{1}{p_2} + \frac{1}{q_2}\right)^{-1} = \left(\frac{1}{0.556 \text{ m}} + \frac{1}{0.150 \text{ m}}\right)^{-1} = 0.118 \text{ m} = \boxed{11.8 \text{ cm}}.$$

    (b) Strategy Use the magnification and total transverse magnification equations.

    Solution Find the total magnification of this lens combination.

    $$m = m_1 \times m_2 = -\frac{q_1}{p_1} \times -\frac{q_2}{p_2} = \frac{(0.3235)(0.150)}{(1.100)(0.5565)} = \boxed{0.0793}$$

5. **Strategy** Draw a ray diagram for the system of lenses. Use the lens equations.

    Solution The ray diagram:

    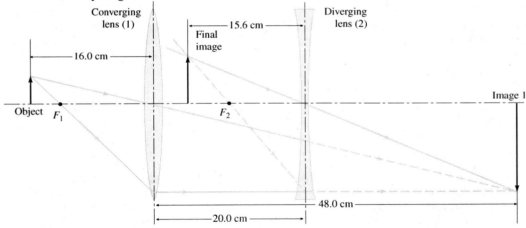

    From the figure, the final image is about 15.6 cm left of lens 2. Verify using the lens equations. Find the image due to the first lens (1).

    $$\frac{1}{p_1} + \frac{1}{q_1} = \frac{1}{f_1}, \text{ so } q_1 = \frac{1}{\frac{1}{f_1} - \frac{1}{p_1}} = \frac{1}{\frac{1}{12.0 \text{ cm}} - \frac{1}{16.0 \text{ cm}}} = 48.0 \text{ cm.}$$

    For the diverging lens (2), the object distance is $p_2 = s - q_1 = 20.0 \text{ cm} - 48.0 \text{ cm} = -28.0 \text{ cm}$.

    Find q_2.

    $$q_2 = \frac{1}{\frac{1}{f_2} - \frac{1}{p_2}} = \frac{1}{\frac{1}{-10.0 \text{ cm}} - \frac{1}{-28.0 \text{ cm}}} = -15.6 \text{ cm}$$

    The final image is located $\boxed{15.6 \text{ cm to the left of the diverging lens}}$.

9. **Strategy** Use the lens equations and transverse and total transverse magnification equations.

    Solution

    (a) Find the image location of the first lens.
    $$\frac{1}{p_1} + \frac{1}{q_1} = \frac{1}{f_1}, \text{ so } q_1 = \left(\frac{1}{f_1} - \frac{1}{p_1}\right)^{-1} = \left(\frac{1}{3.70 \text{ cm}} - \frac{1}{6.00 \text{ cm}}\right)^{-1} = 9.65 \text{ cm}.$$
    So, the object distance for the second lens is $p_2 = s - q_1 = (24.65 \text{ cm} - 6.00 \text{ cm}) - 9.65 \text{ cm} = 9.00 \text{ cm}$.
    Find the focal length of the second lens.
    $$\frac{1}{p_2} + \frac{1}{q_2} = \frac{1}{f_2}, \text{ so } f_2 = \left(\frac{1}{p_2} + \frac{1}{q_2}\right)^{-1} = \left(\frac{1}{9.00 \text{ cm}} + \frac{1}{32.0 \text{ cm} - 24.65 \text{ cm}}\right)^{-1} = \boxed{4.05 \text{ cm}}.$$

    (b) Since its focal length is positive, the lens is $\boxed{\text{converging}}$.

    (c) Find the total magnification of this system.
    $$m = m_1 \times m_2 = -\frac{q_1}{p_1} \times -\frac{q_2}{p_2} = \frac{(9.65)(32.0 - 24.65)}{(6.00)(9.00)} = \boxed{1.31}$$

    (d) Find the image height.
    $$m = \frac{h'}{h}, \text{ so } h' = mh = 1.314(12.0 \text{ cm}) = \boxed{15.8 \text{ cm}}.$$

13. **(a) Strategy and Solution** Referring to Figure 24.6a, we see that rays from the top of the object are incident at the bottom of the image and rays from the bottom of the object are incident at the top of the image, so the image is $\boxed{\text{inverted}}$.

    (b) Strategy Use the transverse magnification equation.

    Solution Compute the magnification.
    $$m = -\frac{q}{p} = -\frac{2.8 \text{ m}}{6.6 \text{ m}} = \boxed{-0.42}$$

    (c) Strategy and Solution The pinhole admits a narrow cone of rays diverging from each point on the object; the cone of rays makes a small circular spot on the film or screen. If the spot is small enough, the image appears clear to the eye. A larger spot results in the spot being spread out and blurry. So, $\boxed{\text{the eye can detect that the rays do not converge to a single point}}$.

    (d) Strategy and Solution The image must be real to expose the film or project the image on a screen and to focus the image. Only $\boxed{\text{converging}}$ lenses form real images.

    (e) Strategy Use $p = 6.6$ m and $q = 2.8$ m in the thin lens equation.

    Solution Compute the focal length.
    $$\frac{1}{p} + \frac{1}{q} = \frac{1}{f}, \text{ so } f = \frac{pq}{p+q} = \frac{(6.6 \text{ m})(2.8 \text{ m})}{6.6 \text{ m} + 2.8 \text{ m}} = \boxed{2.0 \text{ m}}.$$

17. **Strategy** Use the thin lens and transverse magnification equations. $h' = -20$ mm since the image is inverted.

    Solution Find the image location.
    $$-\frac{q}{p} = \frac{h'}{h}, \text{ so } q = -\frac{h'p}{h} = -\frac{(-20 \text{ mm})(300 \text{ m})}{300 \text{ m}} = 20 \text{ mm}.$$
    Find the focal length.
    $$\frac{1}{f} = \frac{1}{p} + \frac{1}{q}, \text{ so } f = \frac{1}{\frac{1}{p} + \frac{1}{q}} = \frac{1}{\frac{1}{300 \text{ m}} + \frac{1}{0.020 \text{ m}}} = 0.020 \text{ m} = \boxed{20 \text{ mm}}.$$
    Since $p \gg q$, $f \approx q$, so the result is reasonable.

19. **Strategy** The slide is inverted with respect to the image, so h is negative. Use the thin lens and transverse magnification equations.

    Solution First find the object distance.
    $$\frac{1}{p} + \frac{1}{q} = \frac{1}{f}, \text{ so } p = \frac{1}{\frac{1}{f} - \frac{1}{q}} = \frac{1}{\frac{1}{12 \text{ cm}} - \frac{1}{5.0 \times 10^2 \text{ cm}}} = 12.3 \text{ cm}.$$
    Find the image height.
    $$\frac{h'}{h} = -\frac{q}{p}, \text{ so } h' = -\frac{qh}{p} = -\frac{(5.0 \text{ m})(-2.4 \text{ cm})}{12.3 \text{ cm}} = 98 \text{ cm}.$$
    Find the image width.
    $$w' = -\frac{(5.0 \text{ m})(-3.6 \text{ cm})}{12.3 \text{ cm}} = 150 \text{ cm}$$
    The screen must be at least $\boxed{98 \text{ cm by } 150 \text{ cm}}$.

21. Strategy Use the mirror/thin lens, magnification, and total transverse magnification equations.

Solution Find the image due to the mirror.

$$\frac{1}{p}+\frac{1}{q}=\frac{1}{f},\ \text{so } q=\left(\frac{1}{f}-\frac{1}{p}\right)^{-1}=\frac{fp}{p-f}=\frac{(4.00\ \text{cm})(6.00\ \text{cm})}{6.00\ \text{cm}-4.00\ \text{cm}}=12.0\ \text{cm}.$$

Find the magnification.

$$m=-\frac{q}{p}=-\frac{12.0}{6.00}=-2.00$$

Since the magnification is negative, the image is inverted. So, an inverted image with magnification −2.00 is formed 12.0 cm to the right of the mirror.

Find the image due to the lens.

$$\frac{1}{p}+\frac{1}{q}=\frac{1}{f},\ \text{so } q=\left(\frac{1}{f}-\frac{1}{p}\right)^{-1}=\frac{fp}{p-f}=\frac{(3.00\ \text{cm})(18.00\ \text{cm})}{18.00\ \text{cm}-3.00\ \text{cm}}=3.60\ \text{cm}.$$

Find the magnification.

$$m=-\frac{q}{p}=-\frac{3.60}{18.00}=-0.200$$

So, an inverted image with magnification −0.200 is formed 3.60 cm to the right of the lens.

Use the image formed by the mirror as the object of the lens. Then, $p_2 = 18.00\ \text{cm} + 6.00\ \text{cm} - 12.0\ \text{cm} = 12.0\ \text{cm}.$

$$\frac{1}{p_2}+\frac{1}{q_2}=\frac{1}{f_2},\ \text{so } q_2=\left(\frac{1}{f_2}-\frac{1}{p_2}\right)^{-1}=\left(\frac{1}{3.00\ \text{cm}}-\frac{1}{12.0\ \text{cm}}\right)^{-1}=4.00\ \text{cm}.$$

Find the magnification.

$$m=m_1\times m_2=m_1\times\left(-\frac{q_2}{p_2}\right)=-2.00\times\left(-\frac{4.00}{12.0}\right)=0.667$$

An upright image (with respect to the original object) with magnification 0.667 (with respect to the original object) is formed 4.00 cm to the right of the lens.

23. (a) Strategy and Solution A focal length range of 1.85 cm to 2.00 cm corresponds to a 2.00 cm lens-retina distance in a normal eye. For a distant object, the focus of the eye can be adjusted to 1.90 cm so that the object is seen clearly, but for close objects, there is not much room for adjustment by the eye muscle to accommodate. Thus, this eye is farsighted.

(b) Strategy Use the thin lens equation.

Solution Solve for the object location.

$$\frac{1}{p}+\frac{1}{q}=\frac{1}{f},\ \text{so } p=\frac{1}{\frac{1}{f}-\frac{1}{q}}.$$

Find p for $q = 1.90$ cm and f between 1.85 cm and 1.90 cm.

$$p=\frac{1}{\frac{1}{1.85\ \text{cm}}-\frac{1}{1.90\ \text{cm}}}=70\ \text{cm}\quad\text{and}\quad p=\frac{1}{\frac{1}{1.90\ \text{cm}}-\frac{1}{1.90\ \text{cm}}}=\infty.$$

The eye can focus from 70 cm to infinity.

25. Strategy The refractive power of a lens is the reciprocal of the focal length. Use the thin lens equation. Let $p = \infty$ for distant objects and $q = -2.0$ m for a virtual image at Colin's far point.

Solution Find the required refractive power.
$$P = \frac{1}{f} = \frac{1}{q} + \frac{1}{p} = \frac{1}{-2.0 \text{ m}} + \frac{1}{\infty} = \boxed{-0.50 \text{ D}}$$

27. Strategy The refractive power of a lens is the reciprocal of the focal length. Use the thin lens equation.

Solution

(a) $p = \infty$ and $q = -2.0$ m $+ 0.020$ m. Find the necessary refractive power of the eyeglass lenses.
$$P = \frac{1}{f} = \frac{1}{q} + \frac{1}{p} = \frac{1}{-2.0 \text{ m} + 0.020 \text{ m}} + \frac{1}{\infty} = \boxed{-0.51 \text{ D}}$$

(b) The refractive power of the eye by itself can be calculated using $p = 2.0$ m and $q = 2.0$ cm $= 0.020$ m.
$$P_{\text{eye}} = \frac{1}{f} = \frac{1}{p} + \frac{1}{q} = \frac{1}{2.0 \text{ m}} + \frac{1}{0.020 \text{ m}} = 50.5 \text{ D}$$
This is the relaxed state. Using the accommodation gives $P_{\text{eye}} = 50.5 \text{ D} + 4.0 \text{ D} = 54.5 \text{ D}$.

This gives an object distance of $p = \dfrac{1}{\frac{1}{f} - \frac{1}{q}} = \dfrac{1}{54.5 \text{ m}^{-1} - \frac{1}{0.020 \text{ m}}} = 22$ cm.

Without his glasses, his near point is 22 cm.
The refractive power of the eye with the glasses is $P_{\text{e-g}} = P_{\text{eye}} + P_{\text{glasses}} = 50.5 - 0.505 = 50 \text{ D}$.
The accommodation gives $P_{\text{e-g}} = 50 + 4.0 = 54 \text{ D}$.

The object distance is $p = \dfrac{1}{\frac{1}{f} - \frac{1}{q}} = \dfrac{1}{54 \text{ m}^{-1} - \frac{1}{0.020 \text{ m}}} = 25$ cm.

With his glasses, his near point is 25 cm.

29. Strategy The refractive power of a lens is the reciprocal of the focal length. Use Eq. (24-6), where $N = 0.25$ m.

Solution Compute the angular magnification.
$$M = \frac{N}{f} = \frac{0.25 \text{ m}}{\frac{1}{5.5 \text{ m}^{-1}}} = \boxed{1.4}$$

31. Strategy Use the thin lens equation and the equation for the angular magnification found in Example 24.6.

Solution

(a) $q = -25$ cm and $f = 5.0$ cm. Find the object distance; that is, the distance between the magnifying glass and the beetle.
$$\frac{1}{f} = \frac{1}{p} + \frac{1}{q}, \text{ so } p = \frac{fq}{q - f} = \frac{(5.0 \text{ cm})(-25 \text{ cm})}{-25 \text{ cm} - 5.0 \text{ cm}} = \boxed{4.2 \text{ cm}}.$$

(b) The angular magnification is $M = \dfrac{N}{p} = \dfrac{25 \text{ cm}}{4.2 \text{ cm}} = \boxed{6.0}$.

33. (a) **Strategy** When the magnifying glass focuses the image to its smallest size, the image distance is equal to the focal length. The mean distance of the Sun from the Earth is 1.50×10^{11} m. The mean radius of the Sun is 6.96×10^8 m. Use the thin lens and transverse magnification equations.

Solution Find the image size.
$$m = -\frac{q}{p} = -\frac{f}{p} = \frac{h'}{h}, \text{ so } h' = -\frac{f}{p}h = -\frac{0.060 \text{ m}}{1.50 \times 10^{11} \text{ m}}(2 \times 6.96 \times 10^8 \text{ m}) = -0.56 \text{ mm}.$$
The size of the image of the Sun is about $\boxed{0.56 \text{ mm in diameter}}$.

(b) **Strategy** The intensity is inversely proportional to the area.

Solution Find the intensity of the image by forming a proportion.
$$\frac{I_{\text{image}}}{I_{\text{glass}}} = \frac{I_2}{I_1} = \frac{A_1}{A_2} = \left(\frac{d_1}{d_2}\right)^2, \text{ so } I_2 = \left(\frac{d_1}{d_2}\right)^2 I_1 = \left(\frac{40 \text{ mm}}{0.557 \text{ mm}}\right)^2 (0.85 \text{ kW/m}^2) = \boxed{4.4 \times 10^3 \text{ kW/m}^2}.$$

(c) **Strategy** Use Equation (24-6) and the equation for the angular magnification found in Example 24.6.

Solution The angular magnification when the image is at the near point is
$$M = \frac{N}{p} = \frac{N}{\frac{Nf}{N+f}} = \frac{N(N+f)}{Nf} = \frac{N+f}{f} = \boxed{\frac{N}{f}+1}.$$

For an image at infinity, the angular magnification is $M_\infty = \dfrac{N}{f}$, so $\boxed{M = M_\infty + 1}$.

37. **Strategy** Use Eq. (24-8).

Solution

(a) Find the angular magnification of the microscope.
$$M_{\text{total}} = -\frac{L}{f_o} \times \frac{N}{f_e} = -\frac{18.0 \text{ cm}}{1.44 \text{ cm}} \times \frac{25 \text{ cm}}{1.25 \text{ cm}} = \boxed{-250}$$

(b) Since the angular magnification is inversely proportional to the objective focal length, the magnification is doubled if the focal length is halved. Therefore, the required focal length is $(1.44 \text{ cm})/2 = \boxed{0.720 \text{ cm}}$.

41. **Strategy** Since the final image is not at infinity, the image from the objective lens is not at the focal point of the eyepiece. The angular magnification due to the eyepiece is equal to the near point divided by the object distance for the eyepiece. Use the thin lens equation and Figure 24.16.

    Solution Find the object distance for the eyepiece.
    $$M_e = \frac{N}{p_e}, \text{ so } p_e = \frac{N}{M_e} = \frac{25.0 \text{ cm}}{5.00} = 5.00 \text{ cm}.$$
    Find the focal length of the eyepiece.
    $$\frac{1}{p_e} + \frac{1}{q_e} = \frac{1}{f_e}, \text{ so } f_e = \frac{1}{\frac{1}{p_e} + \frac{1}{q_e}} = \frac{1}{\frac{1}{5.00 \text{ cm}} + \frac{1}{-25.0 \text{ cm}}} = 6.25 \text{ cm}.$$
    Find the image distance for the objective lens.
    $$q_o + p_e = f_o + L + f_e, \text{ so } q_o = f_o + L + f_e - p_e = 1.50 \text{ cm} + 16.0 \text{ cm} + 6.25 \text{ cm} - 5.00 \text{ cm} = 18.8 \text{ cm}.$$
    Find the object distance for the objective lens.
    $$\frac{1}{p_o} + \frac{1}{q_o} = \frac{1}{f_o}, \text{ so } p_o = \frac{1}{\frac{1}{f_o} - \frac{1}{q_o}} = \frac{1}{\frac{1}{1.50 \text{ cm}} - \frac{1}{18.8 \text{ cm}}} = \boxed{1.63 \text{ cm}}.$$

43. **(a) Strategy** Use the thin lens equation.

    Solution Find the object distance for the eyepiece.
    $$\frac{1}{p_e} + \frac{1}{q_e} = \frac{1}{f_e}, \text{ so } p_e = \frac{f_e q_e}{q_e - f_e} = \frac{(2.80 \text{ cm})(-25.0 \text{ cm})}{-25.0 \text{ cm} - 2.80 \text{ cm}} = 2.52 \text{ cm}.$$
    The distance between the lenses is $q_o + p_e = 16.5 \text{ cm} + 2.52 \text{ cm} = \boxed{19.0 \text{ cm}}$.

    (b) Strategy The angular magnification of the eyepiece is $M_e = N/p_e$. Use the thin lens and the transverse magnification equations and the fact that the total magnification is equal to the product of magnifications due to the eyepiece and the objective.

    Solution The transverse magnification for the objective is
    $$m_o = -\frac{q_o}{p_o} = -q_o\left(\frac{1}{p_o}\right) = -q_o\left(\frac{1}{f_o} - \frac{1}{q_o}\right).$$
    The total magnification is
    $$M_{total} = m_o M_e = -q_o\left(\frac{1}{f_o} - \frac{1}{q_o}\right) \times \frac{N}{p_e} = -(16.5 \text{ cm})\left(\frac{1}{0.500 \text{ cm}} - \frac{1}{16.5 \text{ cm}}\right) \times \frac{25.0 \text{ cm}}{2.518 \text{ cm}} = \boxed{-318}.$$

    (c) Strategy Use the thin lens equation.

    Solution Find the object distance for the objective lens.
    $$\frac{1}{p_o} + \frac{1}{q_o} = \frac{1}{f_o}, \text{ so } p_o = \frac{1}{\frac{1}{f_o} - \frac{1}{q_o}} = \frac{1}{\frac{1}{5.00 \text{ mm}} - \frac{1}{165 \text{ mm}}} = \boxed{5.16 \text{ mm}}.$$

45. **(a) Strategy and Solution** Since the angular magnification due to a telescope is equal to the ratio of the objective and eyepiece focal lengths, the magnification would be 1; that is, there would be no magnification, so you can't really make a telescope.

   (b) Strategy Since the lenses have different focal lengths, there will be a magnification. The lens with the smaller strength should be the objective lens, since it has the longer focal length. Use Eqs. (24-9) and (24-10).

   Solution Compute the magnification.

   $$M = -\frac{f_o}{f_e} = -\frac{\frac{1}{1.3 \text{ m}^{-1}}}{\frac{1}{3.5 \text{ m}^{-1}}} = -2.7$$

   Compute the barrel length.

   $$f_o + f_e = \frac{1}{1.3 \text{ m}^{-1}} + \frac{1}{3.5 \text{ m}^{-1}} = 1.05 \text{ m}$$

   Using a lens from each pair of glasses, the telescope would be 1.05 m long and have an angular magnification of -2.7.

49. **(a) Strategy** Use Eq. (24-10).

   Solution Since the magnifying power is equal to the ratio of the focal length of the objective lens to that of the eyepiece, we choose the largest lens for the objective and the smallest lens for the eyepiece. Thus, the two lenses are 80.0 cm and 1.00 cm. Compute the angular magnification.

   $$M = -\frac{f_o}{f_e} = -\frac{80.0 \text{ cm}}{1.00 \text{ cm}} = \boxed{-80.0}$$

   (b) Strategy and Solution Since the magnifying power is equal to the ratio of the focal length of the objective lens to that of the eyepiece, we choose the largest lens for the objective and the smallest lens for the eyepiece. Thus, the 80.0-cm lens is the objective and the 1.00-cm lens is the eyepiece.

   (c) Strategy Use the equation for the barrel length of a telescope, Eq. (24-9), for the distance between the lenses.

   Solution Find the distance between the objective and the eyepiece—the barrel length.

   barrel length $= f_o + f_e = 80.0 \text{ cm} + 1.00 \text{ cm} = \boxed{81.0 \text{ cm}}$

51. **Strategy** Use Eq. (24-9) for the barrel length of a telescope. Use Eq. (24-10) for the angular magnification.

   Solution Find the focal length of the objective.

   $f_o = $ barrel length $- f_e = 45.0 \text{ cm} - 5.0 \text{ cm} = 40.0 \text{ cm}$

   Compute the angular magnification.

   $$M = -\frac{f_o}{f_e} = -\frac{40.0 \text{ cm}}{5.0 \text{ cm}} = \boxed{-8.0}$$

53. (a) Strategy Since $p_o \gg f_o = 36$ cm, $q_o \approx f_o$. Refer to the figure and use the fact that the eyepiece is a diverging lens.

Solution From the figure, $f_o = |f_e| + d$, so $|f_e| = 36$ cm $- 32$ cm $= 4$ cm. Since the eyepiece is a diverging lens, $f_e = \boxed{-4 \text{ cm}}$.

(b) Strategy and Solution The intermediate image is a virtual object for the eyepiece, since it is located at the coincident F_o and F'_e, which is beyond the eyepiece. So, $p_e = f_e = -4$ cm. For an object at the focal point, the image is at $\boxed{\text{infinity}}$.

(c) Strategy Use the transverse magnification equation. The final magnification is the product of the magnifications of the objective and the eyepiece.

Solution Since no light rays actually pass through the final image, it is $\boxed{\text{virtual}}$.

$p_o > 0$ and $q_o > 0$, so according to $m = -\dfrac{q}{p}$, $m_o < 0$. Both p_e and q_e are virtual, so $m_e < 0$. Therefore, $m_{\text{final}} = m_o m_e > 0$ and the final image is $\boxed{\text{upright}}$.

(d) Strategy The angular magnification is $M = \beta / \alpha$. Since the rays are considered paraxial, the small angle approximation $\tan \theta \approx \theta$ can be used. Let the height of the intermediate image be h, then $\tan \alpha = h / f_o \approx \alpha$ and $\tan \beta = h / |f_e| \approx \beta$.

Solution Find the angular magnification.
$$M = \frac{h / |f_e|}{h / f_o} = \frac{f_o}{|f_e|} = \frac{36 \text{ cm}}{4 \text{ cm}} = \boxed{9}$$

57. Strategy Use the angular magnification for an astronomical telescope and the fact that the distance separating the lenses is the sum of the focal lengths.

Solution

(a) Find the focal length of the objective lens in terms of that of the eyepiece.
$$M = -\frac{f_o}{f_e}, \text{ so } f_o = -M f_e = 5.0 f_e.$$
The focal length of the objective must be 5 times the focal length of the eyepiece. The focal length of lens 1 is 5 times the focal length of lens 2, so $\boxed{\text{lens 1 is the objective and lens 2 is the eyepiece}}$.

(b) Find the distance between the lenses.
$$25.0 \text{ cm} + 5.0 \text{ cm} = \boxed{30.0 \text{ cm}}$$

61. (a) Strategy and Solution A large magnitude magnification is desired. Since $M = -f_o/f_e$, the objective lens should have the longer focal length. So, $\boxed{\text{the lens with the 30.0-cm focal length}}$ should be the objective.

(b) Strategy Use Eq. (24-10).

Solution Compute the angular magnification.

$$M = -\frac{f_o}{f_e} = -\frac{30.0 \text{ cm}}{3.0 \text{ cm}} = \boxed{-10}$$

(c) Strategy The distance between the lenses is the sum of the focal lengths.

Solution Compute the distance between the lenses.

$$30.0 \text{ cm} + 3.0 \text{ cm} = \boxed{33.0 \text{ cm}}$$

65. Strategy Use the thin lens and transverse magnification equations.

Solution

(a) Light must be incident on the film to expose it and create a photograph, so the image must be $\boxed{\text{real}}$.

(b) Diverging lenses only form virtual images, so the lens must be $\boxed{\text{converging}}$.

(c) Find the image distance; that is, the distance from the lens to the film.

$$\frac{1}{f} = \frac{1}{p} + \frac{1}{q}, \text{ so } q = \frac{fp}{p-f} = \frac{(50.0 \text{ mm})(3.0 \times 10^3 \text{ mm})}{3.0 \times 10^3 \text{ mm} - 50.0 \text{ mm}} = \boxed{51 \text{ mm}}.$$

(d) Find the image height.

$$\frac{h'}{h} = -\frac{q}{p}, \text{ so } h' = -\frac{qh}{p} = -\frac{(51 \text{ mm})(1.0 \text{ m})}{3.0 \text{ m}} = -17 \text{ mm}.$$

The image is $\boxed{17 \text{ mm}}$ tall.

(e) For objects at infinity, the lens-film distance must be 50.0 mm since the parallel rays converge at the focal point of the lens. For an object distance of 1.00 m, the image distance is

$$q = \frac{fp}{p-f} = \frac{(50.0 \text{ mm})(1.00 \times 10^3 \text{ mm})}{1.00 \times 10^3 \text{ mm} - 50.0 \text{ mm}} = 52.6 \text{ mm}.$$

The lens must move a distance of $52.6 \text{ mm} - 50.0 \text{ mm} = \boxed{2.6 \text{ mm}}$.

69. (a) Strategy The refractive power of a lens is the reciprocal of the focal length. The distance between the lenses equals the sum of the focal lengths of the lenses and the tube length L.

Solution Compute the focal lengths of the lenses.

$$f_o = f_e = \frac{1}{18 \text{ D}} = 0.056 \text{ m} = 5.6 \text{ cm}$$

Find the tube length.

$$\text{lens distance} = f_o + f_e + L$$
$$28 \text{ cm} = 5.6 \text{ cm} + 5.6 \text{ cm} + L$$
$$L = \boxed{17 \text{ cm}}$$

(b) Strategy Use Eq. (24-8).

Solution Calculate the angular magnification for the microscope.

$$M_{\text{total}} = -\frac{L}{f_o} \times \frac{N}{f_e} = -L \times P_o \times N \times P_e = -0.17 \text{ m} \times 18 \text{ D} \times 0.25 \text{ m} \times 18 \text{ D} = \boxed{-14}$$

(c) Strategy Use the thin lens equation with $q_o = L + f_o$ and $1/f_o = P_o = 18$ D.

Solution Find the object distance for the objective.

$$\frac{1}{p_o} + \frac{1}{q_o} = \frac{1}{f_o}, \text{ so } p_o = \left(\frac{1}{f_o} - \frac{1}{q_o}\right)^{-1} = \left(P_o + \frac{1}{L + f_o}\right)^{-1} = \left(18 \text{ m}^{-1} - \frac{1}{0.17 \text{ m} + 0.056 \text{ m}}\right)^{-1} = \boxed{7.4 \text{ cm}}.$$

73. Strategy Use the thin lens and transverse magnification equations and the relationship between the radius of curvature of a mirror and its focal length.

Solution

(a) Find the intermediate image; that is, the image formed by the converging lens.

$$\frac{1}{p_1} + \frac{1}{q_1} = \frac{1}{f_1}, \text{ so } q_1 = \left(\frac{1}{f_1} - \frac{1}{p_1}\right)^{-1} = \left(\frac{1}{3.00 \text{ cm}} - \frac{1}{7.00 \text{ cm}}\right)^{-1} = 5.25 \text{ cm}.$$

The intermediate image is located $\boxed{5.25 \text{ cm to the right of the lens}}$.

(b) The intermediate image is on the opposite side of the lens with respect to the object, so the image is $\boxed{\text{real}}$—the light rays from a point on the object converge to the corresponding point on the image. Since real images of real objects are always inverted, and since both the object and image are real, the image is $\boxed{\text{inverted}}$.

(c) The magnification of the image is $m = -\dfrac{q_1}{p_1} = -\dfrac{5.25 \text{ cm}}{7.00 \text{ cm}} = \boxed{-0.750}$.

(d) The image of the lens is the object for the mirror. The object distance is
$p_2 = s - q_1 = 12.00 \text{ cm} - 5.25 \text{ cm} = 6.75 \text{ cm}.$
A convex mirror is a diverging mirror, so f is negative.
$$f = -\frac{1}{2}R = -\frac{1}{2}(4.00 \text{ cm}) = -2.00 \text{ cm}$$
Find the image distance for the mirror.
$$q_2 = \left(\frac{1}{f_2} - \frac{1}{p_2}\right)^{-1} = \left(\frac{1}{-2.00 \text{ cm}} - \frac{1}{6.75 \text{ cm}}\right)^{-1} = -1.54 \text{ cm}$$
Since $q_2 < 0$, the image formed by the mirror is $\boxed{1.54 \text{ cm to the right of the mirror}}$.

(e) Since the second image, the image formed by the mirror, is behind the mirror, it is a $\boxed{\text{virtual}}$ image. Since the light rays diverge from the real image of the lens, it is a real object for the mirror, and since virtual images of real objects are always upright (and the image formed by the lens is inverted with respect to the original object), the image formed by the mirror is $\boxed{\text{inverted with respect to the original object}}$.

(f) The magnification due to the mirror is $m = -\dfrac{q_2}{p_2} = -\dfrac{-1.543 \text{ cm}}{6.75 \text{ cm}} = \boxed{0.229}$.

(g) The total magnification is $m = m_1 \times m_2 = -0.750 \times 0.2286 = \boxed{-0.171}$.

77. **Strategy** Use the thin lens equation. The total angular magnification is equal to the product of the magnifications due to each lens.

Solution

(a) The image from the objective is the object for the eyepiece. Solve the thin lens equation for p_e using $q_e = -25.0$ cm and $f_e = 2.00$ cm .

$$p_e = \frac{f_e q_e}{q_e - f_e} = \frac{(2.00 \text{ cm})(-25.0 \text{ cm})}{-25.0 \text{ cm} - 2.00 \text{ cm}} = \boxed{1.85 \text{ cm}}$$

(b) The distance between the lenses is equal to the image distance for the objective plus the object distance for the eyepiece, so d = distance between lenses = $p_e + q_o$. Find the object distance p_o.

$$\frac{1}{p_o} + \frac{1}{q_o} = \frac{1}{f_o}, \text{ so } p_o = \left(\frac{1}{f_o} - \frac{1}{q_o}\right)^{-1} = \left(\frac{1}{f_o} - \frac{1}{d - p_e}\right)^{-1} = \left(\frac{1}{3.00 \text{ cm}} - \frac{1}{20.0 \text{ cm} - 1.85 \text{ cm}}\right)^{-1} = \boxed{3.59 \text{ cm}}.$$

(c) Calculate the angular magnification of the microscope.

$m_o = -\dfrac{L}{f_o}$ for the objective lens. $M_e = \dfrac{N}{p_e}$ since the intermediate image is not at F_e. So,

$$M_{\text{total}} = m_o M_e = -\frac{LN}{f_o p_e} = -\frac{(q_o - f_o)N}{f_o p_e} = -\left(\frac{q_o}{f_o} - 1\right)\frac{N}{p_e} = -\left(\frac{20.0 \text{ cm} - 1.85 \text{ cm}}{3.00 \text{ cm}} - 1\right)\frac{25.0 \text{ cm}}{1.85 \text{ cm}} = \boxed{-68.2}.$$

Chapter 25

INTERFERENCE AND DIFFRACTION

Conceptual Questions

1. Coherent waves have a definite fixed phase relationship. Two waves of different frequencies do not have a fixed phase relationship, and are therefore incoherent. Nevertheless, if the two frequencies are nearly identical, the waves can be nearly coherent, and maintain an approximately fixed phase relationship over a large distance. If the two frequencies are significantly different, the phase relationship is not even approximately constant over any appreciable distance.

5. The magnitude of diffractive effects is proportional to the wavelength of the diffracted wave. The wavelengths of sound waves are several orders of magnitude larger than those of visible light, so diffractive effects are much more pronounced for sound.

9. The smallest distance that a microscope can resolve is about the same size as the wavelength of light used to illuminate the object. For visible light, the wavelength is about 400–700 nm, which is far too large to be able to resolve an object that is 0.1 nm in width.

13. As the width of the slit decreases, the diffraction pattern spreads out and becomes dimmer.

17. The first ray incident from the air ($n = 1.00$) reflects from the layer of MgF_2 ($n = 1.38$), while the second ray travels into the MgF_2 and is reflected from the lens ($n = 1.51$). Both of these rays are incident in a material with lower n and reflect from a material with higher n, so both reflected rays undergo phase shifts of 180°. If the antireflective coating had $n = 1.62$ instead, then the first ray would be phase shifted while the second would not.

Problems

1. (a) **Strategy** The wavelength and frequency are related by $\lambda f = c$.

   Solution Compute the wavelength of a 60-kHz EM wave.
   $$\lambda = \frac{c}{f} = \frac{3.00 \times 10^8 \text{ m/s}}{6.0 \times 10^4 \text{ Hz}} = \boxed{5.0 \text{ km}}$$

   (b) **Strategy** Compute the path difference to determine the phase difference.

   Solution The path difference is $\Delta l = (19 \text{ km} + 12 \text{ km}) - 21 \text{ km} = 10 \text{ km}$, which is equal to two wavelengths. So, the path difference results in constructive interference, but the reflection of the signal from the helicopter results in a half-wavelength phase difference, which cause destructive interference. To summarize: $\boxed{\text{Destructive interference occurs, since the path difference is 10 km and there is a } \lambda/2 \text{ phase shift}}$.

3. **Strategy** Since the radio waves interfere destructively and the reflections at the planes introduce 180° phase shifts, the path differences must be equal to integral numbers of wavelengths.

Solution The path differences are $d + h - 102$ km, where $d = \sqrt{(102 \text{ km})^2 + h^2}$. The wavelength is $\lambda = c/f$. Find the wavelength in terms of the path differences.

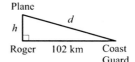

$$\Delta l = \sqrt{(102 \text{ km})^2 + h^2} + h - 102 \text{ km} = m\lambda, \text{ so}$$

$\lambda = \dfrac{\sqrt{(102 \text{ km})^2 + h^2} + h - 102 \text{ km}}{m}$. Now, the heights of the planes are $h = 780$ m, 975 m, and 1170 m, so we

have $\lambda = \dfrac{783 \text{ m}}{m_1} = \dfrac{980 \text{ m}}{m_2} = \dfrac{1177}{m_3}$ for the respective heights. Form ratios of the integers m.

$\dfrac{m_2}{m_1} = \dfrac{980}{783} = 1.25 = \dfrac{5}{4}$ and $\dfrac{m_3}{m_1} = \dfrac{1177}{783} = 1.50 = \dfrac{6}{4}$. So, $m_1 = 4$ and $\lambda = \dfrac{783 \text{ m}}{4} = 196$ m. Therefore, the

frequency of the signal is $f = \dfrac{c}{\lambda} = \dfrac{3.00 \times 10^8 \text{ m/s}}{196 \text{ m}} = \boxed{1530 \text{ kHz}}$.

5. **(a) Strategy** Sketch four sinusoidal waves, each with wavelength 6 cm. From top to bottom, the amplitudes are 2 cm, 2 cm, 3 cm, and 1 cm. The third wave is 180° out of phase with the other three.

Solution The waves are shown.

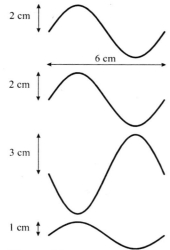

Three of the waves are in phase, so their amplitudes add. The amplitude of the third wave is 180° out of phase, so its amplitude is subtracted. The amplitude of the sum is $2 \text{ cm} + 2 \text{ cm} - 3 \text{ cm} + 1 \text{ cm} = \boxed{2 \text{ cm}}$.

(b) Strategy The first two waves interfere constructively, so use Eq. (25-3) to find the intensity of the new wave. The second two waves interfere destructively, so use Eq. (25-6) to find the intensity of the new wave; this wave will also be 180° out of phase with the sum of the first two, since the third wave's amplitude is greater than the fourth's, so use Eq. (25-6) again to find the final intensity. The first two waves each have intensity I_0. Use the fact that intensity is directly proportional to amplitude to find the intensity of the third and fourth waves in terms of I_0.

Solution Find the intensity of the combined first two waves.

$$I = I_1 + I_2 + 2\sqrt{I_1 I_2} = I_0 + I_0 + 2\sqrt{I_0 I_0} = 4I_0$$

Find the intensities of the third and fourth waves in terms of I_0. Form proportions.

$$\frac{I_3}{I_0} = \frac{3^2}{2^2} = \frac{9}{4}, \text{ so } I_3 = \frac{9}{4}I_0. \quad \frac{I_4}{I_0} = \frac{1^2}{2^2} = \frac{1}{4}, \text{ so } I_4 = \frac{1}{4}I_0.$$

Find the intensity of the combined third and fourth waves.

$$I = I_3 + I_4 - 2\sqrt{I_3 I_4} = \frac{9}{4}I_0 + \frac{1}{4}I_0 - 2\sqrt{\frac{9}{4}I_0 \frac{1}{4}I_0} = \frac{5}{2}I_0 - 2\sqrt{\frac{9}{16}I_0^2} = I_0$$

Find the intensity of the combination of all four waves.

$$I = 4I_0 + I_0 - 2\sqrt{4I_0 I_0} = 5I_0 - 4I_0 = \boxed{I_0}$$

(c) Strategy Three of the waves are in phase, so their amplitudes add. The amplitude of the fourth wave is 180° out of phase, so its amplitude is subtracted.

Solution The amplitude of the sum is $2 \text{ cm} + 2 \text{ cm} + 3 \text{ cm} - 1 \text{ cm} = \boxed{6 \text{ cm}}$.

(d) Strategy Repeat the process of part (b), but this time, the combination of the second two waves will be in phase with the combination of the first two.

Solution Find the intensity of the combined first two waves.

$$I = I_1 + I_2 + 2\sqrt{I_1 I_2} = I_0 + I_0 + 2\sqrt{I_0 I_0} = 4I_0$$

Find the intensities of the third and fourth waves in terms of I_0. Form proportions.

$$\frac{I_3}{I_0} = \frac{3^2}{2^2} = \frac{9}{4}, \text{ so } I_3 = \frac{9}{4}I_0. \quad \frac{I_4}{I_0} = \frac{1^2}{2^2} = \frac{1}{4}, \text{ so } I_4 = \frac{1}{4}I_0.$$

Find the intensity of the combined third and fourth waves.

$$I = I_3 + I_4 - 2\sqrt{I_3 I_4} = \frac{9}{4}I_0 + \frac{1}{4}I_0 - 2\sqrt{\frac{9}{4}I_0 \frac{1}{4}I_0} = \frac{5}{2}I_0 - 2\sqrt{\frac{9}{16}I_0^2} = I_0$$

Find the intensity of the combination of all four waves.

$$I = 4I_0 + I_0 + 2\sqrt{4I_0 I_0} = 5I_0 + 4I_0 = \boxed{9I_0}$$

7. Strategy Since the light from each lamp is incoherent, the intensity of the lamps together is just the sum of the intensities.

Solution Find the combined intensity of the lamps.

$$I = I_0 + 4I_0 = \boxed{5I_0}$$

9. **(a) Strategy** As in Example 25.1, the formula relating the path length difference to the wavelength is $2\Delta x = \lambda$.

   Solution The maxima in the figure are at $x = 0.7$ cm and $x = 2.3$ cm, so $\Delta x = 1.6$ cm and the wavelength is
   $\lambda = 2 \cdot 1.6$ cm $= \boxed{3.2 \text{ cm}}$.

   (b) Strategy Amplitude is proportional to the square root of intensity and intensity is proportional to power, so amplitude is proportional to the square root of power.

   Solution The maxima have power readings of 3.4 units and 2.6 units. The ratio of the amplitudes is
   $\sqrt{3.4/2.6} = \boxed{1.1}$.

13. **(a) Strategy and Solution** Since the wavelength of light is shorter in glass than in vacuum, it will take a larger number of wavelengths to pass through the glass. To increase the number of wavelengths in the other arm, the mirror must be moved $\boxed{\text{out}}$.

    (b) Strategy Let the original length of each arm be D. Let T be the thickness of the glass. Then, the total distance traveled by the light in the glass is $2T$, and the distance traveled in the same arm but outside of the glass is $2(D-T)$. If the vacuum wavelength of one of the components of the white light is λ_0, then the wavelength in the glass is $\lambda = \lambda_0/n$.

    Solution Find the thickness of the slab of glass.
    The number of wavelengths traveled in the glass is $\dfrac{2T}{\lambda} = \dfrac{2T}{\lambda_0/n} = \dfrac{2Tn}{\lambda_0}$.

    The number of wavelengths traveled in the same arm but outside of the glass is $\dfrac{2(D-T)}{\lambda_0}$.

    The total number of wavelengths traveled in this arm is $\dfrac{2D-2T}{\lambda_0} + \dfrac{2Tn}{\lambda_0} = \dfrac{2D + 2T(n-1)}{\lambda_0} = \dfrac{2D + 2T(0.46)}{\lambda_0}$.

    Let d be the distance the mirror moved. Then, the number of wavelengths traveled in this arm is $\dfrac{2D+2d}{\lambda_0}$.

    The number of wavelengths in both cases must be equal.
    $\dfrac{2D + 2T(0.46)}{\lambda_0} = \dfrac{2D+2d}{\lambda_0}$, so $T = \dfrac{d}{0.46} = \dfrac{6.73 \text{ cm}}{0.46} = \boxed{15 \text{ cm}}$.

15. **(a) Strategy** When light reflects from a boundary with a medium with a higher index of refraction, it is inverted (180° phase change). Inversion alone will result in destructive interference.

    Solution If there is no gap between the plates of glass—they are touching—the light is inverted at the boundary and destructive interference occurs. Therefore, the minimum distance between the two glass plates for one of the dark regions is zero.

    (b) Strategy The light passes through oil, so its wavelength is reduced by a factor of $1/n_{oil}$. There must be some nonzero thickness of oil between the plates for constructive interference to occur. Let this distance between the plates be t.

    Solution The light is inverted after is reflects of the bottom layer of glass after it has passed through the thin film of oil between the plates, so the path difference due to the light traveling through the thin film of oil must be equal to one half wavelength to compensate. The light passes through the thickness of the oil twice, before and after reflection. Find t.

    $$2t = \frac{\lambda/n_{oil}}{2}, \text{ so } t = \frac{\lambda}{4n_{oil}} = \frac{550 \text{ nm}}{4(1.50)} = \boxed{92 \text{ nm}}.$$

    (c) Strategy Refer to parts (a) and (b). For destructive interference, the path difference due to the light traveling through the thin film of oil must be equal to one full wavelength.

    Solution Find the next largest distance between the plates resulting in a dark region.

    $$2t = \frac{\lambda}{n_{oil}}, \text{ so } t = \frac{\lambda}{2n_{oil}} = \frac{550 \text{ nm}}{2(1.50)} = \boxed{180 \text{ nm}}.$$

17. **Strategy** At the air-oil boundary, reflected light is 180° out of phase, since $n_{oil} > n_{air}$. For transmitted light reflected at the oil-water boundary, the reflective ray is not phase shifted since $n_{water} < n_{oil}$. The two reflected rays are 180° out of phase, so constructive interference occurs when the relative path length difference, $2t$, is an odd multiple of one half the wavelength in the oil, $\lambda = \lambda_0 / n_{oil}$.

    Solution Find the wavelength.

    $$2t = \left(m + \frac{1}{2}\right)\lambda = \left(m + \frac{1}{2}\right)\frac{\lambda_0}{n_{oil}}$$

    Solving for λ_0 yields $\lambda_0 = \dfrac{2tn_{oil}}{m + \frac{1}{2}}$.

    $n_{air} = 1.00$
    $n_{oil} = 1.50$ $t = 0.40 \text{ μm}$
    $n_{water} = 1.33$
    (Rays are normally incident)

    Substituting $t = 0.40 \text{ μm}$ and $n_{oil} = 1.50$ gives $\lambda_0 = \dfrac{1200 \text{ nm}}{m + \frac{1}{2}}$.

    Evaluating this for $m = 0, 1, 2$, and 3 gives $\lambda_0 = 2400 \text{ nm}$, 800 nm, 480 nm, and 343 nm, respectively. Of these values, only $\boxed{480 \text{ nm}}$ is in the visible spectrum.

19. **Strategy** The wavelength most strongly transmitted will be the wavelength whose reflected rays combine destructively (minimum reflection). At both boundaries, the reflected rays are inverted, since $n_{air} < n_{film}$ and $n_{film} < n_{glass}$. So, the phase difference between the reflected rays is due only to the thickness of the film. Destructive interference occurs when the path length difference, $2t$, is equal to an odd multiple of half the wavelength in the film, $\lambda = \lambda_0 / n_{film}$.

Solution Find the wavelength.

$$2t = \left(m + \frac{1}{2}\right)\lambda = \left(m + \frac{1}{2}\right)\frac{\lambda_0}{n_{film}}$$

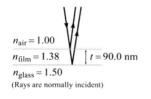

$n_{air} = 1.00$
$n_{film} = 1.38$ $t = 90.0$ nm
$n_{glass} = 1.50$
(Rays are normally incident)

Solving for λ_0, and substituting $n_{film} = 1.38$, $t = 90.0$ nm, and $m = 0$ yields

$$\lambda_0 = \frac{2tn_{film}}{m + \frac{1}{2}} = \frac{2 \cdot 90.0 \text{ nm} \cdot 1.38}{0 + \frac{1}{2}} = \boxed{497 \text{ nm}}.$$

All other values of m give wavelengths not in the visible spectrum.

21. **(a) Strategy** The weakest wavelengths in transmitted light correspond to the strongest wavelengths in reflected light. Rays reflected off the front of the film will be inverted since $n_{air} < n_{film}$, but rays reflected off the back of the film are not inverted, so the reflected rays will be 180° out of phase plus the phase shift caused by the path length difference. The condition for constructive interference is used to find the strongest wavelengths in reflected light. This occurs when the path length difference, $2t$, is an odd multiple of half the wavelength in the film.

Solution Find the weakest wavelengths.

$$2t = \left(m + \frac{1}{2}\right)\lambda = \left(m + \frac{1}{2}\right)\frac{\lambda_0}{n_{film}}$$

$n_{air} = 1.00$
$n_{film} = 1.50$ $t = 910.0$ nm
$n_{air} = 1.00$

(Rays are normally incident)

Solving for λ_0 with $t = 910.0$ nm and $n_{film} = 1.50$ gives

$$\lambda_0 = \frac{2tn_{film}}{m + \frac{1}{2}} = \frac{2730 \text{ nm}}{m + \frac{1}{2}}.$$

Substituting $m = 4$, 5, and 6 gives wavelengths in the visible spectrum:
$\lambda_0 = \boxed{607 \text{ nm, } 496 \text{ nm, and } 420 \text{ nm}}$.

(b) Strategy The strongest wavelengths is transmitted light correspond to the weakest wavelengths in reflected light. Destructive interference in the reflected light occurs when the path length difference, $2t$, is equal to an integral number of wavelengths in the film, $\lambda = \lambda_0 / n_{film}$.

Solution Find the strongest wavelengths.

$$2t = m\lambda = m\frac{\lambda_0}{n_{film}}$$

Solving for λ_0 with $t = 910.0$ nm and $n_{film} = 1.50$ gives

$$\lambda_0 = \frac{2tn_{film}}{m} = \frac{2730 \text{ nm}}{m}.$$

Substituting $m = 4$, 5, and 6 gives wavelengths in the visible spectrum: $\lambda_0 = \boxed{683 \text{ nm, } 546 \text{ nm, and } 455 \text{ nm}}$.

25. **Strategy** Constructive interference occurs when the rays are in phase. Destructive interference occurs when they are 180° out of phase. Consider phase shifts due to reflections and path-length differences.

    Solution Both rays 1 and 2 are inverted. For constructive interference, the path length difference must be equal to an integral number of wavelengths in the film, so $2t = \dfrac{m\lambda_0}{n_2}$.

    For destructive interference, the path length difference must be an odd multiple of the wavelength in the film, so $2t = \left(m + \dfrac{1}{2}\right)\dfrac{\lambda_0}{n_2}$.

    Ray 4 is inverted by its first reflection, so the path length difference between rays 3 and 4 must be equal to an odd multiple of the wavelength in the film for constructive interference, $2t = \left(m + \dfrac{1}{2}\right)\dfrac{\lambda_0}{n_2}$, and for destructive interference, $2t = \dfrac{m\lambda_0}{n_2}$. So, when rays 1 and 2 interfere constructively/destructively, rays 3 and 4 interfere destructively/constructively.

29. **Strategy** Sketch the diagram according to the instructions in the Problem statement. Compare the results to Eq. (25-10).

    Solution

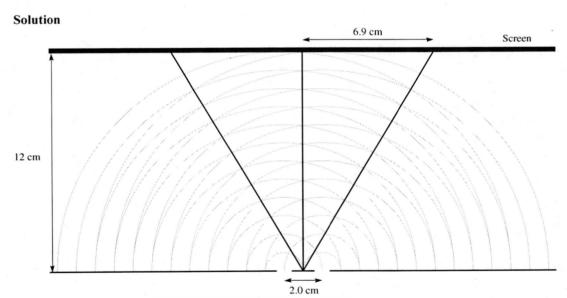

    The $m = \pm 1$ maxima are $\boxed{\pm 6.9 \text{ cm from the central maximum}}$. As measured by a protractor, the angles between the $m = \pm 1$ maxima and the central maximum are $\boxed{\pm 30°}$. Solve Eq. (25-10) for θ.

    $d\sin\theta = m\lambda$, so $\theta = \sin^{-1}\dfrac{m\lambda}{d} = \sin^{-1}\left(m\dfrac{1.0 \text{ cm}}{2.0 \text{ cm}}\right) = \sin^{-1}(0.50m)$.

    So, $\theta_0 = 0$ and $\theta_{\pm 1} = \pm 30°$ for $m = 0$ and $m = \pm 1$, respectively. $\boxed{\text{Yes, the angles agree}}$.

31. **Strategy** Show that the small-angle approximation is justified. Then, use the result for the slit separation obtained in Example 25.4.

    Solution Compare x and D.
    $$\frac{D}{x} = \frac{2.50 \text{ m}}{0.760 \times 10^{-2} \text{ m}} = 329, \text{ so } x \ll D \text{ and the small angle approximation is justified.}$$
    Find the wavelength of the light.
    $$d = \frac{\lambda D}{x}, \text{ so } \lambda = \frac{dx}{D} = \frac{(0.0150 \times 10^{-2} \text{ m})(0.760 \times 10^{-2} \text{ m})}{2.50 \text{ m}} = \boxed{456 \text{ nm}}.$$

33. **Strategy** Check to see if the small angle approximation is justified. If so, use the result of Example 25.4 for the distance d between the slits.

    Solution Compare x and D.
    $$\frac{D}{x} = \frac{1.5 \text{ m}}{0.0135 \text{ m}} = 111, \text{ so } x \ll D \text{ and the small angle approximation is justified}$$
    Compute the slit separation.
    $$d = \frac{\lambda D}{x} = \frac{(630 \times 10^{-9} \text{ m})(1.5 \text{ m})}{0.0135 \text{ m}} = \boxed{7.0 \times 10^{-5} \text{ m}}$$

37. **Strategy** Apply Eq. (25-10) for each line.

    Solution The third-order red line gives
    $d \sin \theta_{R3} = m\lambda = 3\lambda_R$.
    For the fourth-order blue line, we have
    $d \sin \theta_{B4} = m\lambda = 4\lambda_B$.
    Since these lines overlap, the angles θ_{R3} and θ_{B4} must be equal, which implies $d \sin \theta_{R3} = d \sin \theta_{B4}$. So, the right-hand sides of the previous equations must also be equal.
    $$3\lambda_R = 4\lambda_B, \text{ so } \lambda_B = \frac{3}{4}\lambda_R = \frac{3}{4}(630 \text{ nm}) = \boxed{470 \text{ nm}}.$$

39. **Strategy** Find the slit separation. Then solve Eq. (25-10) for m using $\lambda = 412$ nm and $\theta = 90°$.

    Solution Compute the slit separation.
    $$d = \frac{1}{5000.0 \text{ slits/cm}} = 2.0000 \times 10^{-6} \text{ m}$$
    Find the number of orders.
    $$d \sin \theta = m\lambda, \text{ so } m = \frac{d \sin \theta}{\lambda} = \frac{(2.0000 \times 10^{-6} \text{ m}) \sin 90°}{412 \times 10^{-9} \text{ m}} = 4.85.$$
    Since m must be an integer, $\boxed{\text{four}}$ orders can be observed.

41. **Strategy** Use Eq. (25-10). Draw a diagram. Set $\theta = 90°$ to find the highest-order spectral line.

   Solution

   (a) In these first-order maxima, the shortest wavelength will appear first and the longest wavelength last, since $\sin\theta \propto \lambda$. Thus, $\boxed{A \text{ is the blue line, } B \text{ is the yellow line, and } C \text{ is the red line}}$.

   (b) Solve for the wavelength.

   $d\sin\theta = m\lambda$, so $\lambda = \dfrac{d\sin\theta}{m}$.

   Here, $d = 1870$ nm and $m = 1$. To

   find θ, use $\tan\theta = \dfrac{\text{Position of line } C}{\text{distance between grating and screen}} = \dfrac{x}{D}$.

   $\theta = \tan^{-1}\left(\dfrac{11.5 \text{ cm}}{30.0 \text{ cm}}\right) = 20.97°$

   Substitute this value for θ and the given values of d and m into the equation for λ.

   $\lambda = \dfrac{(1870 \text{ nm})\sin 20.97°}{1} = \boxed{669 \text{ nm}}$

   (c) To find the highest-order of spectral line C, solve for m using $\lambda = 669$ nm, $d = 1870$ nm, and $\theta = 90°$.

   $d\sin\theta = m\lambda$, so $m = \dfrac{d\sin\theta}{\lambda} = \dfrac{(1870 \text{ nm})\sin 90°}{669 \text{ nm}} = 2.80$.

   Since m must be an integer, the highest order of spectral line C that it is possible to see is $\boxed{2}$.

45. (a) **Strategy** Use Eq. (25-10). The slit separation is given by $d = 1/N$.

   Solution Find the number of slits per cm of the grating.

   $d\sin\theta = m\lambda$, so $\dfrac{\sin\theta}{m\lambda} = \dfrac{1}{d} = N$.

   Since the higher orders are at greater angles than the shorter ones, set $\theta = 90°$; then for the second order, $N = 1/(2\lambda)$. To get the maximum N allowed, the larger wavelength is chosen. An N larger than this maximum value excludes all wavelengths greater than 661.4 nm, but includes smaller wavelengths.

   $N_{\text{max}} = \dfrac{1}{2(661.4 \times 10^{-7} \text{ cm})} = \boxed{7560 \text{ slits/cm}}$

   (b) **Strategy** Find the average and difference of the wavelengths; then use the given equation for N.

   Solution The average of the two wavelengths is

   $\lambda = \dfrac{\lambda_1 + \lambda_2}{2} = \dfrac{660.0 \text{ nm} + 661.4 \text{ nm}}{2} = 660.7 \text{ nm}$.

   The difference between the two wavelengths is

   $\Delta\lambda = \lambda_2 - \lambda_1 = 661.4 \text{ nm} - 660.0 \text{ nm} = 1.4 \text{ nm}$.

   Substituting these values and $m = 2$ into the given equation yields $N = \dfrac{\lambda}{m\Delta\lambda} = \dfrac{660.7 \text{ nm}}{2 \cdot 1.4 \text{ nm}} = \boxed{240}$.

47. Strategy Let $x = 1.0$ mm, which is the same as the distance from the center to the first minimum. Since $x = 1.0$ mm $\ll D = 1.0$ m, using the small angle approximation is justified, so use the result of Example 25.8 for half the width of the central maximum.

Solution Find the width of the slit.

$$x = \frac{\lambda D}{a}, \text{ so } a = \frac{\lambda D}{x} = \frac{(610 \times 10^{-9})(1.0 \text{ m})}{0.0010 \text{ m}} = \boxed{0.61 \text{ mm}}.$$

49. Strategy Call the width of the central maximum W. Referring to Figure 25.33, we have $W = 2x$, and by trigonometry, $x = D \tan \theta$, so $W = 2x = 2D \tan \theta$. Assume θ is a small angle. Use Eq. (25-12).

Solution Since θ is assumed to be small, we have $W = 2D \tan \theta \approx 2D \sin \theta$.

From Eq. (25-12), we have $\sin \theta = \dfrac{\lambda}{a}$ since $m = 1$, so W becomes $W = \dfrac{2L\lambda}{a}$.

If a is replaced with $2a$, the new width is $W_{\text{new}} = \dfrac{2L\lambda}{2a} = \dfrac{1}{2} \dfrac{2L\lambda}{a} = \dfrac{1}{2} W.$

Thus, the new width is half the old width.

53. Strategy The angular width of the central diffraction maximum is two times the angular position of the first minimum. Solve Eq. (25-13) for θ, using $\lambda = 590$ nm and $a = 7.0$ mm.

Solution Find the angular distance to the first minimum.

$$a \sin \theta = 1.22\lambda, \text{ so } \theta = \sin^{-1} \frac{1.22\lambda}{a} = \sin^{-1} \frac{1.22(590 \times 10^{-9} \text{ m})}{7.0 \times 10^{-3} \text{ m}} = 0.0059°.$$

The angular width is twice this or $\boxed{0.012°}$.

57. (a) Strategy Use Snell's law to relate the angular separation of the two sources and the angular separation of the two images.

Solution By Snell's law, we have $n_{\text{air}} \sin \Delta\theta = n \sin \beta$ for ray 2.

Using $n_{\text{air}} = 1$ yields $\boxed{\sin \Delta\theta = n \sin \beta}$.

(b) Strategy If $\beta \geq \phi$ then $\sin \beta \geq \sin \phi$. In the fluid, $\lambda = \lambda_0 / n$. Apply Eq. (25-13) to image 1 and solve for $\sin \phi$.

Solution

$$a \sin \phi = 1.22\lambda = 1.22\frac{\lambda_0}{n}, \text{ so } \sin \phi = \frac{1.22\lambda_0}{an}.$$

Taking the result from part (a) and solving for $\sin \beta$ gives $\sin \beta = \sin \Delta\theta / n$. Substitute this expression for $\sin \beta$ and the previous expression for $\sin \phi$ into the inequality $\sin \beta \geq \sin \phi$.

$$\frac{\sin \Delta\theta}{n} \geq \frac{1.22\lambda_0}{an}$$

$$a \sin \Delta\theta \geq 1.22\lambda_0$$

This is equivalent to Eq. (25-14), Rayleigh's criterion.

61. Strategy Find the smallest $\Delta\theta$ the eye can discern using Eq. (25-14) with $a = 7$ mm (diameter of the pupil) and $\lambda_0 = 550$ nm (middle of the visible spectrum). Two objects separated by a distance d, at a distance L from the observer, will have an angular separation $\Delta\theta$ given by $\tan\Delta\theta = d/L$.

Solution Find $\Delta\theta$.

$$a\sin\Delta\theta = 1.22\lambda_0, \text{ so } \Delta\theta = \sin^{-1}\frac{1.22\lambda_0}{a} = \sin^{-1}\frac{1.22(550\times10^{-9}\text{ m})}{7\times10^{-3}\text{ m}} = 0.0055°.$$

Estimate the maximum distance at which the headlights can be resolved. Let the distance between the headlights be $d = 2$ m.

$$L = \frac{d}{\tan\Delta\theta} = \frac{2\text{ m}}{\tan 0.0055°} = \boxed{20\text{ km}}$$

20 km is rather large—clearly diffraction is not the only limitation!

65. Strategy The first single-slit diffraction minimum is coincident with the fifth-order double-slit interference maximum. Use the equations for the double-slit interference maxima and single-slit diffraction minima. The distance to the first diffraction minimum is half the width of the central maximum.

Solution

(a) Find the width of the slits.

$$a\sin\theta = m\lambda = (1)\lambda = \lambda, \text{ so } a = \frac{\lambda}{\sin\theta}.$$

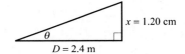

Since $x = (2.40\text{ cm})/2 = 0.0120\text{ m} \ll D = 2.4\text{ m}$, $\sin\theta \approx \tan\theta = \dfrac{x}{D}$ and

$$a = \frac{\lambda D}{x} = \frac{(510\times10^{-9}\text{ m})(2.4\text{ m})}{0.0120\text{ m}} = \boxed{0.10\text{ mm}}.$$

(b) Find the relationship between the slit width a and the distance between the slits d using the fifth-order double-slit interference maximum and the first single-slit diffraction minimum.

$$d\sin\theta = 5\lambda \text{ and } a\sin\theta = \lambda, \text{ so } \frac{d\sin\theta}{a\sin\theta} = \frac{d}{a} = \frac{5\lambda}{\lambda} = 5 \text{ and } d = 5a = 5(0.102\text{ mm}) = \boxed{0.51\text{ mm}}.$$

69. Strategy Rays reflected at both boundaries will be inverted, since $n_{air} < n_{coating} < n_{glass}$, so the phase difference in the two reflected rays depends only on the thickness of the coating. For enhanced reflection, constructive interference is needed. This occurs when the path length difference in the coating, $2t$, is equal to the wavelength in the coating, $\lambda_{coating} = \lambda/n_{coating}$ (for minimum thickness).

Solution Find the minimum thickness of the coating.

$$2t = \lambda_{coating} = \frac{\lambda}{n_{coating}}, \text{ so } \boxed{t = \frac{\lambda}{2n_{coating}}}.$$

73. **Strategy** Use Rayleigh's criterion, Eq. (25-14). The equation relating the angular separation, the distance to the moon D, and the distance x between two objects on the moon is $\tan \Delta\theta = x/D$.

    Solution Find the minimum angular separation.

    $a \sin \Delta\theta \geq 1.22\lambda_0$, so $\Delta\theta \geq \sin^{-1} \dfrac{1.22\lambda_0}{a} = \sin^{-1} \dfrac{1.22(520 \times 10^{-9} \text{ m})}{5.08 \text{ m}} = 1.249 \times 10^{-7}$ rad.

    Solve for x.

    $x = D \tan \Delta\theta = (3.845 \times 10^8 \text{ m}) \tan(1.249 \times 10^{-7} \text{ rad}) = \boxed{48 \text{ m}}$

77. **(a) Strategy and Solution** All points on the line $\theta = 0$ are equidistant from the towers. Since the radio waves started out in phase, they will arrive in phase and combine constructively at all points on $\theta = 0$. The power will be a $\boxed{\text{maximum}}$.

    (b) Strategy This can be treated as a double-slit problem with the antennas playing the role of the slits. Solving Eq. (25-10) for θ, and using $d = \lambda$ and $m = 0$ and ± 1 yields the maxima.

    Solution Solve for the angle.

    $d \sin \theta = \lambda \sin \theta = m\lambda$, so $\theta = \sin^{-1} m$.
    Substitute the values of m.

    $\theta_0 = \sin^{-1}(0) = 0$ and $\theta_{\pm 1} = \sin^{-1}(\pm 1) = \pm 90°$.

    The situation for $90° < \theta < 180°$ and $-180° < \theta < -90°$ is the mirror image of the maxima found above. All together, the maxima occur at $\theta = -180°, -90°, 0°, 90°,$ and $180°$, where $\pm 180°$ is the same location. The minima are midway between the maxima: $\pm 45°$ and $\pm 135°$. A qualitative graph is shown.

    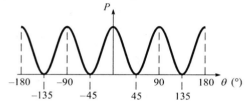

    (c) Strategy Solve Eq. (25-10) for θ using $d = \lambda/2$ and $m = 0$.

    Solution Solve for the angle.

    $d \sin \theta = \dfrac{\lambda}{2} \sin \theta = m\lambda$, so $\theta = \sin^{-1}(2m)$.

    For $m = 0$, we have $\theta_0 = \sin^{-1}(0) = 0$.

    By symmetry, $\theta = \pm 180°$ will also be a maximum. Since $d = \lambda/2$, destructive interference will occur at $\theta = \pm 90°$.

    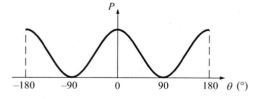

REVIEW AND SYNTHESIS: CHAPTERS 22–25

Review Exercises

1. **Strategy** The sound of the bat hitting the ball travels 22 m from the bat to the microphone in a time Δt_1. Then, the EM wave travels 4500 km from the stadium to the television set in a time Δt_2. Finally, the sound travels 2.0 m from the television set to your ears in a time Δt_3. Compute the time it takes for each wave, sound and EM, to travel the stated distances using $\Delta x = v\Delta t$.

   Solution Compute the minimum time it takes for you to hear the crack of the bat after the batter hit the ball.
   $$\Delta t_1 + \Delta t_2 + \Delta t_3 = \frac{22\ \text{m}}{343\ \text{m/s}} + \frac{4.50 \times 10^6\ \text{m}}{3.00 \times 10^8\ \text{m/s}} + \frac{2.0\ \text{m}}{343\ \text{m/s}} = \boxed{85\ \text{ms}}.$$

5. **Strategy** Since the radio waves interfere destructively and the reflection at the plane introduces a 180° phase shift, the minimum path difference must be equal to one wavelength.

   Solution

   (a) Find the wavelength of the radio waves.
   $$\lambda = \frac{c}{f} = \frac{2.998 \times 10^8\ \text{m/s}}{1408 \times 10^3\ \text{Hz}} = 212.93\ \text{m}$$
   $2\lambda = 425.9\ \text{m}$ and $3\lambda = 638.8\ \text{m}$. If Juanita is correct that the plane is at least 500 m over her head, $\boxed{\text{the plane must be at least } 3\lambda = 638.8\ \text{m over Juanita's head}}$.

   (b) The lower estimate is $2\lambda = \boxed{425.9\ \text{m}}$ and the higher estimate is $4\lambda = \boxed{851.7\ \text{m}}$.

9. **Strategy and Solution** You don't see an interference pattern on your desk when light from two different lamps are illuminating the surface because $\boxed{\text{the lamps are not emitting coherent light}}$.

13. **Strategy** The distance between adjacent slits is the reciprocal of the number of slits per unit length. Use Eq. (25-10).

    Solution

    (a) Compute the distance between adjacent slits.
    $$d = \frac{1}{5550\ \text{slits/cm}} = \frac{1\ \text{cm}}{5550\ \text{slits}} = \frac{10^{-2}\ \text{m}}{5550\ \text{slits}} = \boxed{1.80 \times 10^{-6}\ \text{m}}$$

    (b) Find the distance between the central bright spot and the first-order maximum. Find θ.

    $d \sin\theta = m\lambda = (1)\lambda = \lambda$, so $\theta = \sin^{-1}\dfrac{\lambda}{d}$. Find x.

    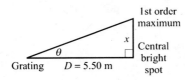

    $\tan\theta = \dfrac{x}{D}$, so

    $$x = D\tan\theta = D\tan\left(\sin^{-1}\frac{\lambda}{d}\right) = (5.50\ \text{m})\tan\left[\sin^{-1}\frac{0.680 \times 10^{-6}\ \text{m}}{1.8018 \times 10^{-6}\ \text{m}}\right] = \boxed{2.24\ \text{m}}.$$

(c) Find the distance between the central bright spot and the second-order maximum. Find θ.

$$d \sin \theta = m\lambda = 2\lambda, \text{ so } \theta = \sin^{-1}\frac{2\lambda}{d}. \text{ Find } x.$$

$$\tan \theta = \frac{x}{D}, \text{ so } x = D\tan\theta = D\tan\left(\sin^{-1}\frac{2\lambda}{d}\right) = (5.50 \text{ m})\tan\left[\sin^{-1}\frac{2(0.680\times10^{-6} \text{ m})}{1.8018\times10^{-6} \text{ m}}\right] = \boxed{6.33 \text{ m}}.$$

(d) $\boxed{\text{The assumption that } \sin\theta = \tan\theta \text{ is not valid because the angles are not small.}}$

17. Strategy Geraldine want to display the zeroth- and first-order interference maxima across the full width of a 20.0-cm screen. To do so, the first-order minima must be at the edge of the screen. Use Eq. (25-11).

Solution Find D, the distance between the double slit and the screen.

$$d \sin\theta = d\frac{x}{\sqrt{x^2 + D^2}} = \left(m + \frac{1}{2}\right)\lambda = \left(1 + \frac{1}{2}\right)\lambda = \frac{3}{2}\lambda$$

Solve for D.

$$\frac{x}{\sqrt{x^2 + D^2}} = \frac{3\lambda}{2d}$$

$$\frac{x^2}{x^2 + D^2} = \left(\frac{3\lambda}{2d}\right)^2$$

$$x^2 + D^2 = \left(\frac{2d}{3\lambda}\right)^2 x^2$$

$$D = x\sqrt{\left(\frac{2d}{3\lambda}\right)^2 - 1} = \frac{0.200 \text{ m}}{2}\sqrt{\left[\frac{2(20.0\times10^{-6} \text{ m})}{3(423\times10^{-9} \text{ m})}\right]^2 - 1} = \boxed{3.15 \text{ m}}$$

21. Strategy Use Eq. (25-10). Let the subscript g stand for the green light and the subscript v stand for the violet light. Assume that the angles are small. Then $\sin\theta \approx \tan\theta = x/D$.

Solution

(a) Since $\sin\theta \propto \lambda$, the green maximum will be farther than the violet maximum from the central maximum. Find the separation $x_g - x_v$.

$$d \sin\theta_g = d\frac{x_g}{D} = \lambda_g \text{ and } d\sin\theta_v = d\frac{x_v}{D} = \lambda_v, \text{ so}$$

$$x_g - x_v = \frac{D}{d}\lambda_g - \frac{D}{d}\lambda_v = \frac{D}{d}(\lambda_g - \lambda_v) = \frac{0.720 \text{ m}}{0.020\times10^{-3} \text{ m}}(520\times10^{-9} \text{ m} - 412\times10^{-9} \text{ m}) = \boxed{3.9 \text{ mm}}.$$

(b) Find the separation.

$$d \sin\theta_g = d\frac{x_g}{D} = 2\lambda_g \text{ and } d\sin\theta_v = d\frac{x_v}{D} = 2\lambda_v, \text{ so}$$

$$x_g - x_v = \frac{D}{d}(2\lambda_g) - \frac{D}{d}(2\lambda_v) = \frac{2D}{d}(\lambda_g - \lambda_v) = \frac{2(0.720 \text{ m})}{0.020\times10^{-3} \text{ m}}(520\times10^{-9} \text{ m} - 412\times10^{-9} \text{ m}) = \boxed{7.8 \text{ mm}}.$$

25. **Strategy** Use Snell's law.

   Solution

   (a) Find the angle.
   $$n_i \sin \theta_i = n_f \sin \theta_f$$
   $$1.00 \sin (90° - 30°) = 1.35 \sin (90° - \theta_1)$$
   $$1.00 \sin (60°) = 1.35 \cos \theta_1$$
   $$\theta_1 = \cos^{-1} \frac{1.00 \sin (60°)}{1.35} = \boxed{50.1°}$$

   (b) Find the critical angle.
   $$n_g \sin \theta_c = n_{air} \sin 90° = 1.00, \text{ so } \theta_c = \sin^{-1} \frac{1.00}{n_g} = \sin^{-1} \frac{1.00}{1.35} = \boxed{47.8°}.$$

   (c) The incident angle is the same as θ_1, which is greater than the critical angle. Therefore, total internal reflection occurs and the light follows $\boxed{\text{path } A}$.

   (d) Find θ_A.
   $$\theta_A = 90° - \theta_1 = 90° - 50.1° = \boxed{39.9°}$$

MCAT Review

1. **Strategy** Use the transverse magnification equation.

   Solution Find the ratio of the height of the image to the height of the object.

   $$m = \frac{h'}{h} = -\frac{q}{p} = -\frac{\frac{3}{2}f}{3f} = -\frac{1}{2}$$

   The correct answer is $\boxed{A}$.

2. **Strategy** The object and image distances are the same. Use the mirror equation and the fact that $f = R/2$.

   Solution Find p, the object location.

   $$\frac{1}{p} + \frac{1}{q} = \frac{1}{p} + \frac{1}{p} = \frac{2}{p} = \frac{1}{f} = \frac{2}{R}, \text{ so } p = R = 50 \text{ cm.}$$

   The correct answer is $\boxed{C}$.

3. **Strategy and Solution** When the telescope is focused on a very distant object, the image is located at the focal point of the mirror. The image is formed in front of the mirror, so it is real. The image and object distances are positive, so according to the magnification equation, $m = -q/p < 0$, therefore, the image is inverted.

   The correct answer is $\boxed{B}$.

4. **Strategy** The magnification is the equal to the ratio of the focal length of the mirror to that of the eyepiece.

   Solution The magnification of the Hubble is approximately

   $$\frac{f_{\text{mirror}}}{f_{\text{eyepiece}}} = \frac{13 \text{ m}}{2.5 \times 10^{-2} \text{ m}} = 520. \text{ The correct answer is } \boxed{C}.$$

5. **Strategy** For a very distant object, p is very large, so $1/p$ is approximately zero. Use the mirror equation.

   Solution Find the image location.

   $$\frac{1}{f} = \frac{1}{q} + \frac{1}{p} \approx \frac{1}{q} + 0 = \frac{1}{q}, \text{ so } q \approx f.$$

   The image location is very close to the focal point of the mirror. The correct answer is $\boxed{C}$.

6. **Strategy and Solution** Visible light has wavelengths greater than ultraviolet light. The wavelengths of visible light are large compared to atoms and molecules, so visible light is not that easily absorbed by the atmosphere. The wavelengths of ultraviolet light are closer to the size of atoms and molecules in the atmosphere, so ultraviolet waves are more readily absorbed by the atmosphere. The correct answer is $\boxed{A}$.

Chapter 26

RELATIVITY

Conceptual Questions

1. Whether something is obvious depends on our common sense about it. In everyday experience we only observe objects in motion with speeds much smaller than the speed of light. We have no experience with, and hence no common sense about, objects moving close to the speed of light. Therefore, it is not obvious that moving clocks don't run slow or that lengths don't shorten when the speeds involved approach the speed of light.

5. The astronaut would measure about 52 beats per minute. He is in his own rest frame, so he measures the proper time between heartbeats—the same as if he were taking his pulse back on Earth.

9. Celia will find that the flash from the left will reach her ship before the flash from the right. According to Abe, this occurs because her ship is moving toward the left flash. According to Celia, however, the two pulses travel equal distances to reach her, so the flash from the left must have been emitted earlier than the flash from the right.

Problems

1. **Strategy** Compute the times required for the optical signal and the train to reach the station and find the difference.

   Solution The time required for the optical signal to reach the station, as measured by an observer at rest relative to the station (the stationmaster) is
   $$\frac{d}{c} = \frac{1.0 \text{ km}}{3.00 \times 10^5 \text{ km/s}} = 3.3 \text{ μs.}$$
   Measured by the same observer, the time needed for the train to arrive is
   $$\frac{d}{v} = \frac{d}{0.60c} = \frac{1.0 \text{ km}}{0.60 \cdot 3.00 \times 10^5 \text{ km/s}} = 5.6 \text{ μs.}$$
   The difference in the arrival times is $5.56 \text{ μs} - 3.33 \text{ μs} = \boxed{2.2 \text{ μs}}$.

5. **Strategy** The proper time interval Δt_0 is the time interval measured by the Rolex. The time interval measured at mission control is $\Delta t = 12.0$ h. The time-dilation equation can be used to find Δt_0.

   Solution
   $$\Delta t_0 = \frac{\Delta t}{\gamma} = \Delta t \sqrt{1 - \frac{v^2}{c^2}} = (12.0 \text{ h}) \sqrt{1 - \frac{(2.0 \times 10^8 \text{ m/s})^2}{(3.00 \times 10^8 \text{ m/s})^2}} = \boxed{8.9 \text{ h}}$$

9. **(a) Strategy** The age of the passenger at the end of the trip is found by adding the elapsed time Δt_0, as measured by the ship's clock, to the passenger's age at the start of the trip. To find the elapsed time Δt_0, use the time dilation equation with $\Delta t = \dfrac{\text{distance relative to Earth}}{\text{speed relative to Earth}} = \dfrac{710 \text{ ly}}{0.9999c} = 710 \text{ yr}.$

Solution Find Δt.

$$\Delta t_0 = \frac{\Delta t}{\gamma} = \Delta t \sqrt{1 - v^2/c^2} = (710 \text{ yr})\sqrt{1 - 0.9999^2} = 10 \text{ yr}$$

The passenger's age at the end of the trip is 20 years old + 10 years = $\boxed{30 \text{ years old}}$.

(b) Strategy and Solution The spaceship takes slightly more than 710 years to reach its destination, as measured by an observer on Earth. The radio signal takes 710 years to reach Earth. The total elapsed time between the departure of the ship and the arrival of the signal back on Earth is 710 years + 710 years = 1420 years. The year will be 2000 + 1420 = $\boxed{3420}$.

11. **Strategy** The proper time interval is 8.0 h. The time interval Δt measured by the clock on the ground is given by the time dilation equation $\Delta t = \gamma \Delta t_0$. Use the binomial approximation.

Solution The time difference is
$$\Delta t - \Delta t_0 = \gamma \Delta t_0 - \Delta t_0 = (\gamma - 1)\Delta t_0 = \left[(1 - v^2/c^2)^{-1/2} - 1 \right] \Delta t_0.$$

Since $v \ll c$, by the binomial approximation, $\gamma = (1 - v^2/c^2)^{-1/2} \approx 1 + \dfrac{v^2}{2c^2}.$

The time difference is $\Delta t - \Delta t_0 \approx \left(1 + \dfrac{v^2}{2c^2} - 1 \right) \Delta t_0 = \dfrac{v^2}{2c^2} \Delta t_0 = \dfrac{(220 \text{ m/s})^2}{2(3.00 \times 10^8 \text{ m/s})^2} (8.0 \text{ h}) = \boxed{7.7 \text{ ns}}.$

13. **(a) Strategy** The observers on Earth see the contracted heights of the occupants of the spaceship, since the heights are along the direction of motion. Use Eq. (26-4).

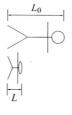

Solution Compute the approximate heights of the occupants as viewed by others on the ship.

$$L = \frac{L_0}{\gamma}, \text{ so } L_0 = \gamma L = \frac{L}{\sqrt{1 - v^2/c^2}} = \frac{0.50 \text{ m}}{\sqrt{1 - 0.97^2}} = \boxed{2 \text{ m}}.$$

(b) Strategy The widths of the occupants are perpendicular to their direction of motion, so the widths are not contracted.

Solution The others on the ship see the same widths as the observers on Earth, $\boxed{0.50 \text{ m}}$.

15. (a) **Strategy** The proper length is $L_0 = 91.5$ m. The length measured in the rest frame of the particle is given by Eq. (26-4).

    Solution
    $$L = \frac{L_0}{\gamma} = L_0\sqrt{1 - v^2/c^2} = (91.5 \text{ m})\sqrt{1 - 0.50^2} = \boxed{79 \text{ m}}$$

    (b) **Strategy** Since the speed of the particle is constant relative to Earth, the time is found using the equation $\Delta x = v\Delta t$.

    Solution
    $$\Delta t = \frac{\Delta x}{v} = \frac{91.5 \text{ m}}{0.50(3.00\times10^8 \text{ m/s})} = 6.1\times10^{-7} \text{ s} = \boxed{610 \text{ ns}}$$

    (c) **Strategy** In the rest frame of the particle, the distance traveled is 79 m and the speed is $0.50c$.

    Solution The time required is
    $$\Delta t = \frac{\Delta x}{v} = \frac{79 \text{ m}}{0.50(3.00\times10^8 \text{ m/s})} = 5.3\times10^{-7} \text{ s} = \boxed{530 \text{ ns}}.$$

17. **Strategy** Use Eqs. (26-2) and (26-4).

    Solution The length measured from the other ship is contracted by a factor
    $$\gamma = (1 - v^2/c^2)^{-1/2} = (1 - 0.90^2)^{-1/2} = 2.3.$$
    The contracted length is $L = \dfrac{L_0}{\gamma} = \dfrac{30.0 \text{ m}}{2.3} = \boxed{13 \text{ m}}.$

19. (a) **Strategy and Solution** Length contraction occurs only along the direction of motion. Since the rod is held perpendicular to the direction of motion, the Earth observer will measure the same length for the rod as the pilot of the ship did, $\boxed{1.0 \text{ m}}$.

    (b) **Strategy** The pilot still measures the proper length of the rod as $L_0 = 1.0$ m. To the Earth observer, the rod is now contracted. Use Eq. (26-4).

    Solution Find the length of the rod according to the Earth observer.
    $$L = \frac{L_0}{\gamma} = L_0\sqrt{1 - v^2/c^2} = (1.0 \text{ m})\sqrt{1 - 0.40^2} = \boxed{0.92 \text{ m}}$$

21. **Strategy** The observer on the ground measures the proper length between the towers as $L_0 = 3.0$ km. The distance is contracted for a passenger on the train. Use Eq. (26-4).

    Solution

    (a) Relative to the passenger on the train, the distance between the towers is

    $$L = \frac{L_0}{\gamma} = L_0\sqrt{1 - v^2/c^2} = (3.0 \text{ km})\sqrt{1 - 0.80^2} = 1.8 \text{ km}.$$

    The time interval measured by the passenger is $\Delta t = \dfrac{L}{v} = \dfrac{1.8 \text{ km}}{0.80(3.00\times10^5 \text{ km/s})} = 7.5\times10^{-6}$ s $= \boxed{7.5 \text{ μs}}$.

    (b) The observer on the ground measures the time interval

    $$\Delta t = \frac{L_0}{v} = \frac{3.0 \text{ km}}{0.80(3.00\times10^5 \text{ km/s})} = 1.3\times10^{-5} \text{ s} = \boxed{13 \text{ μs}}.$$

25. **Strategy** According to the postulates of special relativity, the speed of light must be the same in all inertial reference frames.

    Solution The Earth and the spaceship are both inertial reference frames (no acceleration), so Siu-Ling will measure a speed of light of $\boxed{3.00\times10^8 \text{ m/s}}$.

27. **Strategy** The velocity of particle B as seen by particle A is written v_{BA}. Let E refer to the Earth observer. Use Eq. (26-5).

    Solution Substituting $v_{BE} = -0.90c$ and $v_{EA} = -v_{AE} = -0.90c$ yields

    $$v_{BA} = \frac{v_{BE} + v_{EA}}{1 + v_{BE}v_{EA}/c^2} = \frac{-0.90c + (-0.90c)}{1 + (-0.90)(-0.90)} = \boxed{-0.994c}.$$

29. **Strategy** The velocity of rocket ship *Baker* relative to the Earth observer (v_{BE}) is given by Eq. (26-5). $v_{AE} = 0.400c$ is the velocity of ship *Able* relative to the Earth observer; $v_{BA} = -v_{AB} = -0.114c$ is the velocity of ship *Baker* relative to ship *Able*.

    Solution

    $$v_{BE} = \frac{v_{BA} + v_{AE}}{1 + v_{BA}v_{AE}/c^2} = \frac{-0.114c + 0.400c}{1 + (-0.114)(0.400)} = 0.300c$$

    The speed of *Baker* relative to the Earth observer is $\boxed{0.300c}$.

31. **Strategy** The velocity of rocket Alpha as measured from rocket Bravo v_{AB}. The speeds of the two rockets relative to the Earth are given as $0.90c$ for rocket *Alpha* and $0.60c$ for rocket *Bravo*. Choose the direction that rocket *Alpha* is traveling as the positive direction. Then $v_{AE} = 0.90c$ and $v_{EB} = -v_{BE} = -(-0.60c) = 0.60c$. Use Eq. (26-5).

    Solution

    $$v_{AB} = \frac{v_{AE} + v_{EB}}{1 + v_{AE}v_{EB}/c^2} = \frac{0.90c + 0.60c}{1 + (0.90)(0.60)} = 0.974c$$

    The speed of rocket *Alpha* as measured from rocket *Bravo* is $\boxed{0.974c}$.

33. Strategy The magnitude of the momentum is given by $p = \gamma mv$.

Solution Find the momentum of the proton.
$$p = \gamma mv = (1 - v^2/c^2)^{-1/2} m_{\mathrm{p}} v$$
$$= (1 - 0.90^2)^{-1/2}(1.673 \times 10^{-27} \text{ kg})(0.90)(3.00 \times 10^8 \text{ m/s}) = \boxed{1.0 \times 10^{-18} \text{ kg} \cdot \text{m/s}}$$

37. (a) Strategy To find the force, use the impulse-momentum equation: $F \Delta t = \Delta p = p_{\mathrm{f}} - p_{\mathrm{i}} = p_{\mathrm{f}} - 0 = p_{\mathrm{f}}$. The final velocity is a large fraction of c, so the relativistic form of the momentum, $p_{\mathrm{f}} = \gamma mv$, must be used.

Solution $F \Delta t = \gamma mv$, so
$$F = \frac{\gamma mv}{\Delta t} = \frac{(1 - v^2/c^2)^{-1/2} mv}{\Delta t} = \frac{(1 - 0.70^2)^{-1/2}(2200 \text{ kg})(0.70 \times 3.00 \times 10^8 \text{ m/s})}{3.6 \times 10^4 \text{ s}} = \boxed{1.8 \times 10^7 \text{ N}}.$$

(b) Strategy Use Newton's second law.

Solution Find the initial acceleration.
$$a = \frac{F}{m} = \frac{1.8 \times 10^7 \text{ N}}{2200 \text{ kg}} = \boxed{8200 \text{ m/s}^2}$$
Since $g \approx 10 \text{ m/s}^2$, $a \approx 820g$. $\boxed{\text{This is much larger than any human could survive}}$.

39. Strategy Since no energy is lost to the environment, the change in the kinetic energy of the two lumps can be equated to a change in the rest energy of the system.

Solution $\Delta E_0 = (\Delta m)c^2$, so
$$\Delta m = \frac{\Delta E_0}{c^2} = \frac{\Delta K_1 + \Delta K_2}{c^2} = \frac{\frac{1}{2} m_1 v_1^2 + \frac{1}{2} m_2 v_2^2}{c^2}.$$
The nonrelativistic form of the kinetic energy is used since the speeds are very small compared to c. Using $m_1 = m_2 = 1.00 \text{ kg}$ and $v_1 = v_2 = 30.0 \text{ m/s}$ gives
$$\Delta m = \frac{\frac{1}{2}(1.00 \text{ kg})(30.0 \text{ m/s})^2 + \frac{1}{2}(1.00 \text{ kg})(30.0 \text{ m/s})^2}{(3.00 \times 10^8 \text{ m/s})^2} = 1.00 \times 10^{-14} \text{ kg}.$$
The mass of the system $\boxed{\text{increased by } 1.00 \times 10^{-14} \text{ kg}}$.

41. Strategy The mass of the Sun is 1.987×10^{30} kg. 80.0% of the limiting mass of the white dwarf is converted to energy. Use Eq. (26-7).

Solution Compute the energy released by the explosion of the white dwarf.
$$E_0 = mc^2 = 0.800(1.4)(1.987 \times 10^{30} \text{ kg})(3.00 \times 10^8 \text{ m/s})^2 = \boxed{2.0 \times 10^{47} \text{ J}}$$

43. Strategy Total energy must be conserved.

Solution Find the energy released in the decay.
$$\text{rest energy before decay} = m_{\mathrm{Rn}} c^2 = \text{rest energy after decay} + \text{energy released} = m_{\mathrm{Po}} c^2 + m_\alpha c^2 + E, \text{ so}$$
$$E = (m_{\mathrm{Rn}} - m_{\mathrm{Po}} - m_\alpha)c^2 = (221.970\,39\,\mathrm{u} - 217.962\,89\,\mathrm{u} - 4.001\,51\,\mathrm{u})(931.494 \text{ MeV/u}) = \boxed{5.58 \text{ MeV}}.$$

45. Strategy Use Eq. (26-9).

Solution Find the electron's speed.
$$E = \gamma mc^2 = (1 - v^2/c^2)^{-1/2}mc^2$$
$$\sqrt{1 - \frac{v^2}{c^2}} = \frac{mc^2}{E}$$
$$\frac{v^2}{c^2} = 1 - \frac{m^2c^4}{E^2}$$
$$v = c\sqrt{1 - \frac{m^2c^4}{E^2}} = c\sqrt{1 - \frac{(9.109\times10^{-31}\ \text{kg})^2(3.00\times10^8\ \text{m/s})^4}{(1.02\times10^{-13}\ \text{J})^2}} = \boxed{0.595c}$$

47. Strategy The object is moving at $1.80/3.00 = 0.600$ times c. Use Eq. (26-8).

Solution Find the kinetic energy of the object.
$$K = (\gamma - 1)mc^2 = [(1 - v^2/c^2)^{-1/2} - 1]mc^2 = [(1 - 0.600^2)^{-1/2} - 1](0.12\ \text{kg})(3.00\times10^8\ \text{m/s})^2 = \boxed{2.7\times10^{15}\ \text{J}}$$

49. Strategy Solve Eq. (26-10) for p using $E = 5.0mc^2$ and $E_0 = mc^2$.

Solution Find the magnitude of the electron's momentum as observed in the laboratory.
$$E^2 = (5.0mc^2)^2 = 25m^2c^4 = E_0^2 + (pc)^2 = m^2c^4 + p^2c^2, \text{ so } p = \sqrt{24}mc = \boxed{4.9mc}.$$

53. Strategy Convert 1 MeV to joules. Then divide the result by c.

Solution
$$1\ \text{MeV} = (1\times10^6\ \text{eV})\frac{1.602\times10^{-19}\ \text{J}}{1\ \text{eV}} = 1.602\times10^{-13}\ \text{J}, \text{ so } \frac{1\ \text{MeV}}{c} = \frac{1.602\times10^{-13}\ \text{J}}{2.998\times10^8\ \text{m/s}} = 5.344\times10^{-22}\ \text{kg·m/s}.$$
The conversion is $\boxed{1\ \text{MeV}/c = 5.344\times10^{-22}\ \text{kg·m/s}}$.

57. Strategy Use Eq. (26-10) and the definition of total energy.

Solution Show that $(pc)^2 = K^2 + 2KE_0$.
The total energy is $E = K + E_0$.
Squaring both sides gives $E^2 = (K + E_0)^2 = K^2 + 2KE_0 + E_0^2$.
Combining this with the energy-momentum relation $E^2 = E_0^2 + (pc)^2$ yields
$E_0^2 + (pc)^2 = K^2 + 2KE_0 + E_0^2$, or $(pc)^2 = K^2 + 2KE_0$, which is Eq. (26-11).

61. **Strategy** Use the length contraction equation to find γ. Then, use the time dilation equation to find how long the astronaut would say that she exercised.

    Solution Find γ.

    $L = \dfrac{L_0}{\gamma}$, so $\gamma = \dfrac{L_0}{L}$.

    Find Δt_0.

    $\Delta t = \gamma \Delta t_0 = \dfrac{L_0}{L} \Delta t_0$, so $\Delta t_0 = \dfrac{L}{L_0} \Delta t = \dfrac{30.5 \text{ m}}{35.2 \text{ m}} (22.2 \text{ min}) = \boxed{19.2 \text{ min}}$.

65. **Strategy** Solve the time dilation equation for v. $\Delta t_0 = 3.0$ days and $\Delta t = 4.0$ days.

    Solution Find the speed of the starship relative to the space station.

    $$\Delta t = \gamma \Delta t_0$$
    $$4.0 \text{ d} = (1 - v^2/c^2)^{-1/2}(3.0 \text{ d})$$
    $$\sqrt{1 - v^2/c^2} = 0.75$$
    $$\dfrac{v^2}{c^2} = 1 - 0.75^2$$
    $$v = c\sqrt{1 - 0.75^2} = \boxed{0.66c}$$

69. **(a) Strategy** Let L_{sy} and L_{sE} be the lengths of the ship in your and the Earth's reference frames, respectively. Find the proper length of the ship. Then, find its length as viewed by observers on Earth. From Problem 71, the speed of the ship relative to the Earth is $0.966c$. Use Eq. (26-4).

    Solution Find the proper length L_0.

    $L_{sy} = L_0\sqrt{1 - v_{sy}^2/c^2}$, so $L_0 = \dfrac{L_{sy}}{\sqrt{1 - v_{sy}^2/c^2}} = \dfrac{24 \text{ m}}{\sqrt{1 - 0.50^2}} = 28$ m.

    Find the length of the ship as measured by observers on Earth.

    $L_{sE} = L_0\sqrt{1 - v_{sE}^2/c^2} = (28 \text{ m})\sqrt{1 - 0.966^2} = \boxed{7.2 \text{ m}}$

    (b) Strategy Your speed relative to the Earth is $0.90c$. Use Eq. (26-4).

    Solution Find the length of the rod as measured by observers on Earth.

    $L = L_0\sqrt{1 - v^2/c^2} = (24 \text{ m})\sqrt{1 - 0.90^2} = \boxed{10 \text{ m}}$

    (c) Strategy Your speed relative to the other ship is $0.50c$. Use Eq. (26-4).

    Solution Find the length of the rod as measured by observers on the other ship.

    $L = L_0\sqrt{1 - v^2/c^2} = (24 \text{ m})\sqrt{1 - 0.50^2} = \boxed{21 \text{ m}}$

73. Strategy Use the definition of total energy and Eq. (26-10).

Solution If K is much larger than E_0, then E is also much larger than E_0, since
$E = K + E_0 > K \gg E_0$. Now, $E^2 = E_0^2 + (pc)^2$, so $E^2 - E_0^2 = (pc)^2$. Since $E \gg E_0$, $E^2 \gg E_0^2$, so
$E^2 - E_0^2 \approx E^2$. Therefore, $E^2 \approx (pc)^2$, or $E \approx pc$.

77. Strategy Use Eq. (26-9).

Solution First find the rest energy of the particle.
$$E_0 = \frac{E}{\gamma} = (0.638 \text{ MeV})\sqrt{1 - 0.600^2} = 0.510 \text{ MeV}$$
The particle's total energy at the new speed is
$$E = \gamma E_0 = (1 - 0.980^2)^{-1/2}(0.510 \text{ MeV}) = \boxed{2.6 \text{ MeV}}.$$

81. Strategy The frequency will be reduced because the rocket ship is moving away from Earth. Use the result derived in Problem 80.

Solution Compute the frequency of the signal received by the astronauts in the rocket ship.
$$f_{\text{r}} = f_{\text{s}}\sqrt{\frac{1 - v/c}{1 + v/c}} = (55 \text{ kHz})\sqrt{\frac{1 - \frac{1.2 \times 10^8 \text{ m/s}}{3.00 \times 10^8 \text{ m/s}}}{1 + \frac{1.2 \times 10^8 \text{ m/s}}{3.00 \times 10^8 \text{ m/s}}}} = \boxed{36 \text{ kHz}}$$

85. Strategy Use conservation of momentum and total energy, as well as Eq. (26-11).

Solution The relativistic energies are given by $(p_{\text{n}}c)^2 = K_{\text{n}}^2 + 2K_{\text{n}}E_{0\text{n}}$ and $(p_{\pi}c)^2 = K_{\pi}^2 + 2K_{\pi}E_{0\pi}$.
By conservation of momentum, $p_{\text{n}} = p_{\pi}$, so $K_{\text{n}}^2 + 2K_{\text{n}}E_{0\text{n}} = K_{\pi}^2 + 2K_{\pi}E_{0\pi}$ (1).
The total energy of the system is conserved, so
$$m_{\Lambda}c^2 = m_{\text{n}}c^2 + m_{\pi}c^2 + K_{\text{n}} + K_{\pi}$$
$$(m_{\Lambda} - m_{\text{n}} - m_{\pi})c^2 = K_{\text{n}} + K_{\pi} = K$$
where K is the total kinetic energy. Substitute $K_{\text{n}} = K - K_{\pi}$ into (1) and simplify.
$$K^2 - 2KK_{\pi} + K_{\pi}^2 + 2KE_{0\text{n}} - 2K_{\pi}E_{0\text{n}} = K_{\pi}^2 + 2K_{\pi}E_{0\pi}$$
$$2KK_{\pi} + 2K_{\pi}E_{0\text{n}} + 2K_{\pi}E_{0\pi} = 2KE_{0\text{n}} + K^2$$
$$2K_{\pi}(K + E_{0\text{n}} + E_{0\pi}) = K(K + 2E_{0\text{n}})$$
Solve for K_{π}.
$$K_{\pi} = \frac{K(K + 2E_{0\text{n}})}{2(K + E_{0\text{n}} + E_{0\pi})} = \frac{(m_{\Lambda} - m_{\text{n}} - m_{\pi})c^2[(m_{\Lambda} - m_{\text{n}} - m_{\pi})c^2 + 2m_{\text{n}}c^2]}{2[(m_{\Lambda} - m_{\text{n}} - m_{\pi})c^2 + m_{\text{n}}c^2 + m_{\pi}c^2]}$$
$$= \frac{(1115.7 \text{ MeV} - 939.6 \text{ MeV} - 135.0 \text{ MeV})[(1115.7 \text{ MeV} - 939.6 \text{ MeV} - 135.0 \text{ MeV}) + 2(939.6 \text{ MeV})]}{2[(1115.7 \text{ MeV} - 939.6 \text{ MeV} - 135.0 \text{ MeV}) + 939.6 \text{ MeV} + 135.0 \text{ MeV}]}$$
$$= \boxed{35.4 \text{ MeV}}$$
Calculate K_{n}.
$$K_{\text{n}} = (1115.7 \text{ MeV} - 939.6 \text{ MeV} - 135.0 \text{ MeV}) - 35.4 \text{ MeV} = \boxed{5.7 \text{ MeV}}$$

Chapter 27

EARLY QUANTUM PHYSICS AND THE PHOTON

Conceptual Questions

1. The photoelectric effect describes the mechanism by which electrons are ejected from materials when certain frequencies of electromagnetic radiation are incident upon them. The first puzzling feature of this effect was that brighter light increases the current of electrons but does not affect their kinetic energy. This is understood with the photon model of light because increasing the intensity of the radiation simply increases the number of photons colliding with the material but does not increase their energy. The second puzzling feature was that the maximum kinetic energy of the electrons depends upon the wavelength of the photons. This too is explained by the photon theory because a shorter wavelength photon has a greater energy, and therefore, will provide a larger "kick" to the photoelectron. The third puzzling feature was that for a given metal there is a threshold frequency below which no electrons will be emitted. This is an understandable phenomenon because some minimum amount of energy is required to set the electrons free from the material to which they are bound, and if the photon energy is too low, an electron will not be ejected after absorption of a photon. The final puzzling feature of the photoelectric effect was that electrons are emitted virtually instantaneously after turning on a photon source. This too makes sense as a result of the photon theory because each individual photon carries enough energy to cause a photoelectron to be emitted—only a single photon is required to start a photocurrent.

5. According to the Bohr model, the difference in energy between two levels of the hydrogen atom is proportional to the value of the electron charge. If the electron charge varied from particle to particle, the photons emitted from a hydrogen atom would not all have the same energy. Furthermore, when the emission spectrum of hydrogen was observed, photons from the same transition would have slightly different wavelengths and would therefore not form sharp lines.

9. The Bohr model contains four assumptions: 1) The electron can exist without radiating only in certain circular orbits; 2) The laws of Newtonian mechanics apply to the motion of the electron in any of these stationary states; 3) The electron can make a transition between stationary states through the emission or absorption of a single photon; 4) The stationary states are those circular orbits in which the electron's angular momentum is quantized in integral multiples of $h/(2\pi)$.

13. The process of pair production becomes especially important for photons with energies in excess of 1.02 MeV.

17. In a hydrogen gas discharge tube, the hydrogen atoms are excited to higher energy states in a random process of collisions with energetic electrons. Once in an excited state, the atoms decay back to the ground state, possibly passing through several intermediate states, and emitting one or more photons. Because of the random nature of the excitations in this process, the atoms in the tube are excited, on average, to every possible state, and pass through every possible intermediate state during the decays. The emission spectrum of hydrogen therefore contains a line for every possible atomic transition. In an absorption spectrum, the atoms absorb incident photons in a similarly random process that puts them on average in every excited state. Once excited, however, the atom quickly decays back to the ground state by emitting one or more photons. The excited states do not live long enough to absorb a significant amount of the incident radiation and undergo transitions from intermediate states to higher energy states. For this reason, an absorption spectrum shows lines corresponding to transitions from the ground state only, while an emission spectrum contains lines from transitions between any two states.

21. Since UV light has a higher frequency than blue, and both metals produce photoelectrons for blue light, both metals will also produce photoelectrons for UV light. Metal 1 produces photoelectrons for red light, so it *might* produce them for infrared as well. Metal 2 will not produce them for infrared light, since it doesn't produce them for red light. Photoelectrons are produced whenever the photon energy is sufficient to overcome the work function of the metal. Therefore, the metal with the higher threshold frequency has the larger work function, which is metal 2.

Problems

1. (a) **Strategy and Solution** The energy of a photon is inversely proportional to its wavelength, so the photons with the shorter wavelength have the greater energy. Therefore, a single ⎡ultraviolet⎤ photon has the greater energy.

    (b) **Strategy** The energy of a photon of EM radiation with frequency f is $E = hf$. The frequency and wavelength are related by $\lambda f = c$.

    Solution Compute the energy of a single infrared photon.
    $$E_{\text{infrared}} = hf = \frac{hc}{\lambda} = \frac{(6.626\times10^{-34}\text{ J}\cdot\text{s})(3.00\times10^{8}\text{ m/s})}{2.0\times10^{-6}\text{ m}} = \boxed{9.9\times10^{-20}\text{ J}}$$
    Compute the energy of a single ultraviolet photon.
    $$E_{\text{ultraviolet}} = hf = \frac{hc}{\lambda} = \frac{(6.626\times10^{-34}\text{ J}\cdot\text{s})(3.00\times10^{8}\text{ m/s})}{7.0\times10^{-8}\text{ m}} = \boxed{2.8\times10^{-18}\text{ J}}$$

    (c) **Strategy** Use the definition of power. Let N be the number of photons. The total energy is NE, where E is the energy of a single photon.

    Solution Find the number of infrared photons emitted per second.
    $$P = \frac{NE}{\Delta t}, \text{ so } \frac{N_{\text{infrared}}}{\Delta t} = \frac{P}{E} = \frac{200\text{ W}}{9.9\times10^{-20}\text{ J/photon}} = \boxed{2.0\times10^{21}\text{ photons/s}}.$$
    Find the number of ultraviolet photons emitted per second.
    $$\frac{N_{\text{ultraviolet}}}{\Delta t} = \frac{P}{E} = \frac{200\text{ W}}{2.84\times10^{-18}\text{ J/photon}} = \boxed{7.0\times10^{19}\text{ photons/s}}$$

3. **Strategy** The energy of a photon of EM radiation with frequency f is $E = hf$. The frequency and wavelength are related by $\lambda f = c$.

    Solution

    (a) Calculate the wavelength of a photon with energy 3.1 eV.
    $$E = hf = \frac{hc}{\lambda}, \text{ so } \lambda = \frac{hc}{E} = \frac{1240\text{ eV}\cdot\text{nm}}{3.1\text{ eV}} = \boxed{400\text{ nm}}.$$

    (b) Calculate the frequency of a photon with energy 3.1 eV.
    $$f = \frac{c}{\lambda} = \frac{3.00\times10^{8}\text{ m/s}}{400\times10^{-9}\text{ m}} = \boxed{7.5\times10^{14}\text{ Hz}}$$

5. (a) **Strategy** Use Einstein's photoelectric equation.

    Solution Calculate the maximum kinetic energy.
    $$K_{\text{max}} = hf - \phi = \frac{hc}{\lambda} - \phi = \frac{1240\text{ eV}\cdot\text{nm}}{413\text{ nm}} - 2.16\text{ eV} = \boxed{0.84\text{ eV}}$$

    (b) **Strategy** The threshold wavelength is related to the threshold frequency by $\lambda_0 f_0 = c$. Use Eq. (27-8).

    Solution Find the threshold wavelength.
    $$hf_0 = \frac{hc}{\lambda_0} = \phi, \text{ so } \lambda_0 = \frac{hc}{\phi} = \frac{1240\text{ eV}\cdot\text{nm}}{2.16\text{ eV}} = \boxed{574\text{ nm}}.$$

7. **Strategy** The work function for the metal is $\phi = 2.60$ eV. Use Eq. (27-8) to find the maximum wavelength of the photons that will eject electrons from the metal. The frequency and wavelength are related by $\lambda f = c$.

Solution Find the longest wavelength.

$$f_0 = \frac{\phi}{h} = \frac{c}{\lambda}, \text{ so } \lambda = \frac{hc}{\phi} = \frac{1240 \text{ eV} \cdot \text{nm}}{2.60 \text{ eV}} = \boxed{477 \text{ nm}}.$$

9. **Strategy** The stopping potential V_s is related to the maximum kinetic energy of the electrons by $K_{max} = eV_s$. Use Einstein's photoelectric equation and $\lambda f = c$.

Solution Find the work function.

$$K_{max} = eV_s = hf - \phi = \frac{hc}{\lambda} - \phi, \text{ so } \phi = \frac{hc}{\lambda} - eV_s = \frac{1240 \text{ eV} \cdot \text{nm}}{220 \text{ nm}} - 1.1 \text{ eV} = \boxed{4.5 \text{ eV}}.$$

13. **Strategy** The frequency intercept gives the threshold frequency, so $f_0 = 43.9 \times 10^{13}$ Hz. The stopping potential is related to the maximum kinetic energy of the ejected electrons by $K_{max} = eV_s$. Einstein's photoelectric equation is

$K_{max} = hf - \phi$, so $V_s = \frac{hf}{e} - \frac{\phi}{e}$. The work function ϕ is given by the stopping potential intercept $V_0 = -\phi / e$,

where $V_s = mf + V_0$. $V_s = 0$ when $f = f_0$.

Solution

(a) The slope of the graph is $m = h/e$. Compute the numerical value.

$$m = \frac{1.50 \text{ V} - 0 \text{ V}}{80.0 \times 10^{13} \text{ Hz} - 43.9 \times 10^{13} \text{ Hz}} = 4.155 \times 10^{-15} \text{ V} \cdot \text{s}$$

Planck's constant is $h = em = (1.602 \times 10^{-19} \text{ C})(4.155 \times 10^{-15} \text{ V} \cdot \text{s}) = \boxed{6.66 \times 10^{-34} \text{ J} \cdot \text{s}}$.

(b) Find the work function.

$$0 = mf_0 - \frac{\phi}{e}, \text{ so } \phi = emf_0 = e(4.155 \times 10^{-15} \text{ V} \cdot \text{s})(43.9 \times 10^{13} \text{ Hz}) = \boxed{1.82 \text{ eV}}.$$

15. **Strategy** The maximum kinetic energy K of an electron is equal to hf_{max}. $\lambda_{min} = c/f_{max}$ and $K = eV$, where V is the applied voltage.

Solution Find the applied voltage.

$$hf_{max} = \frac{hc}{\lambda_{min}} = K = eV, \text{ so } V = \frac{hc}{\lambda_{min}e} = \frac{1240 \text{ eV} \cdot \text{nm}}{0.46 \text{ nm} \cdot e} = \boxed{2.7 \text{ kV}}.$$

17. **Strategy** The cutoff (maximum) frequency occurs when the photon's energy and the electron's kinetic energy are equal. The kinetic energy of the electron is equal to the electric potential energy of the electron, eV. Use Eq. (27-9).

Solution Find the cutoff frequency.

$$hf_{max} = K = eV, \text{ so } f_{max} = \frac{eV}{h} = \frac{46 \times 10^3 \text{ eV}}{4.136 \times 10^{-15} \text{ eV} \cdot \text{s}} = \boxed{1.1 \times 10^{19} \text{ Hz}}.$$

21. Strategy Use the Compton shift, Eq. (27-14).

Solution

(a) Find the wavelength of the scattered x-rays.

$$\lambda' - \lambda = \frac{h}{m_e c}(1 - \cos\theta), \text{ so } \lambda' = \lambda + \frac{h}{m_e c}(1 - \cos\theta) = 10.0 \text{ pm} + (2.426 \text{ pm})(1 - \cos 45.0°) = \boxed{10.7 \text{ pm}}.$$

(b) $\lambda' = \lambda + \dfrac{h}{m_e c}(1 - \cos\theta) = 10.0 \text{ pm} + (2.426 \text{ pm})(1 - \cos 90.0°) = \boxed{12.4 \text{ pm}}$

25. Strategy Use conservation of momentum to find the velocity of the electron. Assume that the velocity of the electron is sufficiently less than the speed of light to use the nonrelativistic relationship between momentum and velocity. From Problem 24, $\theta = 54.0°$.

Solution Find the speed of the electron.

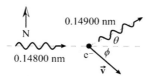

$$p_{e,\text{south}} = m_e v_s = p'_{\text{north}} = \frac{h}{\lambda'}\sin\theta, \text{ so}$$

$$v_s = \frac{h}{m_e \lambda'}\sin\theta = \frac{6.626\times10^{-34} \text{ J}\cdot\text{s}}{(9.109\times10^{-31} \text{ kg})(0.14900\times10^{-9} \text{ m})}\sin 54.0° = 3.95\times10^6 \text{ m/s}.$$

$$p_{e,\text{east}} = m_e v_e = p - p'_{\text{east}} = \frac{h}{\lambda} - \frac{h}{\lambda'}\cos\theta, \text{ so}$$

$$v_e = \frac{h}{m_e}\left(\frac{1}{\lambda} - \frac{\cos\theta}{\lambda'}\right) = \frac{6.626\times10^{-34} \text{ J}\cdot\text{s}}{9.109\times10^{-31} \text{ kg}}\left(\frac{1}{0.14800\times10^{-9} \text{ m}} - \frac{\cos 54.0°}{0.14900\times10^{-9} \text{ m}}\right) = 2.05\times10^6 \text{ m/s}.$$

$$v = \sqrt{v_e^2 + v_s^2} = \sqrt{(2.05\times10^6 \text{ m/s})^2 + (3.95\times10^6 \text{ m/s})^2} = 4.45\times10^6 \text{ m/s}$$

Find the angle.

$$\theta = \tan^{-1}\frac{-v_s}{v_e} = \tan^{-1}\frac{-3.95}{2.05} = -62.6°$$

Thus, the velocity of the electron is $\boxed{4.45\times10^6 \text{ m/s at } 62.6° \text{ south of east}}$.

Since $\gamma \approx 1.000110$, our initial assumption was reasonable.

29. Strategy The change in kinetic energy of the electron is the difference in the energies of the incident photon and the scattered photon.

Solution Find the change in kinetic energy.

$$\Delta K = E - E' = \frac{hc}{\lambda} - \frac{hc}{\lambda'} = \frac{1240 \text{ eV}\cdot\text{nm}}{0.0100 \text{ nm}} - \frac{1240 \text{ eV}\cdot\text{nm}}{0.0124 \text{ nm}} = \boxed{2.4\times10^4 \text{ eV}}$$

31. Strategy The orbital radius in the nth state is given by $r_n = n^2 a_0$.

Solution Compute the orbital radius.

$$r_3 = 3^2 \cdot 52.9 \text{ pm} = 476 \text{ pm} = \boxed{0.476 \text{ nm}}$$

33. Strategy The orbital radius in the nth state is given by $r_n = n^2 a_0$.

Solution

(a) Find the difference in the radii of the $n = 1$ and $n = 2$ states for hydrogen.

$$r_n = n^2 a_0, \text{ so } r_2 - r_1 = (2^2 - 1^2)a_0 = 3a_0 = 3(52.9 \times 10^{-12} \text{ m}) = \boxed{1.59 \times 10^{-10} \text{ m}}.$$

(b) Find the difference in the radii of the $n = 100$ and $n = 101$ states for hydrogen.

$$r_n = n^2 a_0, \text{ so } r_2 - r_1 = (101^2 - 100^2)a_0 = 201a_0 = 201(52.9 \times 10^{-12} \text{ m}) = \boxed{1.06 \times 10^{-8} \text{ m}}.$$

The result for (b) is much larger than that for (a), so $\boxed{\text{the orbital separations are much larger for larger } n \text{ values}}$.

37. Strategy The energy supplied by the photon raises the atom from the ground state to a higher energy level. The energy for a hydrogen atom in the nth stationary state is given by $E_n = (-13.6 \text{ eV})/n^2$.

Solution Find the energy level to which the atom is excited.

$$E_n - E_1 = \frac{E_1}{n^2} - E_1 = 12.1 \text{ eV}, \text{ so } n = \sqrt{\frac{E_1}{12.1 \text{ eV} + E_1}} = \sqrt{\frac{-13.6 \text{ eV}}{12.1 \text{ eV} - 13.6 \text{ eV}}} = 3.$$

The atom is excited to the $\boxed{n = 3}$ energy level.

39. Strategy The minimum energy for an ionized atom is $E_{\text{ionized}} = 0$. Use Eq. (27-24).

Solution The energy needed to ionize a hydrogen atom initially in the $n = 2$ state is

$$E = E_{\text{ionized}} - E_2 = E_{\text{ionized}} - \frac{E_1}{2^2} = 0 - \left(\frac{-13.6 \text{ eV}}{4}\right) = \boxed{3.40 \text{ eV}}.$$

41. Strategy The energy of a photon is related to its wavelength by $E = hc/\lambda$. Use Eq. (27-24) to find the energy of the transition.

Solution The amount of energy released when a hydrogen atom makes a transition from the $n = 6$ to the $n = 3$ state is $E = E_6 - E_3 = \dfrac{-13.6 \text{ eV}}{6^2} - \dfrac{-13.6 \text{ eV}}{3^2} = 1.13 \text{ eV}$.

The wavelength of the radiation emitted is $\lambda = \dfrac{hc}{E} = \dfrac{1240 \text{ eV} \cdot \text{nm}}{1.133 \text{ eV}} = \boxed{1.09 \text{ μm}}$.

45. Strategy The energy of a photon of EM radiation with wavelength λ is $E = hc/\lambda$.

Solution The energy of the UV photon is $E_1 = hc/\lambda_1$. The solid dissipates 0.500 eV, so the energy available for the emitted photon is $E_2 = E_1 - 0.500 \text{ eV} = \dfrac{hc}{\lambda_1} - 0.500 \text{ eV}$.

Calculate the wavelength of the emitted photon.

$$\lambda = \frac{hc}{E_2} = \frac{hc}{\frac{hc}{\lambda_1} - 0.500 \text{ eV}} = \frac{1}{\frac{1}{\lambda_1} - \frac{0.500 \text{ eV}}{hc}} = \frac{1}{\frac{1}{320 \text{ nm}} - \frac{0.500 \text{ eV}}{1240 \text{ eV} \cdot \text{nm}}} = \boxed{370 \text{ nm}}$$

49. **Strategy** Use Eq. (27-20) and the values of the fundamental constants.

Solution Compute the numerical value of the Bohr radius.
$$a_0 = \frac{\hbar^2}{m_e k e^2} = \frac{h^2}{4\pi^2 m_e k e^2} = \frac{(6.626\times10^{-34}\ \text{J}\cdot\text{s})^2}{4\pi^2(9.109\times10^{-31}\ \text{kg})(8.988\times10^9\ \text{N}\cdot\text{m}^2/\text{C}^2)(1.602\times10^{-19}\ \text{C})^2} = 5.29\times10^{-11}\ \text{m}$$

53. **Strategy** For a photon, the maximum wavelength corresponds to the minimum energy. The minimum energy needed is the rest energy of an electron-positron pair. Use $E_0 = mc^2$ and $E = hc/\lambda$.

Solution Compute the required energy.
$$E = 2m_e c^2 = 2(9.109\times10^{-31}\ \text{kg})(3.00\times10^8\ \text{m/s})^2 = 1.64\times10^{-13}\ \text{J}$$
Compute the wavelength.
$$\lambda = \frac{hc}{E} = \frac{(6.626\times10^{-34}\ \text{J}\cdot\text{s})(3.00\times10^8\ \text{m/s})}{1.64\times10^{-13}\ \text{J}} = \boxed{1.21\ \text{pm}}$$

57. **Strategy** The rest energy of the muon and antimuon are converted into the energy of the two photons of equal energy. Since the photons have equal energy, their wavelengths are the same. The energy of a photon of EM radiation with wavelength λ is $E = hc/\lambda$. Use $E_0 = mc^2$.

Solution Find the wavelength.
$$2E_{\text{photon}} = E_{\text{muon}} + E_{\text{antimuon}} = 207 m_e c^2 + 207 m_e c^2 = 414 m_e c^2 = 2\frac{hc}{\lambda},\ \text{so}$$
$$\lambda = \frac{hc}{207 m_e c^2} = \frac{1240\times10^{-15}\ \text{MeV}\cdot\text{m}}{207(0.511\ \text{MeV})} = \boxed{1.17\times10^{-14}\ \text{m}}.$$

61. (a) **Strategy** The energy of a photon of EM radiation with frequency f is $E = hf$.

Solution Compute the energy of each photon.
$$E = (6.626\times10^{-34}\ \text{J}\cdot\text{s})(89.3\times10^6\ \text{Hz})\left(\frac{1\ \text{eV}}{1.602\times10^{-19}\ \text{J}}\right) = \boxed{3.69\times10^{-7}\ \text{eV}}$$

(b) **Strategy** The ratio of the total radiated power to the energy per photon is the rate of photons emitted.

Solution Compute the rate of photon emission.
$$\frac{P}{E} = \frac{50.0\times10^3\ \text{W}}{\frac{(6.626\times10^{-34}\ \text{J}\cdot\text{s})(89.3\times10^6\ \text{Hz})}{1\ \text{photon}}} = \boxed{8.45\times10^{29}\ \text{photons/s}}$$

65. (a) **Strategy** The energy of a photon of EM radiation with wavelength λ is $E = hc/\lambda$. The momentum of a photon is related to its energy by $E = pc$.

Solution Calculate the energy of a single photon.
$$E = \frac{hc}{\lambda} = \frac{1240\ \text{eV}\cdot\text{nm}}{670\ \text{nm}} = \boxed{1.9\ \text{eV}}$$
Calculate the momentum.
$$p = \frac{E}{c} = \frac{(1.85\ \text{eV})(1.6\times10^{-19}\ \text{J/eV})}{3.00\times10^8\ \text{m/s}} = \boxed{9.9\times10^{-28}\ \text{kg}\cdot\text{m/s}}$$

(b) **Strategy** The power of the laser divided by the energy per photon gives the rate of photons emitted.

Solution Compute the rate of photon emission.

$$\frac{P}{E} = \frac{1 \times 10^{-3} \text{ J/s}}{(1.85 \text{ eV})(1.6 \times 10^{-19} \text{ J/eV})} = \boxed{3 \times 10^{15} \text{ photons/s}}$$

(c) **Strategy** Use the impulse-momentum theorem, $F_{av} \Delta t = \Delta p$.

Solution Find the average force.

$$F_{av} = \frac{\Delta p}{\Delta t} = \frac{E/c}{E/P} = \frac{P}{c} = \frac{1 \times 10^{-3} \text{ W}}{3 \times 10^8 \text{ m/s}} = \boxed{3 \times 10^{-12} \text{ N}}$$

69. **Strategy** The minimum wavelength corresponds to the maximum frequency, which occurs when the energy of the photon is equal to the electron's kinetic energy.

Solution Find the potential difference.

$$hf_{max} = \frac{hc}{\lambda_{min}} = K_e = eV, \text{ so } V = \frac{hc}{e\lambda_{min}} = \frac{1240 \text{ eV} \cdot \text{nm}}{e \times 45.0 \times 10^{-3} \text{ nm}} = \boxed{27.6 \text{ kV}}.$$

73. (a) **Strategy** The energy per second incident is the power incident, which is equal to the intensity times the area.

Solution Compute the energy per second falling on the atom.

$$P = IA = (0.01 \text{ W/m}^2)(0.1 \times 10^{-9} \text{ m})^2 = 1 \times 10^{-22} \text{ W} = \boxed{1 \times 10^{-22} \text{ J/s}}$$

(b) **Strategy** Since $P = \Delta E / \Delta t$, and $\Delta E = \phi$ in this case, $\Delta t = \phi / P$.

Solution Compute the time lag.

$$\Delta t = \frac{\phi}{P} = \frac{(2.0 \text{ eV})(1.6 \times 10^{-19} \text{ J/eV})}{1 \times 10^{-22} \text{ J/s}} = \boxed{3.2 \times 10^3 \text{ s}}$$

(c) **Strategy and Solution** Whether or not an electron is ejected has nothing to do with the intensity of the light; ejection of the electron depends upon the energy of individual photons. $\boxed{\text{An electron is ejected immediately when a single photon of sufficient energy strikes it.}}$

77. **Strategy** The minimum wavelength corresponds to the maximum frequency, which occurs when the photon's energy equals the kinetic energy of the electron. Use Eq. (27-9).

Solution Find the minimum wavelength.

$$hf_{max} = \frac{hc}{\lambda_{min}} = K_e, \text{ so } \lambda_{min} = \frac{hc}{K_e} = \frac{1240 \text{ eV} \cdot \text{nm}}{2.0 \times 10^3 \text{ eV}} = \boxed{0.62 \text{ nm}}.$$

81. Strategy Use Eqs. (27-21), (27-24), (27-25), and (27-26).

Solution Equate the energy levels in He^+ and H atoms.

$$E_{nH} = E_{mHe^+}$$

$$\frac{-13.6 \text{ eV}}{n^2} = \frac{-13.6 \text{ eV} \times 2^2}{m^2}$$

$$m^2 = 2^2 n^2$$

$$m = 2n$$

The ratio of the orbital radii of H and He^+ is $\dfrac{r_{nH}}{r_{mHe^+}} = \dfrac{n^2 a_0}{\frac{m^2 a_0}{2}} = \dfrac{2n^2}{m^2}$.

Now substitute $m = 2n$.

$$\frac{r_{nH}}{r_{mHe^+}} = \frac{2n^2}{(2n)^2} = \frac{2n^2}{4n^2} = \boxed{\frac{1}{2}} = \frac{Z_H}{Z_{He^+}}$$

$\boxed{\text{For levels of equal energy, the ratio of orbital radii appears to equal the ratio of atomic numbers.}}$

85. (a) Strategy The *Balmer series* ($n_f = 2$) gives visible wavelengths. The wavelengths are given by

$$\lambda = \frac{1}{(1.097 \times 10^7 \text{ m}^{-1})\left(\frac{1}{2^2} - \frac{1}{n_i^2}\right)}.$$

Solution $n_i = 3$ gives 656.3 nm and $n_i = 4$ gives 486.2 nm, both of which are visible. So, the incident radiation excites the ground-state atoms into the $n = 4$ state. The energy difference between these states is equal to the energy of the incident radiation.

$$\Delta E = E_4 - E_1 = E_1\left(\frac{1}{4^2} - 1\right) = (-13.6 \text{ eV})\left(\frac{1}{16} - 1\right) = 12.75 \text{ eV}$$

Calculate the wavelength.

$$\lambda = \frac{hc}{E} = \frac{1240 \text{ eV} \cdot \text{nm}}{12.75 \text{ eV}} = \boxed{97.3 \text{ nm}}$$

(b) Strategy Since $\lambda \propto 1/E_{photon}$, the longest wavelength corresponds to the smallest energy for the incident radiation. So, the incident radiation must excite the ground-state atom to its $n = 3$ state for visible light to be emitted.

Solution Calculate the wavelength of the incident radiation.

$$\lambda = \frac{1}{R\left(1 - \frac{1}{3^2}\right)} = \frac{1}{(1.097 \times 10^7 \text{ m}^{-1})\left(1 - \frac{1}{9}\right)} = \boxed{102.6 \text{ nm}}$$

As found in part (a), the wavelength of the emitted radiation for the $n = 3$ to the $n = 2$ state is $\boxed{656.3 \text{ nm}}$.

(c) Strategy The incident photon must be energetic enough to excite an electron from any state to $n_f = \infty$. The case where the electron is in the ground state, $n_i = 1$, represents the minimum energy required, 13.6 eV.

Solution Find the range of wavelengths.

$E \geq 13.6$ eV, so $\dfrac{hc}{\lambda} \geq 13.6$ eV; thus, $\lambda \leq \dfrac{1240 \text{ eV} \cdot \text{nm}}{13.6 \text{ eV}}$ or $\boxed{\lambda \leq 91.2 \text{ nm}}$.

Chapter 28

QUANTUM PHYSICS

Conceptual Questions

1. We know the wavelengths of the electrons and x-rays are the same because a diffraction pattern depends only on the wavelength of the incident particles or radiation and the interatomic spacing of the sample. They would not give the same pattern if their energies were the same because then they would have different wavelengths. For a photon and an electron with the same energy E, $\lambda_{\text{photon}} = hc / E$, while $\lambda_{\text{electron}} = hc / \sqrt{E^2 - (mc^2)^2}$.

5. If we are only concerned with the energy of the state, then we need only specify the principal quantum number, n. This would be the case, for example, if we were talking about the energies of photons emitted during atomic transitions. If we are interested in describing the state more completely, then we need to include more quantum numbers such as the spin magnetic number, m_s.

9. The light that causes optical pumping must excite the atoms in the laser to a higher energy state than the one from which they decay when they emit the laser light. Therefore, the energy of the photons of the optical pumping light must be greater than the energy of the photons of the emitted laser light. This means the wavelength of the laser beam must be greater than the wavelength of the optical pumping light.

13. Photons incident on the photoconducting material transfer their kinetic energy to atoms in the semiconductor. Electrons in these atoms jump into the conduction band, and therefore, increase the conductivity of the material. The band gap for a good photoconductor should be approximately the same size as the energy of the incident photons. If the drum gets hot, the contrast between light and dark areas of the image is degraded because some of the thermal energy can cause electrons to jump into the conduction band even if no photons were incident upon them.

17. Increasing the temperature of a semiconductor increases the number of electrons excited into the conduction band. The increased number of conduction electrons increases the semiconductor's ability to conduct electricity, and therefore, decreases its resistivity.

Problems

1. **Strategy** Find the de Broglie wavelength using Eq. (28-1). Compare the result to the diameter of the hoop.

   Solution
   $$\lambda = \frac{h}{p} = \frac{h}{mv} = \frac{6.626 \times 10^{-34} \text{ J} \cdot \text{s}}{(0.50 \text{ kg})(10 \text{ m/s})} = \boxed{1.3 \times 10^{-34} \text{ m}}$$
   The wavelength is much too small compared to the diameter of the hoop for any appreciable diffraction to occur—for a diameter of ~1 m, it's a factor of 10^{-34} smaller!

5. **Strategy** The electron is relativistic, so use $p = \gamma m v$ to find the magnitude of its momentum. Find the de Broglie wavelength using Eq. (28-1).

   Solution
   $$\lambda = \frac{h}{p} = \frac{h}{\gamma m v} = \frac{h\sqrt{1 - v^2/c^2}}{mv} = \frac{(6.626 \times 10^{-34} \text{ J} \cdot \text{s})\sqrt{1 - (3/5)^2}}{(9.109 \times 10^{-31} \text{ kg})(\frac{3}{5} \times 3.00 \times 10^8 \text{ m/s})} = \boxed{3.23 \text{ pm}}$$

9. **Strategy** The electron's kinetic energy is small compared to its rest energy, so the electron is nonrelativistic and we can use $p = mv$ and $K = \frac{1}{2}mv^2$ to find the momentum. Find the de Broglie wavelength using Eq. (28-1) and the wavelength of the photon using $E = hc/\lambda$.

   Solution First solve for p in terms of K.

   $K = \frac{1}{2}mv^2$, so $v = \sqrt{\dfrac{2K}{m}}$.

   The momentum is then $p = mv = m\sqrt{\dfrac{2K}{m}} = \sqrt{2Km}$.

   Find the de Broglie wavelength of the electron.

   $\lambda_e = \dfrac{h}{p} = \dfrac{h}{\sqrt{2Km_e}}$

   The wavelength of a photon is $\lambda_p = \dfrac{hc}{E_p}$. Compute the ratio of the wavelengths.

   $$\dfrac{\lambda_p}{\lambda_e} = \dfrac{hc/E_p}{h/\sqrt{2Km}} = \dfrac{c\sqrt{2Km}}{E_p} = \dfrac{(3.00\times10^8 \text{ m/s})\sqrt{2(0.100\times10^3 \text{ eV})(1.602\times10^{-19} \text{ J/eV})(9.109\times10^{-31} \text{ kg})}}{(0.100\times10^3 \text{ eV})(1.602\times10^{-19} \text{ J/eV})} = \boxed{101}$$

13. **(a) Strategy** The energy of a photon is given by $E = hc/\lambda$. The minimum energy corresponds to the maximum wavelength of a photon that is capable of resolving the details of the molecule.

    Solution Find the minimum photon energy.

    $E = \dfrac{hc}{\lambda} = \dfrac{1240 \text{ eV}\cdot\text{nm}}{0.1000 \text{ nm}} = \boxed{12.4 \text{ keV}}$

    (b) Strategy The minimum kinetic energy corresponds to the maximum wavelength of an electron that is capable of resolving the details of the molecule. Assume that the electrons are nonrelativistic. Use the de Broglie wavelength, $K = \frac{1}{2}mv^2$, and $p = mv$.

    Solution Find the kinetic energy of the electrons.

    $$K = \frac{1}{2}mv^2 = \dfrac{(mv)^2}{2m} = \dfrac{p^2}{2m} = \dfrac{(h/\lambda)^2}{2m} = \dfrac{h^2}{2m\lambda^2} = \dfrac{(6.626\times10^{-34} \text{ J}\cdot\text{s})^2}{2(9.109\times10^{-31} \text{ kg})(1.000\times10^{-10} \text{ m})^2} \times \dfrac{1}{1.602\times10^{-19} \text{ J/eV}}$$

    $$= \boxed{150 \text{ eV}}$$

    The kinetic energy is small compared to the rest energy of an electron, so the use of the nonrelativistic equations for momentum and kinetic energy was valid.

    (c) Strategy Use conservation of energy.

    Solution Find the required potential difference.

    $\Delta U = -e\Delta V = -\Delta K = -K$, so $\Delta V = \dfrac{K}{e} = \dfrac{150 \text{ eV}}{e} = \boxed{150 \text{ V}}$.

17. **Strategy** Use the position-momentum uncertainty principle, Eq. (28-2).

    Solution Find the uncertainty in the position of the basketball.

    $\Delta x\Delta p \geq \dfrac{1}{2}\hbar$, so $\Delta x \geq \dfrac{\hbar}{2\Delta p} = \dfrac{\hbar}{2p\cdot10^{-6}} = \dfrac{10^6 \hbar}{4\pi mv} = \dfrac{10^6(6.626\times10^{-34} \text{ J}\cdot\text{s})}{4\pi(0.50 \text{ kg})(10 \text{ m/s})} = \boxed{1\times10^{-29} \text{ m}}$.

21. **Strategy** Use the position-momentum uncertainty principle, Eq. (28-2), and $p = mv$.

    Solution

    (a) Estimate the minimum uncertainty in the position of the bullet.
    $$\Delta x \Delta p_x \geq \frac{1}{2}\hbar, \text{ so } \Delta x \geq \frac{\hbar}{2\Delta p_x} = \frac{\hbar}{2m\Delta v_x}. \text{ Thus,}$$
    $$(\Delta x)_{min} = \frac{\hbar}{2m\Delta v_x} = \frac{\hbar}{2m\left(\frac{\Delta v_x}{v_x}\right)v_x} = \frac{6.626 \times 10^{-34} \text{ J} \cdot \text{s}}{4\pi(10.000 \times 10^{-3} \text{ kg})(0.0004)(300.00 \text{ m/s})} = \boxed{4 \times 10^{-32} \text{ m}}.$$

    (b) $(\Delta x)_{min} = \dfrac{6.626 \times 10^{-34} \text{ J} \cdot \text{s}}{4\pi(9.109 \times 10^{-31} \text{ kg})(0.0004)(300.00 \text{ m/s})} = \boxed{0.5 \text{ mm}}$

    (c) The uncertainty in the bullet's position is far too small to be observable. The uncertainty in the electron's position is huge on the atomic scale. Thus, the uncertainty principle can be neglected in the macroscopic world, but not on the atomic scale.

23. **Strategy** The single-slit diffraction minima are given by $a\sin\theta = m\lambda$. The edge of the central fringe corresponds to $m = 1$. Since the width of the central fringe is small compared to the slit-screen distance, use small angle approximations. Use the de Broglie wavelength with $p = \sqrt{2Km_e}$ for the electrons.

    Solution Since $\theta \approx \tan\theta = x/D$ and $\theta \approx \sin\theta = \lambda/a$, we have $\lambda = ax/D$, where x is half the width of the central fringe and D is the distance from the slit to the screen. Solve for p.

    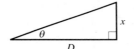

    $$\lambda = \frac{h}{p} = \frac{ax}{D}, \text{ so } p = \frac{Dh}{ax}.$$
    Find the kinetic energy.
    $$p = \frac{Dh}{ax} = \sqrt{2Km_e}, \text{ so}$$
    $$K = \frac{D^2 h^2}{2a^2 x^2 m_e} = \frac{(1.0 \text{ m})^2 (6.626 \times 10^{-34} \text{ J} \cdot \text{s})^2}{2(40.0 \times 10^{-9} \text{ m})^2 (0.031 \text{ m})^2 (9.109 \times 10^{-31} \text{ kg})} \times \frac{1 \text{ eV}}{1.602 \times 10^{-19} \text{ J}} = \boxed{0.98 \text{ eV}}.$$

25. **Strategy** Use the energy-time uncertainty principle, Eq. (28-3).

    Solution Find the uncertainty in the energy.
    $$\Delta E \Delta t \geq \frac{1}{2}\hbar, \text{ so } \Delta E \geq \frac{\hbar}{2\Delta t} \approx \frac{1 \times 10^{-34} \text{ J} \cdot \text{s}}{2(0.1 \times 10^{-9} \text{ s})} = 5 \times 10^{-25} \text{ J}.$$
    Calculate the fractional uncertainty.
    $$\frac{\Delta E}{E} \approx \frac{5 \times 10^{-25} \text{ J}}{(1672 \times 10^6 \text{ eV})(1.602 \times 10^{-19} \text{ J/eV})} = \boxed{2 \times 10^{-15}}$$

27. **Strategy** The kinetic energy of the electron is given by $K = p_n^2/(2m)$, where $p_n = nh/(2L)$. The minimum kinetic energy is the $n = 1$ state.

Solution Find the minimum kinetic energy.
$$K = \frac{p_n^2}{2m} = \frac{n^2 h^2}{8mL^2} = \frac{1^2(6.626\times10^{-34}\ \text{J}\cdot\text{s})^2}{8(9.109\times10^{-31}\ \text{kg})(1.0\times10^{-15}\ \text{m})^2} \times \frac{1\ \text{eV}}{1.602\times10^{-19}\ \text{J}} = \boxed{380\ \text{GeV}}$$

29. **(a) Strategy** The momentum of the marble is given by $p_n = h/\lambda_n = nh/(2L)$.

Solution Find n.
$$p_n = \frac{nh}{2L},\ \text{so}\ n = \frac{2p_n L}{h} = \frac{2mvL}{h} = \frac{2(10\times10^{-3}\ \text{kg})(0.02\ \text{m/s})(0.10\ \text{m})}{6.626\times10^{-34}\ \text{J}\cdot\text{s}} = \boxed{6\times10^{28}}.$$

(b) Strategy Use Eqs. (28-8) and (28-9).

Solution Calculate the energy difference between the n and $n+1$ states.
$$\Delta E = E_{n+1} - E_n = \frac{(n+1)^2 h^2}{8mL^2} - \frac{n^2 h^2}{8mL^2} = \frac{h^2}{8mL^2}[(n+1)^2 - n^2] = \frac{(2n+1)h^2}{8mL^2} = \frac{(12\times10^{28}+1)(6.626\times10^{-34}\ \text{J}\cdot\text{s})^2}{8(0.010\ \text{kg})(0.10\ \text{m})^2}$$
$$= 7\times10^{-35}\ \text{J}$$
We cannot observe the quantization of the marble's energy because the energy difference between levels is too small to observe.

33. **Strategy** The energy of a photon is related to its wavelength by $E = hc/\lambda$. Use Eqs. (28-8), (28-9), and (28-10).

Solution

(a) Calculate the wavelength.
$$\lambda = \frac{hc}{E} = \frac{hc}{E_2 - E_1} = \frac{hc}{2^2 E_1 - E_1} = \frac{hc}{3E_1} = \frac{1240\ \text{eV}\cdot\text{nm}}{3(40.0\ \text{eV})} = \boxed{10.3\ \text{nm}}$$

(b) According to Eq. (28-8), $E \propto L^{-2}$, so doubling the length of the box would reduce the energy to one fourth as much as before.

37. **Strategy** Use Table 28.1 and Eq. (28-13).

Solution Since $n = 3$, ℓ can be 0, 1, or 2, and m_ℓ can have values $-2, -1, 0, 1,$ and 2 (for $\ell = 2$). $L_z = m_\ell \hbar$, so it can have values: $\boxed{-2\hbar, -\hbar, 0, \hbar,\ \text{and}\ 2\hbar}$.

39. **Strategy and Solution** Since $n = 7$, ℓ can have values from 0 to 6. For each value of ℓ, there are $2(2\ell+1)$ states. So, for $\ell = 0, 1, 2, 3, 4, 5,$ and 6, there are 2, 6, 10, 14, 18, 22, and 26 electron states, respectively. The total is 98.

41. **Strategy** Bromine (Br) has atomic number 35, so it has 35 electrons in a neutral atom. Bromine is not an exception to the subshell order, so subshells are filled in the order listed in Eq. (28-16).

Solution The ground-state electron configuration of bromine is
$$\boxed{1s^2 2s^2 2p^6 3s^2 3p^6 4s^2 3d^{10} 4p^5}.$$

43. **Strategy** Lithium, sodium, and potassium have 3, 11, and 19 electrons in a neutral atom, respectively. None of these appear in the list of exceptions to the subshell order, so subshells are filled in the order listed in Eq. (28-16). The Periodic Table of the elements is arranged in columns by electron configuration.

    Solution

    (a) The ground-state electron configurations are the following:

    $$\boxed{\text{Li: } 1s^2 2s^1; \text{ Na: } 1s^2 2s^2 2p^6 3s^1; \text{ K: } 1s^2 2s^2 2p^6 3s^2 3p^6 4s^1}.$$

    (b) All three neutral atoms have valence +1. Their outermost electron is in the $\boxed{s^1 \text{ subshell}}$. This is why they are placed in the same column.

45. (a) **Strategy** Carbon ($Z = 6$) has 6 electrons in its neutral state. It is not an exception to the subshell order, so subshells are filled in the order listed in Eq. (28-16).

    Solution The ground-state electron configuration of carbon is

    $$\boxed{1s^2 2s^2 2p^2}.$$

    (b) **Strategy** Use Table 28.1.

    Solution For carbon in its ground state, $n = 2$, so $\ell = 0$ or 1 and m_ℓ can have values -1, 0, and -1 (for $\ell = 1$). For each m_ℓ, $m_s = \pm 1/2$. The states are given in the table.

n	ℓ	m_ℓ	m_s
1	0	0	$\pm\frac{1}{2}$
2	0	0	$\pm\frac{1}{2}$
2	1	-1	$\pm\frac{1}{2}$
2	1	0	$\pm\frac{1}{2}$
2	1	1	$\pm\frac{1}{2}$

49. **Strategy** Since the energy of a photon is inversely proportional to its wavelength, 640 nm is the longest wavelength of photon that can supply an electron with enough energy to jump into the conduction band across the band gap; therefore, the band gap equals the energy of a 640-nm photon.

    Solution Compute the band gap.

    $$E_{\text{gap}} = \frac{hc}{\lambda} = \frac{1240 \text{ eV} \cdot \text{nm}}{640 \text{ nm}} = \boxed{1.9 \text{ eV}}$$

51. **Strategy** The wavelength of a photon is related to its energy by $E = hc/\lambda$.

    Solution The wavelength for this transition is

    $$\lambda = \frac{hc}{\Delta E} = \frac{1240 \text{ eV} \cdot \text{nm}}{20.66 \text{ eV} - 18.38 \text{ eV}} = \boxed{544 \text{ nm}}.$$

    Light of wavelength 544 nm appears $\boxed{\text{green}}$ in color.

53. **(a) Strategy** The location of the first diffraction minimum for a beam passing through a circular aperture is given by $a\sin\Delta\theta = 1.22\lambda$. The wavelength is 694.3 nm.

    Solution Find the angular spread of the beam.

    $$\Delta\theta = \sin^{-1}\frac{1.22\lambda}{a} = \sin^{-1}\frac{1.22(694.3\times10^{-9}\ \text{m})}{0.0050\ \text{m}} = \boxed{0.0097°}$$

    (b) Strategy If the diameter Δx of the spot on the Moon is approximately equal to the arc length of a circle of radius D subtended by $\Delta\theta$, then $\Delta x \approx D\Delta\theta$, where $\Delta\theta$ is now twice the angle found in part (a) and D is the Earth-Moon distance.

    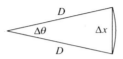

    Solution Find the diameter of the spot on the Moon.

    $$\Delta x \approx D\Delta\theta = (3.845\times10^{8}\ \text{m})(2)\sin^{-1}\frac{1.22(694.3\times10^{-9}\ \text{m})}{0.0050\ \text{m}} = \boxed{130\ \text{km}}$$

57. **Strategy** The kinetic energy of the magnesium ion is equal to $2e\Delta V$ due to the potential difference. Use the relationship between momentum and kinetic energy to find the momentum of the ion. Then, find the de Broglie wavelength of the ion. The mass of the magnesium ion is approximately 24.3 u.

    Solution Find the momentum.

    $$K = \frac{1}{2}mv^2 = \frac{(mv)^2}{2m} = \frac{p^2}{2m} = 2e\Delta V,\ \text{so}\ p = \sqrt{4me\Delta V} = 2\sqrt{me\Delta V}.$$

    Find the de Broglie wavelength.

    $$\lambda = \frac{h}{p} = \frac{h}{2\sqrt{me\Delta V}} = \frac{6.626\times10^{-34}\ \text{J}\cdot\text{s}}{2\sqrt{(24.3\ \text{u})(1.6605\times10^{-27}\ \text{kg/u})(1.602\times10^{-19}\ \text{C})(22\times10^{3}\ \text{V})}} = \boxed{2.8\times10^{-14}\ \text{m}}$$

61. **Strategy** The average kinetic energy of the thermal neutrons is equal to $\frac{3}{2}k_\text{B}T$. Find the de Broglie wavelength using $p = \sqrt{2Km}$, since the neutrons are nonrelativistic [$K = p^2/(2m)$].

    Solution

    (a) Find the average kinetic energy of the neutrons at $T = 400.0$ K.

    $$K_\text{av} = \frac{3}{2}k_\text{B}T = \frac{3}{2}(1.381\times10^{-23}\ \text{J/K})(400.0\ \text{K}) = \boxed{8.286\times10^{-21}\ \text{J}}$$

    (b) Find the de Broglie wavelength.

    $$\lambda = \frac{h}{p} = \frac{h}{\sqrt{2Km}} = \frac{6.626\times10^{-34}\ \text{J}\cdot\text{s}}{\sqrt{2(8.286\times10^{-21}\ \text{J})(1.675\times10^{-27}\ \text{kg})}} = \boxed{125.8\ \text{pm}}$$

65. Strategy Tellurium (Te) has atomic number 52, so there are 52 electrons in the neutral atom. It is not an exception to the subshell order, so subshells are filled in the order listed in Eq. (28-16).

Solution The ground-state electron configuration for Tellurium is

$$\boxed{1s^2 2s^2 2p^6 3s^2 3p^6 4s^2 3d^{10} 4p^6 5s^2 4d^{10} 5p^4}.$$

69. Strategy The neutrons travel a distance $\Delta x = 16.4$ m in a time $\Delta t = 13.0$ ms. The neutron speed is classical, so calculate the de Broglie wavelength using $p = mv$.

Solution

(a) Find the speed of the neutrons selected.

$$v = \frac{\Delta x}{\Delta t} = \frac{16.4 \text{ m}}{13.0 \times 10^{-3} \text{ s}} = \boxed{1.26 \text{ km/s}}$$

(b) Find the de Broglie wavelength.

$$\lambda = \frac{h}{p} = \frac{h}{mv} = \frac{6.626 \times 10^{-34} \text{ J} \cdot \text{s}}{(1.675 \times 10^{-27} \text{ kg})(1.26 \times 10^3 \text{ m/s})} = \boxed{314 \text{ pm}}$$

(c) For the slowest neutrons that get through the two shutters, we have

$$v_s = \frac{16.4 \text{ m}}{13.0 \times 10^{-3} \text{ s} + 0.45 \times 10^{-3} \text{ s}} = 1220 \text{ m/s and } \lambda_s = \frac{6.626 \times 10^{-34} \text{ J} \cdot \text{s}}{(1.675 \times 10^{-27} \text{ kg})(1220 \text{ m/s})} = 324 \text{ pm}.$$

For the fastest neutrons that get through the two shutters, we have

$$v_f = \frac{16.4 \text{ m}}{13.0 \times 10^{-3} \text{ s} - 0.45 \times 10^{-3} \text{ s}} = 1307 \text{ m/s and } \lambda_f = \frac{6.626 \times 10^{-34} \text{ J} \cdot \text{s}}{(1.675 \times 10^{-27} \text{ kg})(1307 \text{ m/s})} = 303 \text{ pm}.$$

The range of wavelengths is $\boxed{303\text{–}324 \text{ pm}}$.

73. **(a) Strategy** The electrons have kinetic energy of 54.0 eV due to the accelerating potential. Since the kinetic energy is so much smaller than the rest energy of the electron, use the nonrelativistic forms of the momentum and kinetic energy. Find the de Broglie wavelength using Eq. (28-1).

    Solution Find the momentum.

    $K = \dfrac{p^2}{2m}$, so $p = \sqrt{2mK}$.

    Calculate the de Broglie wavelength.

    $$\lambda = \frac{h}{p} = \frac{h}{\sqrt{2mK}} = \frac{6.626\times10^{-34}\ \text{J}\cdot\text{s}}{\sqrt{2(9.109\times10^{-31}\ \text{kg})(54.0\ \text{eV})(1.602\times10^{-19}\ \text{J/eV})}} = \boxed{167\ \text{pm}}$$

    (b) Strategy Use Bragg's law.

    Solution Find the Bragg angle for $m = 1$.

    $$2d\sin\theta = m\lambda = (1)\lambda,\ \text{so}\ \theta = \sin^{-1}\frac{\lambda}{2d} = \sin^{-1}\frac{1.669\times10^{-10}\ \text{m}}{2(0.091\times10^{-9}\ \text{m})} = \boxed{66.5^\circ}.$$

    (c) Strategy Make a sketch to show the relationship between the Bragg angle and the scattering angle.

    Solution In the figure to the right, $\theta_i = \theta_r$ and $\theta_{\text{Bragg}} = \theta_r + \alpha$; but $\theta_i = \alpha$, since they are opposite angles, so $\theta_{\text{Bragg}} = \theta_r + \theta_i = 2\theta_i$, or $\theta_{\text{Bragg}} = 2(66.5^\circ) = 133^\circ$, which equals 130° to two significant figures. $\boxed{\text{Yes}}$, the results agree.

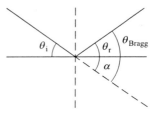

75. **Strategy** Use Figure 28.13 as a guide to make a qualitative sketch of the wave function for the $n = 5$ state. Estimate the number of bound states for the *finite* box by using the energy states for an *infinite* box.

    Solution

    (a) The figure shows the wave function for the $n = 5$ state of an electron in a finite box of length L.

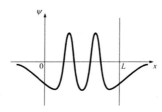

    (b) The energy states for an infinite box are given by $E_n = n^2h^2/(8mL^2)$. Set $E_n = U_0$ and solve for n.

    $$n = \sqrt{\frac{8mL^2U_0}{h^2}} = \sqrt{\frac{8(9.109\times10^{-31}\ \text{kg})(1.0\times10^{-9}\ \text{m})^2(1.0\times10^3\ \text{eV})(1.602\times10^{-19}\ \text{J/eV})}{(6.626\times10^{-34}\ \text{J}\cdot\text{s})^2}} = 51.6$$

    $n = 52$ would give an energy greater than U_0, so there are approximately $\boxed{51}$ bound states.

77. (a) **Strategy** The length of the box can be found by using the equation for the ground-state energy, Eq. (28-8).

   Solution Find the length.

   $$E_1 = \frac{h^2}{8mL^2}, \text{ so } L = \frac{h}{2\sqrt{2mE_1}} = \frac{6.626\times10^{-34}\text{ J}\cdot\text{s}}{2\sqrt{2(9.109\times10^{-31}\text{ kg})(0.010\text{ eV})(1.602\times10^{-19}\text{ J/eV})}} = \boxed{6.1\text{ nm}}.$$

   (b) **Strategy** Use Figure 28.10 to sketch the wave functions for the three lowest energy states of the electron confined to a one-dimensional box of length $L = 6.1$ nm.

   Solution The wave functions:

   Lowest energy state: Second lowest energy state:

   Third lowest energy state:

   (c) **Strategy** The kinetic energy of the electron in the second excited state ($n = 3$) is given by $E_3 = 3^2 h^2/(8mL^2)$. The momentum is related to the kinetic energy by $p = \sqrt{2mK} = \sqrt{2mE_3}$, so the wavelength is given by $\lambda = h/p = h/\sqrt{2mE_3}$.

   Solution Find the wavelength.

   $$\lambda = \frac{h}{\sqrt{2mE_3}} = \frac{h}{\sqrt{2m\frac{3^2 h^2}{8mL^2}}} = \frac{2L}{3} = \frac{2}{3}(6.1\text{ nm}) = \boxed{4.1\text{ nm}}$$

   (d) **Strategy** The energy of a photon is related to its wavelength by $E = hc/\lambda$. Use Eqs. (28-9) and (28-10) to find the energies of the transitions.

   Solution Find the energy of the photon.

   $$E = \frac{hc}{\lambda} = \frac{1240\text{ eV}\cdot\text{nm}}{15{,}500\text{ nm}} = 0.0800\text{ eV}$$

   Find the energy state of the electron after absorbing the photon.

   $$E = E_n - E_1 = n^2 E_1 - E_1, \text{ so } n = \sqrt{\frac{E + E_1}{E_1}} = \sqrt{\frac{0.0800\text{ eV} + 0.010\text{ eV}}{0.010\text{ eV}}} = 3.$$

   The possible transitions (and wavelengths) are

   $3 \rightarrow 1$: one photon; $\lambda_{31} = \boxed{15.5\text{ µm}}$

   $3 \rightarrow 2 \rightarrow 1$: two photons;

   $$\lambda_{32} = \frac{hc}{E_3 - E_2} = \frac{1240\text{ eV}\cdot\text{nm}}{9(0.010\text{ eV}) - 4(0.010\text{ eV})} = \boxed{25\text{ µm}}$$

   $$\lambda_{21} = \frac{hc}{E_2 - E_1} = \frac{1240\text{ eV}\cdot\text{nm}}{4(0.010\text{ eV}) - 0.010\text{ eV}} = \boxed{41\text{ µm}}$$

Chapter 29

NUCLEAR PHYSICS

Conceptual Questions

1. Alpha, beta, and gamma rays were distinguished by their varying abilities to penetrate matter before their masses and electric charges were determined.

5. There is a mass defect due to the binding energy of atomic electrons but it is several orders of magnitude smaller than the binding energy of the nucleus and may thus be omitted without significant loss of accuracy.

9. If none of the iodine were eliminated from the body by biological processes, it would take 8 days for only half as much ^{131}I to be left in the body. However, at least some of the iodine is eliminated from the body by biological means before it has decayed, so it should take less than 8 days for there to be only half as much ^{131}I left in the body.

13. The products of a typical fusion reaction are light, non-radioactive elements. During a fusion reaction, electrons or positrons and photons are typically emitted. These by-products have little tendency to induce radioactivity in materials they come into contact with. A fission reaction, on the other hand, typically has radioactive daughter nuclei as the end products, constituting radioactive waste. Furthermore, fission reactions usually give off neutrons, which tend to induce radioactivity (neutron activation) in materials they come into contact with. The material of the fission reactor itself, the moderator substance, and the control rods all become radioactive as a result of neutron bombardment. As such, they must eventually be discarded as nuclear waste.

Problems

1. **Strategy** Nucleons have masses of approximately 1 u, so dividing the mass of the person by 1 u will give an estimate of the total number of nucleons in the person's body.

   Solution Estimate the number of nucleons.
   $$\frac{75 \text{ kg}}{1 \text{ u}} = \frac{75 \text{ kg}}{1.660\,539 \times 10^{-27} \text{ kg}} = \boxed{4.5 \times 10^{28}}$$

3. **Strategy** For a spherical object, the density equation is $\rho = M/V = M/(\frac{4}{3}\pi r^3)$.

   Solution Solving for r and using $M = M_{\text{Sun}} = 1.99 \times 10^{30}$ kg and $\rho = 2.3 \times 10^{17}$ kg/m^3 (the density of nuclear matter found in Problem 2) yields the result.
   $$r^3 = \frac{M}{\frac{4}{3}\pi\rho}, \text{ so } r = \left(\frac{M}{\frac{4}{3}\pi\rho}\right)^{1/3} = \left[\frac{1.99 \times 10^{30} \text{ kg}}{\frac{4}{3}\pi(2.3 \times 10^{17} \text{ kg/m}^3)}\right]^{1/3} = \boxed{13 \text{ km}}.$$

5. **Strategy** The nucleon number A is the sum of the total number of protons Z and neutrons N. Use the Periodic Table of the elements to find the number of protons.

   Solution Find the number of protons.
   Potassium has atomic number 19, so $Z = 19$.
   $N = $ # of neutrons $= 21$
   Find the nucleon number.
   $A = Z + N = 19 + 21 = 40$
   So, the symbol is $\boxed{^{40}_{19}\text{K}}$.

7. **Strategy** Use the Periodic Table of the elements to find the number of protons.

   Solution Xe has atomic number 54, so the number of protons is $\boxed{54}$.

9. **Strategy** $A = 107$ for technetium-107. Find the radius of the nucleus using Eq. (29-4). Use the volume of a sphere to find the volume of the nucleus.

   Solution Find the radius.
   $$r = r_0 A^{1/3} = (1.2 \text{ fm})(107)^{1/3} = \boxed{5.7 \text{ fm}}$$
   Find the volume.
   $$V = \frac{4}{3}\pi r^3 = \frac{4}{3}\pi r_0^3 A = \frac{4}{3}\pi (1.2 \times 10^{-15} \text{ m})^3 (107) = \boxed{7.7 \times 10^{-43} \text{ m}^3}$$

11. **Strategy** A deuteron has 1 proton and 1 neutron. Use Eqs. (29-7) and (29-8) to find the mass defect and binding energy.

    Solution Find the mass defect.
    $\Delta m = 1 \times 1.007\ 276\ 5 \text{ u} + 1 \times 1.008\ 664\ 9 \text{ u} - 2.013\ 553 \text{ u} = 0.002\ 388 \text{ u}$

    The binding energy is $E_B = (\Delta m)c^2 = 0.002\ 388\ 4 \text{ u} \times 931.494 \text{ MeV/u} = \boxed{2.225 \text{ MeV}}$.

13. **Strategy** The nucleon number A is the sum of the total number of protons Z and neutrons N. Use Eqs. (29-7) and (29-8) to find the mass defect and binding energy. The binding energy per nucleon is the binding energy divided by the total number of nucleons in the nucleus.

    Solution

    (a) The ^{16}O atom has 8 protons, 8 neutrons, and 8 electrons. Its mass is 15.994 914 6 u. Find the mass defect.
    $$\Delta m = (\text{mass of 8 } ^1\text{H atoms and 8 neutrons}) - (\text{mass of } ^{16}\text{O atom})$$
    $$= 8 \times 1.007\ 825\ 0 \text{ u} + 8 \times 1.008\ 664\ 9 \text{ u} - 15.994\ 914\ 6 \text{ u} = 0.137\ 004\ 6 \text{ u}$$

    The binding energy is $E_B = (\Delta m)c^2 = 0.137\ 004\ 6 \text{ u} \times 931.494 \text{ MeV/u} = \boxed{127.619 \text{ MeV}}$.

    (b) Calculate the average binding energy per nucleon.
    $$\frac{E_B}{A} = \frac{127.619 \text{ MeV}}{16 \text{ nucleons}} = \boxed{7.976\ 19 \text{ MeV/nucleon}}$$
    This result matches the value given in Figure 29.2.

17. Strategy Use Eq. (29-8) to find the mass defect. Compare the mass defect to the mass of an electron.

Solution

(a) Calculate the mass defect.

$$\Delta m = \frac{E_B}{c^2} = \frac{13.6 \text{ eV}}{931.494 \times 10^6 \text{ eV/u}} = \boxed{1.46 \times 10^{-8} \text{ u}}$$

(b) The ratio of this mass defect to the mass of the electron is

$$\frac{1.46 \times 10^{-8} \text{ u}}{5.486 \times 10^{-4} \text{ u}} = 2.66 \times 10^{-5}.$$

This is small enough to be ignored. There is $\boxed{\text{no}}$ reason to worry when calculating the mass of the ^1H nucleus, especially since its mass is 1836 times that of the electron.

21. Strategy In beta-minus decay, the atomic number Z increases by 1 while the mass number A remains constant. Use Eq. (29-11).

Solution

For the parent $\left({}_{19}^{40}\text{K}\right)$ $Z = 19$, so the daughter nuclide will have $Z = 19 + 1 = 20$, which is the element Ca. The

symbol for the daughter is $\boxed{{}_{20}^{40}\text{Ca}}$.

23. Strategy In electron-capture decay, the atomic number Z is decreased by 1 while the mass number A stays the same.

Solution In this case, the parent nuclide has $Z = 11$ and $A = 22$, so the daughter nuclide will have $Z = 10$ and $A = 22$, which is the element neon. Write out the reaction.

$${}_{Z}^{A}\text{P} + {}_{-1}^{0}\text{e} \rightarrow {}_{Z-1}^{A}\text{D} + {}_{0}^{0}\nu, \text{ so } \boxed{{}_{11}^{22}\text{Na} + {}_{-1}^{0}\text{e} \rightarrow {}_{10}^{22}\text{Ne} + {}_{0}^{0}\nu}. \text{ The daughter nuclide is } \boxed{{}_{10}^{22}\text{Ne}}.$$

25. Strategy The kinetic energy of the decay products equals the energy associated with the change in mass.

Solution Find the change in mass during the decay using atomic masses.

$\Delta m = (\text{mass of } {}_{86}^{222}\text{Rn} + \text{mass of } {}_2^4\text{He}) - (\text{mass of } {}_{88}^{226}\text{Ra}) = (222.017\ 570\ 5 \text{ u} + 4.002\ 603\ 2 \text{ u}) - 226.025\ 402\ 6 \text{ u}$
$= -0.005\ 228\ 9 \text{ u}$

Find the kinetic energy.

$$K = E = |\Delta m|c^2 = 0.005\ 228\ 9 \text{ u} \times 931.494 \text{ MeV/u} = 4.8707 \text{ MeV}$$

Assuming the ${}_{86}^{222}\text{Rn}$ nucleus takes away an insignificant fraction of the kinetic energy, the alpha particle's

kinetic energy will be $\boxed{4.8707 \text{ MeV}}$.

27. **Strategy** In Problem 21, the daughter nuclide in this decay was found to be $^{40}_{20}$Ca. The maximum kinetic energy of the beta particle is equal to the disintegration energy.

    Solution The reaction is
    $$^{40}_{19}\text{K} \rightarrow {}^{40}_{20}\text{Ca} + {}^{0}_{-1}\text{e} + {}^{0}_{0}\bar{\nu}$$

    The atomic masses of $^{40}_{19}$K and $^{40}_{20}$Ca are 39.963 998 7 u and 39.962 591 2 u, respectively. To get the masses of the nuclei, we subtract Zm_e from each. The mass of the electron is 0.000 548 6 u, and the neutrino's mass is negligible. The mass difference is
    $$\Delta m = [(M_{\text{Ca}} - 20m_e) + m_e] - (M_K - 19m_e) = M_{\text{Ca}} - M_K = 39.962\ 591\ 2\ \text{u} - 39.963\ 998\ 7\ \text{u} = -0.001\ 407\ 5\ \text{u}.$$

    The disintegration energy is $E = |\Delta m|c^2 = 0.001\ 407\ 5\ \text{u} \times 931.494\ \text{MeV/u} = 1.3111\ \text{MeV}.$

    The maximum kinetic energy of the β^- particle is $\boxed{1.3111\ \text{MeV}}$.

29. **Strategy** In a spontaneous decay, the mass of the decay products must be less than the mass of the parent.

    Solution Alpha decay of $^{19}_{8}$O would have a daughter nuclide of $^{15}_{6}$C.
    $$^{19}_{8}\text{O} \rightarrow {}^{15}_{6}\text{C} + {}^{4}_{2}\text{He}$$

    The mass of atomic $^{19}_{8}$O is 19.003 578 7 u. The combined mass of $^{15}_{6}$C and $^{4}_{2}$He is

    15.010 599 3 u + 4.002 603 2 u = 19.013 202 5 u.

    Since 19.013 202 5 u > 19.003 578 7 u, the spontaneous alpha decay of $^{19}_{8}$O is not possible.

33. **Strategy** The activity is reduced by a factor of two for each half-life. Use Eqs. (29-18), (29-20), and (29-22) to find the initial number of nuclei and the probability of decay per second.

    Solution

    (a) Find the number of half-lives.
    $$600.0\ \text{s} = \frac{600.0\ \text{s}}{200.0\ \text{s/half-life}} = 3.000\ \text{half-lives}$$

    The activity after 3.000 half-lives will be $R = \left(\frac{1}{2}\right)^{3.000} \times R_0 = \frac{1}{8.000} \times 80,000.0\ \text{s}^{-1} = \boxed{10,000\ \text{s}^{-1}}$.

    (b) Find the initial number of nuclei.
    $$N_0 = \frac{1}{\lambda} R_0 = \tau R_0 = \frac{R_0 T_{1/2}}{\ln 2} = 80,000.0\ \text{s}^{-1} \times \frac{200.0\ \text{s}}{\ln 2} = \boxed{2.308 \times 10^7}$$

    (c) The probability per second is $\lambda = \frac{1}{\tau} = \frac{\ln 2}{T_{1/2}} = \frac{\ln 2}{200.0\ \text{s}} = \boxed{3.466 \times 10^{-3}\ \text{s}^{-1}}$.

35. **Strategy** The activity as a function of time is given by $R = R_0 e^{-t/\tau}$. Use Eq. (29-22) to find the time constant. Assume that the original activity was 0.25 Bq per g of carbon.

    Solution Find the age of the bones.
    $$e^{-t/\tau} = \frac{R}{R_0}, \text{ so } -\frac{t}{\tau} = \ln\frac{R}{R_0} \text{ or } t = -\tau \ln\frac{R}{R_0} = -\frac{5730\ \text{yr}}{\ln 2} \times \ln\frac{0.242}{0.25} = \boxed{270\ \text{yr}}.$$

37. Strategy The half-life of $^{214}_{83}Bi$ is 19.9 min. The activity is given by $R = R_0(2^{-t/T_{1/2}})$.

Solution Find the activity.

$$R = R_0(2^{-t/T_{1/2}}) = 0.058 \text{ Ci} \times \left(\frac{1}{2}\right)^{60/19.9} = \boxed{0.0072 \text{ Ci}}$$

39. Strategy Use Eq. (29-22) to find the time constant. The number of nuclei is related to the mass by $N = mN_A/M$, where N_A is Avogadro's number and M is the molar mass. The activity is equal to the number of nuclei divided by the time constant.

Solution Convert the half-life of radium-226 to seconds.

$$1600 \text{ yr} \cdot \frac{3.156 \times 10^7 \text{ s}}{\text{yr}} = 5.0496 \times 10^{10} \text{ s}$$

The time constant τ is $\tau = \frac{T_{1/2}}{\ln 2} = \frac{5.0496 \times 10^{10} \text{ s}}{\ln 2} = 7.285 \times 10^{10}$ s.

Find the number of nuclei in 1.0 g of radium.

$$N = \frac{mN_A}{M} = \frac{1.0 \text{ g}}{226.0254 \text{ g/mol}} \times 6.022 \times 10^{23} \text{ nuclei/mol} = 2.6643 \times 10^{21} \text{ nuclei}$$

The activity is $R = \dfrac{N}{\tau} = \dfrac{2.6643 \times 10^{21} \text{ nuclei}}{7.285 \times 10^{10} \text{ s}} \times \dfrac{1 \text{ Ci}}{3.7 \times 10^{10} \text{ Bq}} = \boxed{0.99 \text{ Ci}}$.

41. Strategy Use Eqs. (29-20) and (29-22) to find the decay constant. The total number of nuclei in 1.00 g of carbon times the relative abundance gives the number of carbon-14 nuclei. Use the results of parts (a) and (b) and Eq. (29-18) to find the activity per gram in a living sample.

Solution

(a) Calculate the decay constant.

$$\lambda = \frac{1}{\tau} = \frac{\ln 2}{T_{1/2}} = \frac{\ln 2}{5730 \text{ yr} \times 3.156 \times 10^7 \text{ s/yr}} = \boxed{3.83 \times 10^{-12} \text{ s}^{-1}}$$

(b) Calculate the number of ^{14}C atoms.

$$N = \frac{\text{mass}}{\text{mass per mol}} \times N_A \times (\text{relative abundance}) = \frac{1.00 \text{ g}}{12.011 \text{ g/mol}} \times 6.022 \times 10^{23} \text{ atoms/mol} \times 1.3 \times 10^{-12}$$

$$= \boxed{6.5 \times 10^{10} \text{ atoms}}$$

(c) Calculate the activity per gram.

$$\frac{R}{1.00 \text{ g}} = \lambda \frac{N}{1.00 \text{ g}} = 3.83 \times 10^{-12} \text{ s}^{-1} \times \frac{6.5 \times 10^{10} \text{ atoms}}{1.00 \text{ g}} = \boxed{0.25 \text{ Bq/g}}$$

45. Strategy The number of ionized molecules is equal to the kinetic energy of the alpha particle divided by the ionization energy per molecule.

Solution

$$\# \text{ of ionized molecules} = \frac{6 \times 10^6 \text{ eV}}{20 \text{ eV/molecule}} = \boxed{3 \times 10^5 \text{ molecules}}$$

49. **Strategy** Write out the reaction using variables for the unknown quantities. Balance the reaction to find the unknowns. The total charge and total number of nucleons must remain the same. The emission of an electron is beta-minus decay.

    Solution

    (a) The reaction is
    $$\,_0^1 n + \,_b^a(?_1) \rightarrow \,_d^c(?_2) \rightarrow \,_f^e(?_3) + \,_{-1}^0 e$$
    $$\,_f^e(?_3) \rightarrow \,_2^4 \alpha + \,_2^4 \alpha$$

    Working backward, $e = 4 + 4 = 8$ and $f = 2 + 2 = 4$, so $(?_3) = \boxed{\,_4^8 Be}$. Next, $c = e = 8$ and

    $d = f - 1 = 4 - 1 = 3$, so $(?_2) = \boxed{\,_3^8 Li}$. Finally, $a + 1 = c = 8$, so $a = 7$ and $b = d = 3$, which means

    $(?_1) = \boxed{\,_3^7 Li}$.

    (b) $\boxed{\text{Yes; the emission of an electron (beta-minus decay) is accompanied by the emission of one antineutrino.}}$

53. **Strategy** The excitation energy is equal to the energy given up by the neutron when it becomes bound to the nucleus. The atomic masses of ^{235}U and ^{236}U are 235.043 923 1 u and 236.045 561 9 u, respectively.

    Solution The reaction is
    $$\,_{92}^{235}U + \,_0^1 n \rightarrow \,_{92}^{236}U.$$

    Find the change in mass.
    $$\Delta m = 236.045\ 561\ 9\ u - 235.043\ 923\ 1\ u - 1.008\ 664\ 9\ u = -0.007\ 026\ 1\ u$$

    The excitation energy is $E = |\Delta m| c^2 = 0.007\ 026\ 1\ u \times 931.494$ MeV/u $= \boxed{6.5448\ \text{MeV}}$.

57. **Strategy** The total charge and total number of nucleons must remain the same. The energy released is equal to the difference between the binding energy of the reaction product and that of the deuteron. Compare the thermal energy to the coulomb repulsion.

    Solution

    (a) The atomic number of X must be $1 + 1 = 2$, so X is helium. The mass number of X must be $1 + 2 = 3$. The reaction product is $\boxed{\,_2^3 He}$.

    (b) The binding energies are as follows:
    $\,_1^1 H$: 0, $\,_1^2 H$: 2×1.1 MeV $= 2.2$ MeV, $\,_2^3 He$: 3×2.6 MeV $= 7.8$ MeV.
    Find the energy released.
    7.8 MeV $- 2.2$ MeV $= \boxed{5.6\ \text{MeV}}$

    (c) $\boxed{\text{At room temperature, the kinetic energies of the proton and the deuteron are much too small to overcome their Coulomb repulsion.}}$

61. **Strategy** Isotopes have the same atomic number.

    Solution Compare the atomic numbers of the unknown nuclides.

    $^{175}_{71}(?)$ and $^{167}_{71}(?)$ have the same atomic number 71, so they are isotopes of each other.

    $^{175}_{74}(?)$ and $^{180}_{74}(?)$ have the same atomic number 74, so they are isotopes of each other.

    Lutetium (Lu) has atomic number 71 and Tungsten (W) has atomic number 74.

    To summarize: $\boxed{^{175}_{71}(\text{Lu}) \text{ and } ^{167}_{71}(\text{Lu})}$ are isotopes of each other and $\boxed{^{175}_{74}(\text{W}) \text{ and } ^{180}_{74}(\text{W})}$ are isotopes of each other.

65. **Strategy** The energies of the photons correspond to the differences in energy levels for the transitions.

    Solution From left to right, the energies are:

    $492 \text{ keV} - 0 = \boxed{492 \text{ keV}}$, $472 \text{ keV} - 0 = \boxed{472 \text{ keV}}$, $40 \text{ keV} - 0 = \boxed{40 \text{ keV}}$, $492 \text{ keV} - 40 \text{ keV} = \boxed{452 \text{ keV}}$,

    $472 \text{ keV} - 40 \text{ keV} = \boxed{432 \text{ keV}}$, and $327 \text{ keV} - 40 \text{ keV} = \boxed{287 \text{ keV}}$.

69. (a) **Strategy** The nucleon number A is the sum of the total number of protons Z and neutrons N.

    Solution The reaction is

    $^{226}_{88}\text{Ra} \rightarrow ^{222}_{86}\text{Rn} + ^{4}_{2}\alpha$.

    There will be $226 - 4 - 86 = \boxed{136 \text{ neutrons}}$ and $88 - 2 = \boxed{86 \text{ protons}}$.

    (b) **Strategy** The activity is related to the number of nuclei N and the time constant τ by $R = N/\tau$. The time constant is related to the half-life by $T_{1/2} = \tau \ln 2$.

    Solution Calculate the activity of the radon.

    $$R = \frac{N}{\tau} = \frac{N \ln 2}{T_{1/2}} = \frac{10^7 \cdot \ln 2}{3.8 \text{ d} \times 24 \text{ h/d} \times 3600 \text{ s/h}} = 20 \text{ decays/s}$$

    So, $\boxed{20 \text{ alpha particles per second}}$ are emitted due to the decaying radon nuclei.

73. **Strategy** Find the number of deuterium nuclei in 1.00 liter of water. Then, find the energy obtained from fusing 87.0% of the deuterium. The number of water molecules in 1.00 L of water is $N_w = m_w N_A / M_w$, where N_A is Avogadro's number and M_w is the molar mass.

    Solution The mass of 1.00 L of water is $m_w = 1.00 \times 10^3$ g. Since there are two hydrogen atoms per water molecule, there are $N_H = 2N_w = 2m_w N_A / M_w$ hydrogen atoms. Of this number, 0.016% are deuterium atoms, so there are $N_D = 0.00016 N_H = 0.00032 N_w = 0.00032 m_w N_A / M_w$ deuterium nuclei. Two nuclei are fused to release 3.65 MeV of energy and 87.0% of the nuclei are available for fusion. Compute the amount of energy that can be obtained from the deuterium in the water.

    $$E = (3.65 \times 10^6 \text{ eV})(1.602 \times 10^{-19} \text{ J/eV}) \frac{1 \text{ kW} \cdot \text{h}}{3.600 \times 10^6 \text{ J}} (0.870) \frac{1}{2} N_D$$

    $$= (3.65 \times 10^6 \text{ eV})(1.602 \times 10^{-19} \text{ J/eV}) \frac{1 \text{ kW} \cdot \text{h}}{3.600 \times 10^6 \text{ J}} (0.870) \frac{1}{2} (0.00032) \frac{(1.00 \times 10^3 \text{ g})(6.022 \times 10^{23} \text{ mol}^{-1})}{15.9994 \text{ g/mol} + 2 \times 1.00794 \text{ g/mol}}$$

    $$= \boxed{760 \text{ kW} \cdot \text{h}}$$

77. **Strategy** The original number of samarium-147 nuclei in the rock is equal to the current number plus the number of neodymium-143. The number of nuclei is related to the mass by $N = mN_A/M$, where N_A is Avogadro's number and M is the molar mass. Use Eq. (29-23).

Solution Find the age of the rock.

$$N = N_0 \times 2^{-t/T_{1/2}}$$

$$N_{Sm} = (N_{Sm} + N_{Nd}) \times 2^{-t/T_{1/2}}$$

$$\frac{N_{Sm}}{N_{Sm} + N_{Nd}} = 2^{-t/T_{1/2}}$$

$$\log_2 \frac{N_{Sm}}{N_{Sm} + N_{Nd}} = \log_2 2^{-t/T_{1/2}} = -\frac{t}{T_{1/2}}$$

$$t = T_{1/2} \log_2 \frac{N_{Sm} + N_{Nd}}{N_{Sm}}$$

$$t = T_{1/2} \log_2 \frac{m_{Sm} N_A/M_{Sm} + m_{Nd} N_A/M_{Nd}}{m_{Sm} N_A/M_{Sm}}$$

$$t = T_{1/2} \log_2 \left(1 + \frac{m_{Nd} M_{Sm}}{m_{Sm} M_{Nd}}\right) = (1.06 \times 10^{11}\ \text{yr}) \log_2 \left[1 + \frac{(0.150)(146.914\,897\,9)}{(3.00)(142.909\,814\,3)}\right] = \boxed{7.67 \times 10^9\ \text{yr}}$$

81. **Strategy** $R = R_0(2^{-t/T_{1/2}}) = \lambda N$, $\lambda = \dfrac{1}{\tau}$, and $\tau = T_{1/2}/\ln 2$. Use these equations to find the activities.

Solution

(a) Form the ratio R_A/R_B.

At $t = 0$, $\dfrac{R_A}{R_B} = \dfrac{R_{A0}}{R_{B0}} = \dfrac{\lambda_A N_A}{\lambda_B N_B} = \dfrac{\lambda_A N_0}{\lambda_B N_0} = \dfrac{\tau_B}{\tau_A} = \dfrac{T_B/\ln 2}{T_A/\ln 2} = \dfrac{12.0\ \text{h}}{3.0\ \text{h}} = \boxed{4.0}$.

(b) At $t = 12.0$ h, $\dfrac{R_A}{R_B} = \dfrac{R_{A0}(2^{-t/T_A})}{R_{B0}(2^{-t/T_B})} = 4.0(2)^{t\left(\frac{1}{T_B} - \frac{1}{T_A}\right)} = 4.0(2)^{(12.0\ \text{h})\left(\frac{1}{12.0\ \text{h}} - \frac{1}{3.0\ \text{h}}\right)} = \boxed{0.50}$.

(c) At $t = 24.0$ h, $\dfrac{R_A}{R_B} = 4.0(2)^{(24.0\ \text{h})\left(\frac{1}{12.0\ \text{h}} - \frac{1}{3.0\ \text{h}}\right)} = \boxed{0.063}$.

Chapter 30

PARTICLE PHYSICS

Conceptual Questions

1. Quarks, electrons, and the exchange particles (gluons, photons, W^+, W^-, Z^0, and possibly gravitons) are the only *fundamental* particles that make up an atom of ordinary matter.

5. There are six particles in the lepton family, including their associated anti-particles. These are the electron, the muon, the tau, and the three corresponding neutrinos.

9. Neutrinos interact with matter extremely weakly. This means that matter is essentially transparent to neutrinos. Although billions of them pass through our body per second, practically none of them interact with particles in our body.

Problems

1. **Strategy** The difference in the rest energy of the particles before and after the decay is related to the kinetic energy of the particles by Einstein's mass-energy relation. Neglect the relatively small rest energy of the neutrino.

   Solution Compute the change in mass.

   $\Delta m = (\text{mass of muon}) - (\text{mass of pion}) = 0.106 \text{ GeV}/c^2 - 0.140 \text{ GeV}/c^2 = -0.034 \text{ GeV}/c^2$

   The total kinetic energy of the two particles is $K = E = |\Delta m|c^2 = \boxed{34 \text{ MeV}}$.

5. **Strategy** Choose the correct fundamental force for each decay. Use Tables 30.2 and 30.3.

   Solution

   (a) The decay products are a muon and a muon neutrino, which are leptons. Leptons are associated with the $\boxed{\text{weak}}$ force.

   (b) The decay products are photons, which are associated with the $\boxed{\text{electromagnetic}}$ force.

   (c) Two of the decay products are an electron and an electron antineutrino, which are leptons. Leptons are associated with the $\boxed{\text{weak}}$ force.

9. **Strategy** The rest energy of an electron is 0.5110 MeV, so an electron with energy 7.0 TeV is extremely relativistic.

   Solution Calculate the de Broglie wavelength.

   $E \approx pc$ and $p = \dfrac{h}{\lambda}$, so $\lambda = \dfrac{hc}{E} = \dfrac{1240 \times 10^{-9} \text{ eV} \cdot \text{m}}{7.0 \times 10^{12} \text{ eV}} = \boxed{1.8 \times 10^{-19} \text{ m}}$.

13. **Strategy** Use Newton's second law. The magnitude of the force on a charged particle moving at speed v perpendicular to a magnetic field of strength B is $F = qvB$. Since the particle is extremely relativistic, use $p \approx E/c$.

Solution Find the magnetic field strength.

$$\Sigma F_r = qvB = ma_r = m\frac{v^2}{r}, \text{ so } B = \frac{mv}{qr} = \frac{p}{qr} \approx \frac{E}{qrc} = \frac{E}{q \times \frac{C}{2\pi} \times c} = \frac{7.0 \times 10^{12} \text{ eV} \times 2\pi}{e \times 27 \times 10^3 \text{ m} \times 3.00 \times 10^8 \text{ m/s}} = \boxed{5.4 \text{ T}}.$$

15. **Strategy** The rest energy of the pion is 0.135 GeV. Conservation of energy requires that the energy of the pion be equal to the energy of the two photons. Conservation of momentum requires that the momenta of the photons are equal and opposite.

Solution Relate the energies.
$$E_\pi = E_{\gamma_1} + E_{\gamma_2}$$
Relate the wavelengths of the photons.
$$p_{\gamma_1} = \frac{h}{\lambda_1} = p_{\gamma_2} = \frac{h}{\lambda_2}, \text{ so } \lambda_1 = \lambda_2.$$
The photons are identical, so let $E_{\gamma_1} = E_{\gamma_2} = E_\gamma$.
Find the energy of each photon.
$$E_\pi = E_\gamma + E_\gamma = 2E_\gamma, \text{ so } E_\gamma = \frac{E_\pi}{2} = \frac{0.135 \text{ GeV}}{2} = \boxed{67.5 \text{ MeV}}.$$

17. **Strategy** Use conservation of energy and momentum.

Solution Conservation of energy requires that
$$E_{p_1} + E_{p_2} = E_{p_3} + E_{p_4} + E_{p_5} + E_{\bar{p}}.$$
Assuming p_1 and p_2 have equal kinetic energies, and that p_3, p_4, p_5, and $\bar{p}$ have zero kinetic energy, we have
$$K + E_0 + K + E_0 = 4E_0$$
where E_0 is the rest energy of a proton (or antiproton) and K is the initial kinetic energy required.
Solve for K.
$$2K = 2E_0, \text{ so } K = E_0 = \boxed{938 \text{ MeV}}.$$

21. **Strategy** Mesons contain one quark and one antiquark. Baryons contain three quarks. Use Table 30.1 to determine the quark content of each particle.

Solution The charge of an up antiquark is $-\frac{2}{3}e$, and that of a down quark is $-\frac{1}{3}e$. Therefore, the quark content of the meson with charge $-e$ is $\boxed{\bar{u}d}$.

23. **Strategy** Replace each particle in the decay reaction with its corresponding antiparticle to write the two decay modes.

Solution Since π^+ is the antiparticle of π^-, the decay products of π^+ must be antiparticles of the decay products of π^-. The decay modes of π^+ are then $\boxed{\pi^+ \rightarrow \mu^+ + \nu_\mu \text{ and } \pi^+ \rightarrow e^+ + \nu_e}$.

25. **Strategy** The pion was initially at rest. Conservation of momentum requires that the momenta of the muon and antineutrino be equal in magnitude and opposite in direction, so $p_\mu = p_{\bar{\nu}}$. For the muon, $K = p^2/(2m)$, so $p_\mu = \sqrt{2mK}$. For the antineutrino, $E \approx pc$, so $p_{\bar{\nu}} \approx E_{\bar{\nu}}/c$.

Solution Equate the momenta and solve for K.
$$p_\mu = \sqrt{2mK} = p_{\bar{\nu}} = \frac{E_{\bar{\nu}}}{c}, \text{ so } K = \frac{E_{\bar{\nu}}^2}{2mc^2} = \frac{E_{\bar{\nu}}^2}{2E_{0\mu}}.$$

Use conservation of energy to find the energy of the antineutrino.
$$E_i = E_{0\pi} = E_f = E_{0\mu} + K + E_{\bar{\nu}} = E_{0\mu} + \frac{E_{\bar{\nu}}^2}{2E_{0\mu}} + E_{\bar{\nu}}, \text{ so } 0 = \frac{1}{2E_{0\mu}} E_{\bar{\nu}}^2 + E_{\bar{\nu}} - (E_{0\pi} - E_{0\mu}).$$

Solve using the quadratic formula.
$$E_{\bar{\nu}} = \frac{-1 \pm \sqrt{1^2 + 4\left(\frac{1}{2E_{0\mu}}\right)(E_{0\pi} - E_{0\mu})}}{E_{0\mu}^{-1}} = (0.106 \text{ GeV})\left[-1 \pm \sqrt{1 + 2\left(\frac{0.140}{0.106} - 1\right)}\right]$$

$$= 0.030 \text{ GeV or } -0.242 \text{ GeV, which is extraneous.}$$

So, the kinetic energy of the muon is $K = \dfrac{(0.030 \text{ GeV})^2}{2(0.106 \text{ GeV})} = 0.0042 \text{ GeV} = \boxed{4.2 \text{ MeV}}$.

REVIEW AND SYNTHESIS: CHAPTERS 26–30

Review Exercises

1. **Strategy** Use Eq. (26-5) to find the velocity v_{pE} of the escape pod relative to Earth. Then, use Eq. (26-4) to find how long the pod appears to the people on Earth. Let the positive direction be toward the Earth.

   Solution The velocity of the starship relative to Earth is $v_{sE} = 0.78c$. The velocity of the pod relative to the starship is $v_{ps} = 0.63c$. Find the velocity of the pod relative to Earth.

   $$v_{pE} = \frac{v_{ps} + v_{sE}}{1 + v_{ps}v_{sE}/c^2} = \frac{0.63c + 0.78c}{1 + (0.63)(0.78)} = 0.945c$$

   Find the length of the pod as it appears to the people on Earth.

   $$L = \frac{L_0}{\gamma} = L_0\sqrt{1 - v^2/c^2} = (12.0\ \text{m})\sqrt{1 - 0.945^2} = \boxed{3.91\ \text{m}}$$

5. **Strategy** Use Eq. (26-8).

   Solution Find the speed of the electron.

   $$K = (\gamma - 1)mc^2$$

   $$\frac{K}{mc^2} = \gamma - 1$$

   $$1 + \frac{K}{mc^2} = \frac{1}{\sqrt{1 - \frac{v^2}{c^2}}}$$

   $$1 - \frac{v^2}{c^2} = \left(1 + \frac{K}{mc^2}\right)^{-2}$$

   $$v = c\sqrt{1 - \left(1 + \frac{K}{mc^2}\right)^{-2}} = c\sqrt{1 - \left[1 + \frac{1.02 \times 10^{-13}\ \text{J}}{(9.109 \times 10^{-31}\ \text{kg})(2.998 \times 10^8\ \text{m/s})^2}\right]^{-2}} = \boxed{0.895c = 2.68 \times 10^8\ \text{m/s}}$$

9. **Strategy** The width of the central maximum is the distance between the first minimum on either side of the central maximum. Use conservation of energy to find the speed of the electrons. Use the de Broglie wavelength for the wavelength of the electrons.

   Solution Find the wavelength of the electrons.

   $$p = mv = \frac{h}{\lambda} \text{ and } K = \frac{1}{2}mv^2 = \frac{p^2}{2m} = e\Delta V, \text{ so } \lambda = \frac{h}{\sqrt{2me\Delta V}}.$$

   The first minimum on either side of the central maximum is given by $a\sin\theta = \lambda$. Solve for θ.

   $$\theta = \sin^{-1}\frac{\lambda}{a} = \sin^{-1}\frac{h}{a\sqrt{2me\Delta V}} = \sin^{-1}\frac{6.626\times10^{-34} \text{ J}\cdot\text{s}}{(6.6\times10^{-10} \text{ m})\sqrt{2(9.109\times10^{-31} \text{ kg})(1.602\times10^{-19} \text{ C})(8950 \text{ V})}} = 1.12555°$$

   If x is the distance to the first minimum on either side of the central maximum and D is the distance from the slit to the screen, then $\tan\theta = x/D$. Find $2x$, the width of the central maximum.

   $$2x = 2D\tan\theta = 2(2.50 \text{ m})\tan 1.12555° = \boxed{9.8 \text{ cm}}$$

13. **Strategy** The rest energy of the lambda particle not used to create the proton and the pion will become kinetic energy of the proton and the pion. Use conservation of momentum and energy.

    Solution Find the energy not used to create the proton and the pion.
    $$K = 1116 \text{ MeV} - (938 \text{ MeV} + 139.6 \text{ MeV}) = 38.4 \text{ MeV}$$

    Let the magnitudes of the momenta of the proton and pion be p_p and p_π, respectively. Then, according to conservation of momentum, $p_p = p_\pi$. Let the kinetic energies of the proton and pion be K_p and K_π, respectively. Then, according to conservation of energy, $K = K_p + K_\pi$. According to Eq. (26-11) and conservation of momentum, $(p_\pi c)^2 = K_\pi^2 + 2K_\pi E_{\pi 0} = (p_p c)^2 = K_p^2 + 2K_p E_{p0}$, where E_{p0} and $E_{\pi 0}$ are the rest energies of the proton and pion, respectively. Substituting for K_π in $(p_p c)^2 = K_\pi^2 + 2K_\pi E_{\pi 0}$ gives

    $$(p_p c)^2 = (K - K_p)^2 + 2(K - K_p)E_{\pi 0}.$$

    Subtracting $(p_p c)^2 = K_p^2 + 2K_p E_{p0}$ from $(p_p c)^2 = (K - K_p)^2 + 2(K - K_p)E_{\pi 0}$ gives

    $$0 = K^2 - 2KK_p + 2KE_{\pi 0} - 2K_p E_{\pi 0} - 2K_p E_{p0}.$$

    Solving this equation for the kinetic energy of the proton K_p gives

    $$K_p = \frac{K^2 + 2KE_{\pi 0}}{2(K + E_{\pi 0} + E_{p0})} = \frac{(38.4 \text{ MeV})^2 + 2(38.4 \text{ MeV})(139.6 \text{ MeV})}{2(38.4 \text{ MeV} + 139.6 \text{ MeV} + 938 \text{ MeV})} = \boxed{5.5 \text{ MeV}}.$$

    Therefore, the kinetic energy of the pion is $K_\pi = K - K_p = 38.4 \text{ MeV} - 5.46 \text{ MeV} = \boxed{33 \text{ MeV}}$.

17. **(a) Strategy** Use Eqs. (29-21) and (29-22).

    Solution Find the activity of the sample of gold-198 after 8.10 days.
    $$R = R_0 e^{-t/\tau} = R_0 e^{-t \ln 2/T_{1/2}} = (1.00 \times 10^{10} \text{ Bq}) e^{-8.10 \ln 2/2.70} = \boxed{1.25 \times 10^9 \text{ Bq}}$$

    (b) Strategy The atomic number Z increases by 1, while the mass number A stays the same, so the isotope undergoes beta-minus decay.

    Solution Since the isotope undergoes beta-minus decay, the particles emitted in this process are $\boxed{\text{an electron and an antineutrino}}$.

21. **Strategy** Use Eq. (28-2).

    Solution Find the order of magnitude of the minimum uncertainty in the momentum of the electron.
    $$\Delta x \Delta p_x \geq \frac{\hbar}{2}, \text{ so } (\Delta p_x)_{min} = \frac{\hbar}{2\Delta x} \sim \frac{10^{-34} \text{ J} \cdot \text{s}}{(10^{-11} \text{ m})(10^{-19} \text{ J/eV})} = \boxed{10^{-4} \text{ eV} \cdot \text{s/m}}.$$

25. **Strategy** According to Eq. (27-9), The cutoff frequency f_{max} of x-rays produced by bremsstrahlung (braking radiation) is directly proportional to the kinetic energy of the incident electrons. Since $\lambda_{min} \propto 1/f_{max}$, the minimum wavelength of the x-rays is inversely proportional to the kinetic energy of the electrons. Now, by conservation of energy, an electron accelerated through a potential difference is given kinetic energy equal to $e\Delta V$. Therefore, the minimum wavelength of the x-rays is inversely proportional to the potential difference through which the electrons are accelerated.

    Solution Find the ratio of the minimum wavelength of x-rays in tube A to the minimum wavelength in tube B.
    $$\frac{\lambda_A}{\lambda_B} = \frac{\Delta V_B}{\Delta V_A} = \frac{40 \text{ kV}}{10 \text{ kV}} = \frac{4}{1}; \text{ the ratio is } \boxed{4{:}1}.$$

29. **Strategy** The absorbed dose of ionizing radiation is the amount of radiation energy absorbed per unit mass of tissue.

    Solution Find the equivalent dose of energy due to the x-rays.
    $$E = (20 \times 10^{-3} \text{ rad}) \times \frac{0.01 \text{ Gy}}{\text{rad}} \times \frac{1 \text{ J/kg}}{\text{Gy}} \times \frac{65 \text{ kg}}{3} = 0.0043 \text{ J}$$
    Find the energy absorbed by the patient's body.
    $$\text{energy absorbed} = \frac{\text{biologically equivalent dose of energy}}{\text{QF}} = \frac{4.3 \text{ mJ}}{0.90} = \boxed{4.8 \text{ mJ}}$$

MCAT Review

1. **Strategy** The rest energy of an alpha particle is approximately 3.7 GeV. This is much greater than the kinetic energy of the alpha particle, so $K = \frac{1}{2}mv^2$ is a reasonable approximation of the kinetic energy of the particle.

   Solution Compute the approximate speed of the alpha particle.

   $$K = \frac{1}{2}mv^2, \text{ so } v = \sqrt{\frac{2K}{m}} = \sqrt{\frac{2(4.8\times10^6 \text{ eV})(1.602\times10^{-19} \text{ J/eV})}{(4 \text{ u})(1.66\times10^{-27} \text{ kg/u})}} = 1.5\times10^7 \text{ m/s}.$$

   The correct answer is $\boxed{C}$.

2. **Strategy** For each alpha emitted, the nucleus loses two protons and two neutrons. For each beta emitted, the nucleus gains a proton and loses a neutron. Count the number of protons, neutrons, and betas emitted.

   Solution A total of $1+4+1=6$ alphas are emitted, so the nucleus loses 12 protons and 12 neutrons. A total of $2+1+1=4$ betas are emitted, so the nucleus gains 4 protons and loses 4 neutrons. The new atomic number is $Z = 90-12+4 = 82$ and the new nucleon number is $A = 232-12-12+4-4 = 208$. The correct answer is $\boxed{A}$.

3. **Strategy** There are three protons and $7-3=4$ neutrons in the nucleus of lithium-7. Use Eq. (29-8) to find the binding energy.

   Solution Compute the approximate binding energy.

   $$E_B = (\Delta m)c^2 = [(3\times1.0073 \text{ u} + 4\times1.0087 \text{ u}) - 7.014 \text{ u}]\times931 \text{ MeV/u} = 40.0 \text{ MeV}$$

   The correct answer is $\boxed{D}$.

4. **Strategy** The external magnetic field only exerts forces on moving charged particles.

   Solution Alphas are positively charged, betas are negatively charged, and gammas have no charge. Therefore, gamma rays travel straight and alpha and beta rays are bent in opposite directions. The correct answer is $\boxed{D}$.

5. **Strategy** The nucleus decreases by $|\Delta m| = E/c^2$, where E is the energy of the gamma ray.

   Solution Compute the decrease in mass of the nucleus.

   $$|\Delta m| = \frac{E}{c^2} = \frac{(2.5\times10^6 \text{ eV})(1.602\times10^{-19} \text{ J/eV})}{(2.998\times10^8 \text{ m/s})^2} = 4.5\times10^{-30} \text{ kg}$$

   The correct answer is $\boxed{C}$.

6. **Strategy** Use Eqs. (29-21) and (29-22).

   Solution Find the original activity R_0 of the sodium-24 sample.

   $$R = R_0 e^{-t/\tau} = R_0 e^{-t \ln 2/T_{1/2}}, \text{ so } R_0 = R e^{t \ln 2/T_{1/2}} = (100 \text{ mCi})e^{24 \ln 2/15} = 300 \text{ mCi}.$$

   The correct answer is $\boxed{B}$.

7. **Strategy** Subtract the mass equivalent of the energy required to break the nucleus of the neon-20 atom into its constituent parts from the masses of $Z = 10$ protons and $N = A - Z = 20 - 10 = 10$ neutrons to find the atomic mass of the atom.

   Solution Find the atomic mass of the atom.
   $10 \times 1.0073 \text{ u} + 10 \times 1.0087 \text{ u} - 0.173 \text{ u} = 19.987 \text{ u}$
   The correct answer is $\boxed{\text{A}}$.

8. **Strategy** The intermediate product uranium-234 has $238 - 234 = 4$ fewer neutrons than does uranium-238. Count the number of neutrons lost in each sequence.

   Solution Beta emission does not effect the nucleon number A, but does increase the atomic number by one, so choice A is not the correct answer. Alpha emission reduces the nucleon number by four and the atomic number by two. Evaluate the remaining three sequences.
   B: $A = 238 - 4 = 234$ and $Z = 92 - 2 + 3 = 93$.
   C: $A = 238 - 4 - 4 = 230$ and $Z = 92 - 2 - 2 + 1 + 1 = 90$.
   D: $A = 238 - 4 = 234$ and $Z = 92 - 2 + 1 + 1 = 92$.
   The correct answer is $\boxed{\text{D}}$.

9. **Strategy** Every 8 months, the sample has been reduced by half.

   Solution Find the fraction of the sample still remaining after 2 years.
   $2 \text{ yr} = 24 \text{ mo}$ and $\dfrac{24}{8} = 3$, so the sample has been reduced to $\left(\dfrac{1}{2}\right)^3 = \dfrac{1}{8}$ of its original amount.
   The correct answer is $\boxed{\text{C}}$.

10. **Strategy and Solution** In gamma decay, the atomic number Z and the mass number A stay the same. The correct answer is $\boxed{\text{B}}$.

11. **Strategy** The nucleon number A is the sum of the total number of protons Z and neutrons N.

    Solution The nucleon number of thallium-201 is $A = 201$, which is equal to the total number of protons and neutrons in the nucleus. The atomic number is $Z = 81$, which is equal to the total number of protons in the nucleus. In a neutral atom, the number of electrons is equal to the number of protons; in this case, there are 81 electrons. There are $N = A - Z = 201 - 81 = 120$ neutrons. The correct answer is $\boxed{\text{D}}$.

12. **Strategy** The activity R is given by $R = R_0 e^{-t/\tau}$, where R_0 is the initial activity, t is the time elapsed, and τ is the time constant.

    Solution The activity decreases exponentially with time. The correct answer is $\boxed{\text{C}}$.

13. **Strategy** The energy of a photon is related to its wavelength by $E = hc/\lambda$.

    Solution Compute the wavelength of the gamma ray photon.
    $$\lambda = \frac{hc}{E} = \frac{(4.15 \times 10^{-15} \text{ eV} \cdot \text{s})(3.0 \times 10^8 \text{ m/s})}{1.35 \times 10^5 \text{ eV}} = \frac{(4.15 \times 10^{-15})(3.0 \times 10^8)}{1.35 \times 10^5} \text{ m}$$
    The correct answer is $\boxed{\text{A}}$.

Notes

Notes

Notes

Notes

Notes

Notes

Notes